The Practice of Spatial Analysis

Helen Briassoulis • Dimitris Kavroudakis
Nikolaos Soulakellis
Editors

The Practice of Spatial Analysis

Essays in memory of Professor Pavlos Kanaroglou

Springer

Editors
Helen Briassoulis
Department of Geography
University of the Aegean
Mytilini, Lesvos, Greece

Dimitris Kavroudakis
Department of Geography
University of the Aegean
Mytilini, Lesvos, Greece

Nikolaos Soulakellis
Department of Geography
University of the Aegean
Mytilini, Lesvos, Greece

ISBN 978-3-030-07863-8 ISBN 978-3-319-89806-3 (eBook)
https://doi.org/10.1007/978-3-319-89806-3

Printed on acid-free paper

This Springer imprint is published by the registered company Springer International Publishing AG part of Springer Nature.
The registered company address is: Gewerbestrasse 11, 6330 Cham, Switzerland

Preface

Spatial analysis answers the human need for acquiring, processing, and making sense of spatial information. The spatial knowledge produced is an indispensable aid for going about daily activities, managing private and common affairs, planning, policy making, and making long-term decisions. In this broad sense, spatial analysis comprises all techniques that have developed over time and are still being developed to process information related to the diverse phenomena that variously constitute and leave their signature on space. Contemporary spatial analysis, dating from the mid-1970s, remains a fast-evolving academic and applied research area that responds to the unending demand for spatial knowledge to address the analysis of complex socio-environmental issues and support management and policy decisions. Building on the valuable research that has accumulated over the years, it highly benefits from technological progress while simultaneously providing impetus for the development of advanced technical and technological means to enhance the potential of available and new techniques.

The present honorary volume in memory of Professor Pavlos Kanaroglou offers a snapshot of the wide variety of spatial analysis techniques in use today to address transport and land use issues, the thematic areas in which Prof. Kanaroglou contributed significantly and which constitute essential research domains in the context of the sustainable spatial development discourse. The initiative to produce this volume came from the Department of Geography, University of the Aegean, Lesvos, Greece, where Prof. Kanaroglou served as a faculty member (1999–2007) and as its first elected head (1999–2002). The Department of Geography, the first Geography department since the contemporary Greek state was established in 1830, accepted the first students in 1994. Prof. Kanaroglou's academic and administrative experience, solid research record, and influential personality were critical for the academic formation of the Department and the organization of its technical research infrastructure in these early years.

The editors invited Prof. Kanaroglou's colleagues in Canada and Greece to contribute to this volume and their enthusiastic response made this collection of essays possible. The precious contribution of a number of reviewers who volunteered to evaluate the draft chapters is gratefully acknowledged.

Our hope is that the readers of this volume will find its content informative, useful, and inspiring for their own research endeavors and will be incited to continue their research efforts to provide spatial analysis techniques responsive to the challenges of complex socio-spatial problem solving.

Mytilini, Lesvos, Greece
March 2018

Helen Briassoulis
Dimitris Kavroudakis
Nikolaos Soulakellis

Contents

Contributors

Matthew Adams Department of Geography, University of Toronto Mississauga, Mississauga, ON, Canada

Joshua Auld Argonne National Laboratory, Argonne, IL, USA

Dimitris Ballas Department of Economic Geography, Faculty of Spatial Sciences, University of Groningen, Groningen, The Netherlands

Helen Briassoulis Department of Geography, University of the Aegean, Mytilini, Lesvos, Greece

Tom Broomhead Department of Automatic Control and System Engineering, University of Sheffield, Sheffield, England

Denis Corr Corr Research, Hamilton, ON, Canada

Athanasia Darra Laboratory of Geography and Spatial Analysis, School of Rural and Surveying Engineering, National Technical University of Athens, Athens, Greece

Adam W. Davis University of California, Santa Barbara, Santa Barbara, CA, USA

André de Palma Ecole Normale Supérieure (ENS) Paris-Saclay, CREST, Cachan, France

Mark R. Ferguson McMaster Institute for Transportation and Logistics, McMaster University, Hamilton, ON, Canada

Milad Ghasri Research Centre for Integrated Transport and Innovation (rCITI), School of Civil and Environmental Engineering, University of New South Wales (UNSW), Sydney, NSW, Australia

Nima Golshani Department of Civil and Materials Engineering, University of Illinois at Chicago, Chicago, IL, USA

Konstadinos G. Goulias Department of Geography and GeoTrans Lab, University of California, Santa Barbara, Santa Barbara, CA, USA

Christopher D. Higgins Department of Land Surveying and Geo-Informatics, The Hong Kong Polytechnic University, Hung Hom, Hong Kong

Department of Building and Real Estate, The Hong Kong Polytechnic University, Hung Hom, Hong Kong

Rahaf Husein Department of Civil and Environmental Engineering, University of Windsor, Windsor, ON, Canada

Polydoropoulos Ioakeim Department of Geography, University of the Aegean, Mytilini, Lesvos, Greece

Christos Iossifidis Institute for Space Applications and Remote Sensing, National Observatory of Athens, Athens, Greece

Phil Mike Jones School of Law, Bartolomé House, University of Sheffield, Sheffield, England

Marinos Kavouras Cartography Laboratory, School of Rural and Surveying Engineering, National Technical University of Athens, Athens, Greece

Dimitris Kavroudakis Department of Geography, University of the Aegean, Mytilini, Lesvos, Greece

Themistoklis Kontos Department of Environment, University of the Aegean, University Hill, Mytilini, Greece

Jae Hyun Lee University of California, Santa Barbara, Santa Barbara, CA, USA

Claude Lefèvre Département de Mathématique, Université Libre de Bruxelles (ULB), Bruxelles, Belgium

Hanna Maoh Department of Civil and Environmental Engineering, University of Windsor, Windsor, ON, Canada

Elizabeth C. McBride University of California, Santa Barbara, Santa Barbara, CA, USA

Eric J. Miller Department of Civil Engineering, University of Toronto, Toronto, ON, Canada

Moataz Mohamed Department of Civil Engineering, McMaster University, Hamilton, ON, Canada

Abolfazl Mohammadian Department of Civil and Materials Engineering, University of Illinois at Chicago, Chicago, IL, USA

Toka S. Mostafa Department of Civil Engineering, University of Toronto, Toronto, ON, Canada

Scott Orford School of Geography and Planning, Cardiff University, Cardiff, Wales, UK

Panagiotis Papaioannou Department of Civil Engineering, Aristotle University of Thessaloniki, Thessaloniki, Greece

Kyriakidis Phaedon Faculty of Engineering and Technology, Department of Civil Engineering and Geomatics, Cyprus University of Technology, Limassol, Cyprus

Ioannis Politis Department of Civil Engineering, Aristotle University of Thessaloniki, Thessaloniki, Greece

Dimitris Potoglou School of Geography and Planning, Cardiff University, Cardiff, Wales, UK

Taha Hossein Rashidi Research Centre for Integrated Transport and Innovation (rCITI), School of Civil and Environmental Engineering, University of New South Wales (UNSW), Sydney, NSW, Australia

Srinath Ravulaparthy University of California, Santa Barbara, Santa Barbara, CA, USA

Weeberb Requia McMaster University, McMaster Institute for Transportation and Logistics, Hamilton, ON, Canada

Matthew J. Roorda Department of Civil Engineering, University of Toronto, Toronto, ON, Canada

Ramin Shabanpour Department of Civil and Materials Engineering, University of Illinois at Chicago, Chicago, IL, USA

Nikolaos Sifakis Institute for Space Applications and Remote Sensing, National Observatory of Athens, Athens, Greece

Nikolaos Soulakellis Department of Geography, University of the Aegean, Mytilini, Lesvos, Greece

Thomas S. Stephens Argonne National Laboratory, Argonne, IL, USA

Kizos Thanassis Department of Geography, University of the Aegean, Mytilini, Lesvos, Greece

Yiming Wang School for Policy Studies, University of Bristol, Bristol, UK

Clarence Woudsma School of Planning, University of Waterloo, Waterloo, ON, Canada

Introduction: The Practice of Spatial Analysis

Helen Briassoulis, Dimitris Kavroudakis, and Nikolaos Soulakellis

Abstract The chapter provides a broad-brush introduction to the subject of spatial analysis as a general backdrop to the discussion of the individual chapters of this volume. It first offers a range of definitions of spatial analysis from the human geography viewpoint mainly, which reflect its historical evolution. Then, it discusses the practice of spatial analysis in terms of broad streams of methods and techniques that have developed over time, in several of which Prof. Kanaroglou has contributed important research. Next, it presents the individual contributions and closes with a discussion of their main features, organized in terms of theory, methods/techniques, applications, and policy relevance, which point to the respective open issues that future research is called to address to enhance the role of spatial analysis in contemporary data-rich, technologically advanced, and decision-intensive contexts.

1 Defining Spatial Analysis

This brief introduction aims to provide a general backdrop to the individual chapters of this volume that tackle spatial analysis issues most of which are related to the late Prof. Kanaroglou's research activity. It first presents selected definitions of spatial analysis, from the human geography point of view mainly, that reflect its historical evolution. It then discusses the practice of spatial analysis in terms of broad streams of methods and techniques that have developed over time. After briefly presenting the individual chapters, it closes with a discussion of their main features, organized in terms of theory, methods/techniques, applications, and policy relevance, which point to the respective open issues that future research is called to address to enhance the role of spatial analysis in contemporary data-rich, technologically advanced, and decision-intensive contexts.

In a broad, general sense, spatial analysis encompasses the methods and analytical techniques employed to study diverse natural, economic, socio-cultural, and

H. Briassoulis (✉) · D. Kavroudakis · N. Soulakellis
Department of Geography, University of the Aegean, Mytilini, Lesvos, Greece
e-mail: e.briassouli@aegean.gr; dimitrisk@aegean.gr; n.soulakellis@aegean.gr

H. Briassoulis et al. (eds.), *The Practice of Spatial Analysis*,
https://doi.org/10.1007/978-3-319-89806-3_1

other phenomena using spatial concepts and data on and across various spatial scales (Fotheringham and Rogerson 2009a, b; Goodchild 2009). Its aim is to describe, explain, and predict their structure (composition, patterns, relationships, and order) and dynamics (processes of change and their outcomes) in order to explore scientific, management, and policy issues of interest, answer practical questions, and support pertinent decision-making in the public and in the private domain. The range of its applications extends over numerous fields of the natural, social, decision sciences, and the humanities (Haining 1993; Anselin 1999; Páez and Scott 2004; Rey and Anselin 2006).

In this broad sense, the roots of spatial analysis as a systematic field of study go back to the nineteenth and early twentieth century when the first attempts to study spatial structure and the location of activities were made by von Thunen and Christaller, among others (von Thünen 1966; Christaller 1933; Berry et al. 1962; Haggett et al. 1966). The contemporary, post-WWII spatial analysis took shape in the late 1950s and early 1960s in the context of the quantitative revolution in geography and the birth of regional science (Isard 1960). The emphasis was on quantitative spatial analysis and modeling especially in land use, transportation, retail geography, and planning (Herbert and Stevens 1960; Harris 1962; Berry and Marble 1968; Haggett et al. 1966, 1977; Abler et al. 1971). Early applications involved the use of mainly nonspatial statistical methods and techniques to analyze the pattern and form of geographical objects (points, lines, areas, and surfaces) depicted on maps or defined by coordinates in two- or three-dimensional space (Fischer 1999; O'Sullivan and Perry 2013), thus emphasizing form over process. Statistical software packages such as Minitab, Systat, S-Plus, and GLIM, based on nonspatial statistical techniques, were developed and are still in use.

Starting in the 1970s and continuing into the 1980s, with the notable contributions of researchers such as Dixon et al. (1987), Openshaw (1984), Openshaw et al. (1987, 1990), pertinent research exposed the inherent traits of spatial data owing to their two- or three-dimensional nature and pointed to the associated analytical challenges (Fischer 1999; Páez and Scott 2004; Goodchild 2009). A wealth of spatial analysis methodologies and techniques were developed and applied in various problem settings (Openshaw et al. 1987, 1990) to investigate the properties of spatial features and the relationships among them, i.e., size, shape, orientation, location, distance, neighborhood, distribution, pattern, topology (contiguity, containment, connectivity), spatial interaction, and scale.

From the 1990s on, the confluence of theoretical, methodological, and analytical developments mainly in geography, economics, operations research and planning, technological progress, such as advances in computing and networking technology and, in particular, the development of Geographical Information Systems (GIS), remote sensing (RS) technology, and the worldwide web precipitated the "spatial turn" in various fields of the natural, social, and the decision sciences. Together with the growing availability of large datasets and the storage, retrieval, and visualization capabilities of GIS, the need and quest for spatial thinking, and spatial literacy strengthened (Goodchild 2009; Duckham et al. 2003; Longley et al. 2015) and fed back to spatial analysis that was called to further improve the suite of methods and techniques in order to address contemporary complex socio-spatial problems.

As a result, definitions of spatial analysis became more focused (but not necessarily narrower), differentiated, and specialized, reflecting their disciplinary origins and traditions and, more specifically, their affinity to technical-oriented fields (GIS and RS). The crucial feature that distinguishes traditional from contemporary versions of spatial analysis is the use of spatial (geo-referenced/geocoded) data, i.e., those containing attribute and locational information about spatial objects—points, lines, areas, or spatial primitives located in geographical space (Fischer 1999; Goodchild 2009). In traditional spatial analysis, space is specified in a simplified way while in contemporary spatial analysis space is explicitly modeled by means of spatial data.

The current conception of spatial analysis agrees with the broad, general definition mentioned before, the main difference being that it concerns methods and techniques used to study socio-spatial phenomena using spatial data to answer important questions and to support pertinent decision-making. For Longley et al. (2001: 31): "...spatial analysis is the process by which we turn raw spatial data into useful spatial information" while for Fotheringham et al. (2009a, b): "Spatial analysis is a general term to describe a technique that uses this locational information in order to better understand the processes generating the observed attribute values."

Contemporary spatial analysis comprises the following functions: data collection, data processing, data analysis, and data visualization. Depending on the purpose of analysis, the questions addressed and the methods used, Fotheringham and Rogerson (2009a, b) suggest four types of spatial analysis (a) techniques for reducing volumes of spatial data—descriptive statistics, visualization means, data reduction techniques, detection and description of patterns in spatial data (including detection of outliers and anomalies); (b) exploratory data analysis—testing hypotheses and theories—and visual display of spatial data linked to a map; (c) statistical models used to infer (deterministic and/or stochastic) process or processes generating the data and drawing inferences; (d) mathematical modeling and prediction of spatial patterns and processes (see also Fischer 1999; Páez and Scott 2004; Grubesic and Nelson 2016).

2 The Practice of Spatial Analysis

Spatial analysis belongs to the field of Quantitative Geography and, thus, its development is inevitably rooted in and has been influenced by the methods and techniques of the latter. Adopting the broad conception of spatial analysis mentioned before, the methods and techniques that have developed over time may be grouped into six main streams, distinguished according to the methodological approach adopted, i.e., according to the methods and techniques used to solve a problem depending on the nature of the problem and the purpose of analysis. These are: spatial interaction models, spatial optimization models, spatial (or, geospatial) statistics, spatial econometrics, geo-computation, and spatial simulation. The boundaries between the streams are fluid, variable, and probably overlapping, as "traditional" blend with "modern" techniques depending on disciplinary traditions and practices. Moreover, depending on the particular application, analysis may combine techniques from more than one stream. It is noted that this classification is revisable as future theoretical, methodological, and technological advances may, first, introduce changes in these streams and, second, give rise to new ones.

Spatial interaction models aim to model the interaction among activities in space such as various types of trips (work, shopping, etc.), flows of goods, services, information, and knowledge. (Briassoulis 2000). They can be considered among the first spatially explicit models as they adopt a primitive, zone-based spatial system of reference to assess activity flows between zones at various spatial levels (Adams et al. 2016; Cliff and Ord 1981; Ghosh and Rushton 1987; Haggett et al. 1977; Openshaw 1984; Wilson 1974). Their contemporary versions account for spatial autocorrelation in the data used (Fischer and Griffith 2008). Their combination with econometric models is known as location-allocation models (Cooper 1964; Scott 1970) that are used to forecast changes in spatial structure that result from changes in the magnitude of activities brought about by exogenous socio-economic, policy, and other changes.

Spatial optimization models, initially known as programming models, are oriented towards producing solutions which optimize, under various constraints, one or more user-defined objectives (Briassoulis 2000) such as optimal allocation of land use types in an area, minimization of environmental impacts, and minimization of travel time (Harris 1962; Herbert and Stevens 1960; Nijkamp and Rietveld 1979). Their traditional versions used a zone-based spatial system of reference as it was the case with spatial interaction models. Their contemporary versions rely on geo-referenced data (Tong and Murray 2012).

Spatial (or, geospatial) statistics comprise univariate or multivariate spatial statistical techniques developed to describe and analyze spatial data (i.e., data that are not spatially independent) and to quantify spatial processes (i.e., processes that are not stationary) in order to identify spatial patterns, assess spatial dependence (spatial autocorrelation), handle issues of spatial heterogeneity, the Modifiable Area Unit problem (MAUP) and ecological fallacy, and to represent spatial choice processes and the spatio-temporal evolution of complex spatial systems (Anselin 1995; Berry and Marble 1968; Cressie 1993; Fischer and Getis 2009; Fotheringham and Rogerson 2009a, b; Gelfand et al. 2010; Getis 2004; Goodchild 2009; Griffith 2012; Maoh and Kanaroglou 2007; Openshaw et al. 1987, 1990; Páez and Scott 2004; Ripley 2004).

Spatial econometrics, founded on the seminal work of Paelinck and Klaasen (1979), refers to the application of econometric models (multivariate statistical models such as regression models) with spatial data (Anselin 1988, 1995, 1999; Getis 2004; Gumprecht 2010; LeSage and Pace 2009). According to Anselin (1988) Spatial statistics is a data-driven approach while spatial econometrics is a model-driven approach as it focuses on the relevance of spatial effects on model specification, estimation, and other inferences.

Geo-computation, a field of geographical analysis that emerged and took shape in the 1980s under the initiative and leadership of Stan Openshaw at the University of

Leeds (Openshaw and Abrahart 2000), refers to the use of computing techniques and technology, such as GIS and RS, to perform mapping operations, spatial analysis, and visualization for the study of socio-spatial and socio-temporal phenomena (Bill 1999; Fotheringham and Rogerson 1993; 2009a, b; 2013; Fotheringham et al. 1997; Longley and Batty 1996; Longley et al. 2001; O'Sullivan and Unwin 2010; Stillwell and Clarke 2003).

Spatial simulation comprises traditional and contemporary simulation methods and techniques applied with spatial data to model and understand spatio-temporal processes underlying observed spatial patterns and their dynamics. Included among them are microsimulation, agent-based models, cellular automata, spatial neural network models, etc. (Bankes 2002; Maoh and Kanaroglou 2013; Moss and Davidsson 2000; Parker et al. 2003; O'Sullivan and Perry 2013; White and Engelen 1993).

It is noted that integrated models and spatial decision support systems are not considered to constitute distinct streams of spatial analysis. The former combine several models in a usually modular structure, some of which may belong to one or more of the six streams mentioned above. The latter include spatial models in spatial decision support systems that place emphasis on the locational characteristics and impacts of decisions (see, e.g., Strager and Rosenberger 2006).

All six streams are actively used in the current practice of spatial analysis although spatial statistics seems to be more widespread and cuts across all streams given the greater demand for its applications, the inclusion of spatial statistical techniques in commercial software packages, and the lower expertise requirements compared to the other streams. Agent-based models are popular among scientists in various fields of the natural and the social sciences. The contributions of this volume offer examples of the current practice of spatial analysis with a focus on land use and transportation issues, the research areas where Prof. Kanaroglou's research has mostly focused, and reflect the diversity of approaches taken to address pertinent management and policy issues arising in socio-spatial decision-making.

3 Introducing the Book Chapters

The contributions of this volume address diverse topics in the thematic areas of land use and transportation mainly. They include reviews of selected techniques and applied research works that illustrate the use of spatial analysis techniques. The chapters are organized in four thematic groups. The first group "Spatial analysis—selected topics" includes four chapters that discuss general topics in spatial analysis. In chapter "Pavlos Kanaroglou, His McMaster Institute for Transportation and Logistics and Research Contributions to the City of Hamilton (2007–2016)," Ferguson, Higgins, and Mohamed offer a concise account of the crucial role and contribution of Prof. Pavlos Kanaroglou to the formation of the McMaster Institute for Transportation and Logistics (MITL) at McMaster University, Hamilton, Ontario. They summarize the main themes of MITL's research that intertwined with aspects of Prof. Kanaroglou's expertise and overview selected projects carried out related to freight transport, transit-oriented development, and other topics, such as impacts of public transit on land values, traffic congestion and urban form, estimation of vehicular emissions, sustainable development strategies, smart growth, freight villages, complete streets, and electric mobility, for the Greater Toronto and Hamilton Area (GTHA) and the City of Hamilton. They discuss the spatial analysis techniques being employed that include integrated urban land use-transport modeling, the IMULATE model developed by Prof. Kanaroglou, microsimulation, structural equation modeling, and scenario analysis. The chapter concludes with an account of MITL's research impacts and lessons learned under the guidance of Prof. Kanaroglou.

In chapter "An Ontological Framework for Spatial Socioeconomic Units," Darra and Kavouras discuss the ontological status of spatial units used in geographic and other applications, an issue of fundamental importance in spatial analysis. Spatial units serve as reference to socio-economic phenomena constituting what is known as spatial socio-economic units (SSEUs) that are either predefined (e.g., administrative subdivisions) or they may emerge from multilevel socio-spatial processes (e.g., service areas). The chapter presents the development of an ontological framework for the systematic definition of SSEUs on the basis of their properties and particularly the spatial ones. Special attention is paid to the most common case of SSEUs, i.e., administrative units. Using specific illustrative cases, the chapter demonstrates their conceptual association and comparison via spatialization tools. SSEUs defined on the basis of this framework can be effectively compared not only at the class level but also at the instance (specific place) level. Furthermore, the restructuring and consolidation of SSEUs is greatly facilitated by the proposed property-based definition.

In chapter "Spatial Microsimulation and Agent-Based Modelling," Ballas, Broomhead and Jones critically review the state-of-the-art in spatial microsimulation and agent-based modeling both of which concern individual level modeling. After introducing microsimulation and spatial microsimulation and their recent versions and applications, they discuss the relationship of agent-based modeling (ABM) to microsimulation and argue that ABM offers a better framework for including behavioral rules into the actions of agents (including random behavior) while allowing for interactions between them. The complementarity of the two approaches suggests that they can be linked to better address applied geography problems; microsimulation can give the agents in ABM their initial characteristics and locations, and ABMs can provide the capacity to model individual adaptive behaviors and emergence of new ones. The chapter considers attempts to combine them and articulates a pertinent research agenda.

In chapter "Spatial Patterns of Accessibility Between Islands of the North and South Aegean Regions and Attica, Greece," Kavroudakis, Polydoropoulos, Kyriakidis, and Kizos present an approach to assessing the accessibility of the islands of the North and South Aegean Sea to Attica where the capital of Greece, Athens, is located. The exercise is predicated on the important role of accessibility to economic development especially in the case of this highly fragmented insular geographic space, where the sea between the islands presents additional serious impediments. The approach developed in this chapter adds to the quantitative measures of accessibility that are frequently used to identify over- and under-served areas and, thus, support development policy-making and spatial planning. After a preliminary spatial analysis of selected indicators, such as transportation route frequency, passengers-to-population ratio and mean ticket cost, two-graph theory metrics and clustering methods, enhanced by introducing local indices of spatial association, are employed to determine possible groupings of islands and delineate spatial clusters based on their accessibility. “Hot spots” of accessibility are identified, suggesting that it is more spatially clustered in the study area than would be expected under a random model.

The next six chapters concern the analysis and modeling of urban transportation, one of the first and major areas where spatial analysis has been developed and practiced and the principal research area in which Prof. Kanaroglou devoted much of his research effort. In chapter “Agent-Based Activity/Travel Microsimulation: What’s Next?”, Miller offers a stimulating, concise, and edifying review of the current generation of operational activity/tour-based model systems that employ an agent-based microsimulation (ABM) framework to varying degrees. He argues that although these model systems represent a sound “first generation” of ABM, they have still to realize the full potential of the powerful ABM concept to model spatio-temporally and socio-economically complex behavior such as travel. The chapter perceptively discusses and details a wide range of issues and challenges in advancing the ABM-based activity/travel modeling state-of-the-art. These include the need to develop more advanced and robust, theory-informed travel demand modeling methods to take advantage of the growing availability of “big data” concerning travel behavior from a variety of sources and the continuing growth in computing capabilities as well as to address issues of dynamics, heterogeneity, uncertainty, multi-dimensionality of activity/travel decisions, inter-agent interactions, mode and route choice modeling, activity episode utility, spatial choice, and intercity travel modeling. Finally, it suggests key “next steps” in model development.

In chapter “Bottleneck Models and Departure Time Problems,” de Palma and Lefevre examine the bottleneck models of congestion, first introduced by Vickrey (1969) and elaborated by Arnott, de Palma, and Lindsey (see, e.g., Arnott et al. 1990), which are spatial optimization models developed to represent the supply side in transportation analysis. After first discussing the case of a set of individual drivers through both deterministic and stochastic approaches, the analysis is extended to a new model that combines small and large agents. The chapter provides a stochastic version of dynamic bottleneck models and elaborates on selected issues such as choice of departure time, endogenously produced arrival rates and equilibrium solutions. The chapter concerns the model building stage while model simulations are planned for the near future.

In chapter “Factors Influencing Journey-to-Work by Public Transit in Mega Canadian Cities,” Husein, Maoh, and Potoglou present a spatial econometric modeling exercise carried out to better understand how urban form and network design, together with socio-economic and demographic factors, influence public transit usage for the journey-to-work in Toronto, Ontario and Montreal, Quebec, Canada. Using 2011 census tract level data, a Simultaneous auto-regressive modeling approach was employed to account for the presence of spatial dependencies in the data and, thus, derive unbiased travel demand estimates. Urban form and transport network design characteristics were operationalized via selected measures to differentiate homogeneous from mixed land uses and gridded from curvilinear networks. Their inclusion, together with population over 55, family size (more than five members), minorities percentage, and population density revealed that mixed land uses and gridded transport networks are generally associated with higher public transit usage for the journey-to-work in both cities. The findings support the need for enhancing public transit usage in large Canadian metropolitan areas as a component of sustainable transport policies.

In chapter "Freight in the City: The Evolving Landscape of Matters and Models," Woudsma argues that "freight matters" and provides a critical synopsis of the major matters that have influenced freight transportation over the past 30 years and the methods and models applied to study those matters in the urban/metropolitan realm, highlighting in the process the research contributions of Prof. Kanaroglou. The chapter first presents the traditional emphases of freight transport research and modeling in the 1980s followed by certain major contemporary issues or "matters" related to freight transportation—deregulation and privatization, sustainability and globalization, and the rise of Logistics—and the modeling efforts devoted to tackle those issues. It ascertains that the importance of freight transportation in the urban/regional context has been now recognized and concludes with a discussion of the trends impacting freight and modeling today, such as environmental externalities, e-commerce, technological advances (e.g., electric cars, new analytical approaches, computing technology, data sources) which will fuel future research and policy.

In chapter "A Structural Equation Model of Commercial Vehicle Ownership," Mostafa and Roorda adopt a behavioral freight transportation modeling framework to examine the little-researched issue of commercial vehicle fleet ownership and composition. A structural equation model (SEM) of vehicle ownership of small/medium-sized business establishments (less than 100 employee) in the Greater Toronto and Hamilton Area (GTHA) is employed that utilizes geocoded (via ARCHGIS) business establishment and census tract data collected by three commercial travel surveys for different regions in three different years. The model estimates the number of owned cars, pickups/vans, and trucks for establishments located in the GTHA according to their location, industrial sector, freight demand, and basic characteristics (number of employees, floor space, and age). Its predictive ability can be further improved by considering the effects of lagged endogenous variables, by means of additional longitudinal data, industry-specific information, and other model structures. The model can serve as a component in a large-scale behavioral freight microsimulation developed by the authors to support planning and policy assessment.

In chapter "Applying Behavior Change Theory to Predict Travel Behavior Changes of University Commuters," Papaioannou and Politis combine behavior change theories from Social Psychology with discrete choice modeling to study the differences in the predictions of travel mode choice change obtained from the conventional aggregate approach of analyzing a total sample of commuters and from a disaggregate approach based on sample segmentation based on the commuters' behavioral change stages. A sample of commuters (academic and administrative staff) to the Aristotle University of Thessaloniki, Greece was used and four travel modes—van-pooling, collective taxi, car-pooling, and metro—were considered as alternatives to the private car. A combined revealed and stated preference survey provided the data to run a total of 32 logit models that tested various hypotheses and scenarios concerning travel mode change and predicted the number of commuters who are willing to change travel mode if the right alterative is offered to them. The model results indicated that travel mode choice behavior predictions differ significantly among the persons belonging to different behavioral change stages. The findings suggest that transport policy-makers might better aim for targeted interventions to maximize the positive socio-economic and environmental impacts of travel mode alternatives.

In chapter "Developing a Data Transferability Platform to Analyze National-Level Impacts of Connected Automated Vehicles," Shabanpour, Golshani, Stephens, Auld, and Mohammadian present the application of the spatial transferability approach to assess the potential impacts of the emerging connected automated vehicles (CAVs) technology on travel behavior at the national level. Based on transferability techniques, a methodological framework is developed which uses data and models from a smaller geographical area—the POLARIS simulation results for the CAVs scenario in the Chicago metropolitan area—to provide a robust platform to generate disaggregate travel data at the national level. The POLARIS modeling suite is an open-source agent-based modeling platform, specifically designed for simulating large-scale transportation systems, which is able to generate disaggregate output, including individuals' record of daily activities and travels, travel mode and other choices, along with socio-demographics, land use, and built-environment attributes. Comparison of the distributions of transferred variables, total daily trip rates and travel time at the regional and the national contexts indicates that the platform is capable of transferring travel behavior indices to the national level with high level of accuracy.

The next three chapters concern another topic of methodological and policy interest, integrated land use/transport analysis, on which Prof. Kanaroglou had devoted an important part of his research. In chapter "The Impact of Public Transport Infrastructure on Land Values: Using Spatial Data Analysis to Uncover Policy-Relevant Processes," Potoglou, Maoh, Wang, and Orford employ exploratory and explanatory spatial data analysis to examine the pecuniary impacts of public transport infrastructure on land values controlling for type of property. Using approximately 9000 property-sale records from central Cardiff for the period 2000–2009, spatial statistics was applied to identify the presence of spatial autocorrelation in the data. Two alternative spatial regression models were employed to produce unbiased estimates of the public transport infrastructure–land values relationships. The results showed that (a) there is an underlying bus stop-oriented spatial dependence of the values of residential land, which varies with the type of housing property across different areas in Cardiff and (b) after controlling for spatial dependence, there is still a significant positive correlation between bus stops and land value change. Different types of property seem to benefit differently from land value uplift due to bus stop locations. The highest benefits are identified for flats and semi-detached homes followed by terraced and detached homes. This analysis is relevant in the context of formulating fair taxation policies and, in particular, of the land value tax that is premised on the benefits properties obtain from adjacent public infrastructure funded by tax payers.

In chapter "California Business Establishment Evolution and Transportation Provision," Davis, Lee, McBride, Ravulaparthy, and Goulias analyze the relationship between the survival and economic success of business establishments in four target classes (retail, manufacturing, professional services, and healthcare) and attributes of their local environment, including access to transportation facilities, for the entire State of California. Panel data analysis methods (stepwise probit regression) were employed using data spanning two decades on external (local context, transportation network) and internal (business type, number of employees, age of the establishment) factors influencing business establishment success and failure. The block level data available were spatialized and various metrics were assessed to operationalize business and locational characteristics. The panel models control for factors influencing business life cycle events, such as closures, formation/birth, and relocation. Focusing on manufacturing and retail business establishments, differences and commonalities in event occurrence and the factors that impact them are illustrated. The analytical approach employed can provide decision support for local economic development planning.

In chapter "Investigating the Internal Compromise Between Wife and Husband's Commute Time Changes in Residential Relocation," Rashidi and Ghasri focus on commute time, an important consideration in transportation and urban studies. They propose a statistical modeling framework to explore the compromise between commute time of husband and wife when deciding to change their place of residence and test the hypothesis of an internal negotiation between household breadwinners. From a longitudinal survey of Household Income and Labour Dynamics in Australia (HILDA), they extract the residential relocation history of dual-worker households and first specify a reciprocal model which is then converted into a linear model solved by ordinary least squares (OLS). The results reveal that two types of relationships develop—complementary and substitution—between husband and wife commute time depending on household structure and job types.

The last two chapters engage with the important public health issue of travel-related air pollution. In chapter "Aerosol Optical Thickness Mapper (AOT-mapper): A Geo-Information Software for AOT Mapping at Urban Scale Using Landsat TM or ETM$^+$ Satellite Images," Soulakellis, Kontos, Sifakis, and Iossifidis present a geo-information software, the aerosol optical thickness mapper (AOT-mapper), to support operational aerosol pollution mapping in urban environments. The AOT-mapper is a multi-module object-oriented software, specifically designed with the use of geo-computation techniques, for aerosol optical thickness monitoring and mapping on the basis of high spatial resolution satellite imagery. The chapter presents its four processing modules—satellite image import, geometric correction of images, radiometric calibration and AOT calculation, and AOT visualization—and its successful application to map aerosol load distribution over the city of Athens (Greece).

In chapter "Mapping Air Pollution Health Risk: Providing Citizens the Capability for Personal Risk Avoidance," Adams, Corr, and Requia deal with the issue of communicating air pollution health risk to citizens. An Air Quality Health Index (AQHI) assesses a person's exposure to a cocktail of air pollutants and subsequent air pollution health risk as a function of location and numerous environmental and anthropogenic factors that vary over time. Available AQHIs at the city level fail to depict the spatial variability of air pollution health risk in real time. The chapter describes the development of a neighborhood level, real time, internet-enabled air pollution map for the city of Hamilton, Ontario, Canada. The approach developed employs geo-computation techniques to process fixed location and mobile air pollution monitoring data to produce the air pollution city map. The authors note that data availability is the main limitation of the approach and they discuss selected alternatives, namely land use regression (LUR) modeling and remote sensing, to generate spatially resolved air pollution data.

4 Closing Remarks: Spatial Analysis in (Future) Practice

Spatial analysis has "revolutionized" the analytical approaches used to study diverse scientific and policy issues because, by explicitly accounting for the spatial properties (dependence and heterogeneity) of the data and information employed, it may offer more accurate, compared to nonspatial analysis, estimates of current and predicted values of environmental, socio-economic, and other variables and of their patterns and relationships in a problem setting. The development of advanced spatial analysis techniques parallels the growth of complex socio-spatial problems, offering more apposite resources to address numerous practical issues, such as the incidence, spatial distribution, spatial processes and spatial trends of environmental and socio-economic problems at various spatial levels, spatial structure/form, facility location, optimal land use allocation, travel behavior and transportation data analysis, environmental and socio-economic impacts of human activities, policy and plan evaluation, and many others. The contemporary popularity of spatial analysis, thus, owes to its usefulness in a wide range of scientific and practical problem and decision settings but also to the availability of easily accessible computational tools with enhanced analytical and visualization capabilities and the greater amounts of available and (relatively) easily accessible spatial data.

The collection of essays included in this volume offers a snapshot of the practice of spatial analysis, illustrating its use in the thematic areas of transport, integrated land use/transport analysis, and transport-related air pollution on which Prof. Kanaroglou's research has mostly focused. These areas remain central academic and applied research foci that continue to feed spatial analysis with elaborate methods, techniques, and geo-informatics tools to better comprehend and respond to constantly emerging complex problems. In closing this Introduction, the main features of the contributions are summarized and discussed, organized around questions of theory, methods/techniques, applications, and policy relevance to suggest a number of important open issues which future research is called to address.

Theory. All contributions implicitly hold a positivist or a post-positivist epistemological orientation that implies adoption of a system ontology, a geometric, or a relative view of space and reductionist analytical approaches that emphasize the observable, quantifiable, and measurable aspects of phenomena, which can be subject to quantitative spatial analysis. In this context, instrumental theory is given precedence over substantive theory (Briassoulis 2000). In other words, selected variables are hypothesized to influence the phenomenon studied (e.g., travel demand), hypotheses are statistically tested and generalizations are proposed. Thus, data-driven are prioritized over theory-driven models.

Methods/techniques. The contributions employ techniques that belong to one or more of the streams of spatial analysis discussed before. It seems that application of spatial analysis proper starts with a preliminary analysis of data to detect the presence and degree of spatial autocorrelation that will partly dictate the appropriate model to be used to explore the research questions of a study. Four trends can be discerned, at least: (a) development of advanced spatial statistical techniques, (b) wider applicability of spatial econometric approaches to a variety of problems, (c) diffusion of ABMs to several thematic areas and their combination with microsimulation techniques, and (d) further development of integrated models and their spatialized versions, in particular, for use in practical applications.

It is noted that, with some exceptions, the contributions do not address the issue of specification error in the models used. This is bound to be significant and affect their results not only in terms of variables omitted (due to lack of data, proper theory, and other reasons) but also in terms of the particular ways the variables are operationalized to mostly conform with available data. Moreover, although several models include validation of their results, it is not evident if they have also performed verification of their results that would obviously reinforce their explanatory and predictive power.

Applications. The contributions offer examples of spatial analysis that concern the case of western societies, in general, and of large urbanized and metropolitan areas, in particular. These are characterized by, or at least assumed by the models to possess, relatively well-functioning and stable institutions, regulated housing and other markets, more or less routinized residential and mobility patterns as well as stable "external" (rest of the world) socio-economic conditions that are amenable to the averaging of socio-spatial features and processes that the modeling exercises perform in order to produce a statistical picture of reality. Under these conditions and assumptions, the models appear to be suitable for the particular problem situations and contexts for which they are built and applied.

Four qualifications are in order here. First, the era of stability is past in the western, developed world; globalization, global environmental change, migration, geopolitical and other events cast doubt on the credibility of these assumptions even in well-ordered societies, making it necessary to account for and work with the inevitable uncertainty of contemporary western societies. The freight transport contributions, at least, appear to be aware of these issues.

Second, even in western societies, the metropolitan differ widely from the nonmetropolitan problem settings on various fronts—environmental and land use conditions, population composition, industrial structure, socio-cultural traits and processes, to name a few. This fact alone introduces incommensurability between the modeling and the real world milieus.

Third, the western examples are far removed from non-western situations. Geographical, socio-cultural, institutional, and other differences render the assumptions underlying several spatial analysis techniques questionable; consequently, their results are expected to be at variance with reality and the statistical "explanations" offered to be inadequate aids for practical decision-making.

Fourth, the case was always that "the times they are a-changin'," people and places are in constant flux; people grow, change stages in their life cycle and change life styles unpredictably; the weather changes, trees grow, water resources become scarce; businesses (and jobs) are created, relocate, quit, new industrial sectors appear, and old ones vanish. The spatial structure (built environment) may possess inertia but the socio-spatial structure remains fluid and volatile. Moreover, a great deal of socio-economic activity goes unrecorded, not to say uncharted, but it does not go unnoticed; informal activities are part and particle of socio-economic structure and dynamics in both developed and developing countries, accounting for up to 50% of real socio-economic activity, as international and other organizations have recognized since the late 1980s (Turnham et al. 1990; ILO 2002; Cling et al. 2015). In the present context, this implies model misspecification (that varies with the case) and lack of pertinent data even if models are correctly specified.

The data issue is not limited to the informal sector data only. The contributions of this volume reveal that, although existing data bases and sources provide large volumes of data to model the metropolitan and urbanized areas of the developed world, the appetite of models for data continues unabated, and they remain insatiable as it has always been the case (Briassoulis 2008). With the exception of real-time data perhaps, the data used to calibrate and run the models reflect the conditions at the time of their collection and not at the time their results are considered in decision-making. Moreover, issues of data conceptual (definitions) and spatio-temporal consistency and comparability may be seriously affecting the transferability of model results in addition to other factors (Briassoulis 2008).

Policy relevance. All contributions aim at directly or indirectly providing support to decision- and policy-making. This takes either the form of providing essential information on current or predicted values of variables of interest, their spatial patterns and relationships, such as the relationship of land values to the provision of public infrastructure, or of ex ante or ex post assessing the effectiveness of particular policy measures. This is a simplified conception of policy-making, dictated by the positivist orientation of mainstream spatial analysis, which is blind to the actors participating in policy-making and the processes by which decisions are made and policy measures are implemented (Briassoulis 2008).

Future research faces, thus, several interesting directions that are summarized below following the theory, methods/techniques, applications, and policy relevance grouping.

Theory. Replacing instrumental with substantive theory in modeling exercises and development of theory-driven models is a long-waited research undertaking, as several contributions also note. To tackle the contemporary complex and highly uncertain socio-spatial issues, integrated theoretical frameworks are necessary that account for the multitude of current and emerging environmental and human, formal, and informal, "actors," at all spatial scales, their characteristics, spatial relationships and dynamics, particularly those associated with the land and the labor markets and technology (in the widest sense).

A more advanced research direction, necessitating broad multi- and interdisciplinary collaborations, concerns the adoption of nonpositivist epistemologies, alternative to system ontologies, and relational space conceptions (see, e.g., Olsson et al. 2015) to conceptualize the dynamic and fluid socio-spatial phenomena and frame their analysis.

Methods/techniques. Extant methods and techniques of spatial analysis may be improved by relaxing restrictive and unrealistic assumptions, addressing the specification error, and refining their operational features. However, the real challenge is to develop analytical methods and techniques that reflect the new theoretical orientations discussed above and provide context-sensitive analytical tools. In this respect, the ABM family of models holds significant promise to help put into practice integrated theoretical frameworks, in the context of mixed methods methodologies, as well as to experiment with alternative spatial ontologies. Moreover, it is questionable if quantitative spatial analysis alone can address real world "wicked" problems (Rittel and Webber 1973; Conklin 2005) comprehensively and holistically. Complementing or combining quantitative with qualitative spatial analysis techniques, which is an emerging field of inquiry, is a fruitful avenue to follow, especially in participatory decision settings, which may provide further insights into their value and use in handling more "evidence-based" (i.e., data-based) decision issues (Baur et al. 2014; Rucks-Ahidiana and Bierbaum 2017). Lastly, model verification, an admittedly demanding endeavor, is necessary to complement future analytical efforts in order to enhance the reliability and validity of their results and meaningfully inform policy and practice.

Applications. Besides undertaking more theoretically informed and analytically refined applications in western settings, spatial analysis applications in non-western, often data-poor, settings, which possess different geographical, environmental, socio-economic, cultural, technical, technological, and institutional characteristics represent an immense and challenging field of applied spatial analysis research which may generate novel theoretical, methodological, and practical research insights and outputs.

The new age of big data and crowd-sourced data confronts spatial analysts with another intriguing research direction. New data sources may help overcome problems of data availability and obsolescence but, at the same time, raise questions of data consistency and comparability as well as issues of research ethics (anonymity, confidentiality, etc.). Moreover, in a data super-rich world, the question of how much detail is really necessary in problem analysis arises as Miller (see chapter "Agent-Based Activity/Travel Microsimulation: What's Next?") ponders.

Policy relevance. Policy questions will continue to provide impetus for theoretical and analytical developments as in the past. Besides the growing field of participatory spatial data management and analysis (see, e.g., Dunn 2007), policy-relevant spatial analysis will have to grapple with the issues of better representing and theorizing the policy context, more specifically the actors involved and the processes operating to produce the spatial features and patterns observed and measured on the ground. Lastly, studies of the use and utility of spatial analysis in actual decision settings will contribute to its further progress and hopefully underscore its importance in contemporary data-rich and decision-intensive contexts.

References

Abler, R. A., Gould, J. S., Abler, P. R., Adams, J. S., & Gould, P. (1971). *Spatial organization. The geographer's view of the world*. Englewood Cliffs, NJ: Prentice Hall.

Adams, M. D., Yiannakoulias, N., & Kanaroglou, P. S. (2016). Air pollution exposure: An activity pattern approach for active transportation. *Atmospheric Environment, 140*, 52–59.

Anselin, L. (1988). *Spatial econometrics: Methods and models*. New York: Springer.

Anselin, L. (1995). Local indicators of spatial association—LISA. *Geographical Analysis, 27*, 93–115.

Anselin, L. (1999). The future of spatial analysis in the social sciences. *Geographic Information Sciences, 5*, 67–76.

Arnott, R., de Palma, A., & Lindsey, R. (1990). Economics of a bottleneck. *Journal of Urban Economics, 27*, 111–130.

Bankes, S. C. (2002). Agent-based models: A revolution? *PNAS, 99*, 7199–7200.

Baur, N., Hering, L., Raschke, L., & Thierbach, C. (2014). Theory and methods in spatial analysis. Towards integrating qualitative, quantitative and cartographic approaches in the social sciences and humanities. *Historical Social Research/Historische Sozialforschung, 39*(2), 7–50.

Berry, B. J. L., Barnum, H. G., & Tennant, R. J. (1962). Retail location and consumer behavior. *Papers in Regional Science, 9*, 65–106.

Berry, B. J. L., & Marble, D. F. (1968). *Spatial analysis: A reader in statistical geography*. Englewood Cliffs, NJ: Prentice-Hall.

Bill, R. (1999). Grundlagen der Geo-Informationssysteme. Band 1: *Hardware, Software und Daten*. 4., *völlig neubearb. und erw.* Auflage. Heidelberg: Herbert Wichmann Verlag

Briassoulis, H. (2000). *Analysis of land use change: Theoretical and modeling approaches*. In S. Loveridge (Ed.), *Web book of regional science*. Morgantown: Regional Research Institute, West Virginia University.

Briassoulis, H. (2008). Land use policy and planning, theorizing, and modeling: Lost in translation, found in complexity? *Environment and Planning B and Design, 35*, 16–33.

Cliff, A. D., & Ord, J. K. (1981). *Spatial processes: Models & applications*. London: Taylor & Francis.

Cling, J. -P., Razafindrakoto, M. Roubaud, F. (2015). *Informality and human development*. UNDP. Human Development Reports. Retrieved February 1, 2018, from http://hdr.undp.org/en/informality-and-hd

Conklin, J. (2005). *Dialogue mapping: Building shared understanding of wicked problems*. London: Wiley.

Cressie, N. (1993). *Statistics for spatial data*. New York: John Wiley & Sons.

Christaller, W. (1933). *Die Zentralen Orte in Süddeutschland*. Jena: G. Fischer. (Translated by C.W. Baskin as Central Places in Southern Germany. Englewood Cliffs, NJ: Prentice-Hall, 1966).

Cooper, L. (1964). Heuristic Methods for Location-Allocation Problems. *SIAM Review, 6*, 37–53.

Dixon, J. S., Openshaw, S., & Wymer, C. (1987). *A proposal for a geographical analysis subroutine library. Northern Regional Research Laboratory, Research Report 3*. Newcastle upon Tyne: Department of Geography, University of Newcastle.

Duckham, M., Goodchild, M. F., & Worboys, M. (Eds.). (2003). *Foundations of Geographic Information Science*. London: CRC Press.

Dunn, C. E. (2007). Participatory GIS – a people's GIS? *Progress in Human Geography, 31*(5), 616–637.

Fischer, M. M. (1999). Spatial analysis: Retrospect and prospect. In P. A. Longley, M. F. Goodchild, D. J. Maguire, & D. W. Rhind (Eds.), *Geographical information systems: Principles, technical issues, management issues and applications* (pp. 283–292). New York: Wiley.

Fischer, M. M., & Griffith, D. A. (2008). Modeling spatial autocorrelation in spatial interaction data: an application to patent citation data in the European Union. *Journal of Regional Science, 48*, 969–989.

Fischer, M. M., & Getis, A. (2009). *Handbook of applied spatial analysis: Software tools, methods and applications* (1st ed.). Berlin: Springer.

Fotheringham, A. S., & Rogerson, P. A. (1993). GIS and spatial analytical problems. *International Journal of Geographical Information Science, 7*, 3–19.

Fotheringham, A. S., Clarke, G., & Abrahart, B. (1997). Guest editorial: Geocomputation and GIS. *Transactions in GIS, 2*, 199–200.

Fotheringham, A. S., & Rogerson, P. A. (Eds.). (2009a). *The SAGE Handbook of spatial analysis*. New York: Sage.

Fotheringham, A. S., & Rogerson, P. A. (2009b). Introduction. In A. S. Fotheringham & P. A. Rogerson (Eds.), *The Sage Handbook of spatial analysis* (pp. 1–4). New York: Sage.

Fotheringham, A. S., & Rogerson, P. A. (2013). *Spatial analysis and GIS*. Boca Raton: CRC Press.

Gelfand, A. E., Diggle, P., Guttorp, P., & Fuentes, M. (2010). *Handbook of spatial statistics*. Boca Raton: CRC Press.

Getis, A. (2004). *Spatial econometrics and spatial statistics*. New York: Macmillan.

Ghosh, A., & Rushton, G. (1987). *Spatial analysis and location-allocation models*. New York: Van Nostrand Reinhold Company.

Goodchild, M. (2009). Challenges in spatial analysis. In A. S. Fotheringham & P. A. Rogerson (Eds.), *The Sage Handbook of spatial analysis* (pp. 465–480). New York: Sage.

Griffith, D. A. (2012). *Advanced spatial statistics: Special topics in the exploration of quantitative spatial data series*. New York: Springer Science & Business Media.

Grubesic, T., & Nelson, J. R. (2016). *Spatial analysis*. Retrieved May 12, 2017, from http://www.oxfordbibliographies.com/view/document/obo-9780199874002/obo-9780199874002-0125.xml

Gumprecht, D. (2010). *Spatial methods in econometrics*. Vienna: VDM Verlag Dr. Müller.

Haggett, P., Cliff, A. D., & Frey, A. (1966). *Locational analysis in human geography*. London: Edward Arnold.

Haggett, P., Cliff, A. D., & Frey, A. (1977). Locational analysis in human geography. *Tijdschrift voor Economische en Sociale Geografie, 68*.

Haining, R. (1993). *Spatial data analysis in the social and environmental sciences*. Cambridge: Cambridge University Press.
Harris, B. (1962). *Linear programming and the protection of land uses*. Urban Transp Plng Abstracts/Prelim.
Herbert, J. D., & Stevens, B. H. (1960). A model for the distribution of residential activity in urban areas. *Journal of Regional Science, 2*, 21–36.
ILO. (2002). *Resolution concerning decent work and the informal economy*. Accessed on February 1, 2018, from http://www.ilo.org/public/english/standards/relm/ilc/ilc90/pdf/pr-25res.pdf
Isard, W. (1960). *Methods of regional analysis; an introduction to regional science*. Cambridge: Published jointly by the Technology Press of the Massachusetts Institute of Technology and Wiley, New York.
LeSage, J., & Pace, R. K. (2009). *Introduction to spatial econometrics*. Boca Raton: Chapman and Hall/CRC.
Longley, P. A., & Batty, M. (1996). *Spatial analysis: Modelling in a GIS environment*. New York: John Wiley & Sons.
Longley, P. A., Goodchild, M. F., Maguire, D. J., & Rhind, D. W. (2001). *Geographic information systems and science*. New York: John Wiley & Sons.
Longley, P. A., Goodchild, M. F., Maguire, D. J., & Rhind, D. W. (2015). *Geographic information science and systems* (4th ed.). New York: John Wiley & Sons.
Maoh, H., & Kanaroglou, P. (2007). Geographic clustering of firms and urban form: a multivariate analysis. *Journal of Geographical Systems, 9*(1), 29–52.
Maoh, H., & Kanaroglou, P. (2013). Modelling firm failure: Towards building a firmographic microsimulation model. In F. Pagliara, M. de Bok, D. Simmonds, & A. Wilson (Eds.), *Employment location in cities and regions* (pp. 243–261). Berlin: Springer.
Moss, S., & Davidsson, P. (Eds.). (2000). *Multi agent based simulations*. Berlin: Springer.
Nijkamp, P., & Rietveld, P. (1979). *Multidimensional spatial data and decision analysis*. New York: John Wiley & Sons.
Olsson, L., Jerneck, A., Thoren, H., Persson, J., & O'Byrne, D. (2015). Why resilience is unappealing to social science: Theoretical and empirical investigations of the scientific use of resilience. *Science Advances, 1*(4), e1400217. (American Association for the Advancement of Science).
Openshaw, S. (1984). The modifiable areal unit problem. Geo Abstracts University of East Anglia.
Openshaw, S., Charlton, M., Wymer, C., & Craft, A. (1987). A mark 1 geographical analysis machine for the automated analysis of point data sets. *International Journal of Geographical Information Systems, 1*, 335–358.
Openshaw, S., Cross, A., & Charlton, M. (1990). Building a prototype geographical correlates exploration machine. *International Journal of Geographical Information Systems, 4*, 297–311.
Openshaw, S., & Abrahart, R. J. (2000). *GeoComputation*. London: Taylor and Francis.
O'Sullivan, P. D., & Unwin, D. (2010). *Geographic information analysis* (2nd ed.). New York: Wiley.
O'Sullivan, D., & Perry, G. L. W. (2013). *Spatial simulation: Exploring pattern and process*. New York: John Wiley & Sons.
Paelinck, J. H. P., & Klaasen, K. H. (1979). *Spatial econometrics; studies in spatial analysis*. Farnborough: Saxon House.
Páez, A., & Scott, D. M. (2004). Spatial statistics for urban analysis: A review of techniques with examples. *GeoJournal, 61*, 53–67.
Parker, D. C., Manson, S. M., Janssen, M. A., Hoffmann, M. J., & Deadman, P. (2003). Multi-agent systems for the simulation of land-use and land-cover change: A review. *Annals of the Association of American Geographers, 93*, 314–337.
Rey, S. J., & Anselin, L. (2006). Recent advances in software for spatial analysis in the social sciences. *Geographical Analysis, 38*, 1–4.
Ripley, B. D. (2004). *Spatial statistics*. New York: Wiley-Interscience.

Rittel, H. W., & Webber, M. M. (1973). Dilemmas in a general theory of planning. *Policy Sciences, 4*, 155–169.
Rucks-Ahidiana, Z., & Bierbaum, A. H. (2017). Qualitative spaces: Integrating spatial analysis for a mixed methods approach. *International Journal of Qualitative Methods*. http://journals.sagepub.com/doi/pdf/10.1177/160940691501400208.
Scott, A. J. (1970). Location-allocation systems: a review. *Geographical Analysis, 2*, 95–119.
Stillwell, J., & Clarke, G. (Eds.). (2003). *Applied GIS and spatial analysis* (1st ed.). New York: John Wiley & Sons.
Strager, M. P., & Rosenberger, R. S. (2006). Incorporating stakeholder preferences for land conservation: Weights and measures in spatial MCA. *Ecological Economics, 57*, 627–639.
Tong, D., & Murray, A. T. (2012). Spatial optimization in geography. *Annals of the Association of American Geographers, 102*(6), 1290–1309.
Turnham, D., Salomé, B., & Schwartz, A. (Eds.). (1990). *The informal sector revisited*. Paris: OECD Development Centre.
Vickrey, W. S. (1969). Congestion theory and transportation investment. *American Economic Review, 59*(2), 251–260.
von Thünen, J. (1966[1842]). *Isolated state: an English translation of Der Isolierte Staat* (C. M. Wartenberg, Trans. with intro. by P.G. Hall). Oxford: Pergamon Press (first volume originally published in 1826).
White, R., & Engelen, G. (1993). Cellular automata and fractal urban form: A cellular modelling approach to the evolution of urban land-use patterns. *Environment and Planning A, 25*(8), 1175–1199.
Wilson, A. G. (1974). *Urban and regional models in geography and planning*. New York: John Wiley & Sons Inc.

Part I
Spatial Analysis: Selected Topics

Pavlos Kanaroglou, his McMaster Institute for Transportation and Logistics and Research Contributions to the City of Hamilton (2007–2016)

Mark R. Ferguson, Christopher D. Higgins, and Moataz Mohamed

Abstract At an age when others might have considered slowing down, Pavlos Kanaroglou was just getting started. Neither his age nor the knowledge that he had chronic leukemia were going to stand in his way. Pavlos had accomplished a lot in his years as an academic (he held, for example, a prestigious Tier 1 Canada Research Chair for two consecutive 7 year terms up to 2015) but he felt the need to do something new and ambitious that would be of additional service to society. Pavlos was integral to the process that led to the formation, in October 2007, of the McMaster Institute for Transportation and Logistics (MITL) and he was the natural choice to take on the role of the Institute's founding director. MITL was launched in Hamilton, Ontario, Canada which had been Pavlos' home from the mid-1970s. During Pavlos' tenure, the Institute completed a substantial body of work, associated with approximately $3.5M in funding. Much of the research that was undertaken intertwined with aspects of Pavlos' expertise and was quite applicable for his adopted hometown. The purpose of this chapter is to provide a retrospective look at the insights that emerged for Hamilton as MITL grew and as Pavlos did all he could for the benefit of both, and for the development of his students and team members.

M. R. Ferguson (✉)
McMaster Institute for Transportation and Logistics, McMaster University, Hamilton, ON, Canada
e-mail: fergumr@mcmaster.ca

C. D. Higgins
Department of Land Surveying and Geo-Informatics, The Hong Kong Polytechnic University, Hung Hom, Hong Kong

Department of Building and Real Estate, The Hong Kong Polytechnic University, Hung Hom, Hong Kong

M. Mohamed
Department of Civil Engineering, McMaster University, Hamilton, ON, Canada

H. Briassoulis et al. (eds.), *The Practice of Spatial Analysis*,
https://doi.org/10.1007/978-3-319-89806-3_2

1 Introduction

One of the defining aspects of the last decade of Pavlos Kanaroglou's professional life was his role as founding director of the McMaster Institute for Transportation and Logistics (MITL). The seeds of the concept of MITL came about when representatives of the goods movement sector in and around Hamilton, Ontario approached the leadership of McMaster University in 2006. At initial meetings, the need to form a University organization that would partner with the private and public sectors to address real-world transportation and logistics issues was suggested. It did not take long before the University identified Pavlos as the ideal McMaster researcher to lead MITL, and the Institute formally came into existence in October 2007. One of the aspects that Pavlos saw as attractive was that MITL was to be based in his adopted hometown of Hamilton where he resided for over 40 years and where he and his wife Vina brought up their two children Niki and Savas.

The story of Pavlos and MITL is essentially a story of leadership, empowerment, and of him creating an environment in which researchers could thrive. For MITL's suite of reports, Pavlos was not typically the primary researcher but he was always there, ready and willing to discuss any aspect of a technical issue at hand, the dynamics surrounding a particular MITL partnership, or to offer a push when needed. His determination was contagious and helped to focus the intensity of the research team. The quality of MITL's research mattered a lot to him. Over his tenure, Pavlos gave literally thousands of hours of his time to MITL, and it was his intent to carry on with the Institute after retiring, which he did until the last weeks of his life.

Of all the ways to pay tribute to Pavlos, one of the best is to report on the progress that MITL made in his critical years as founding director. Accordingly, this chapter will explore connections between Pavlos' work and research carried out by the Institute. In connecting the dots of his career and MITL's development, we find that there are examples where Pavlos' early and mid-career work were influential in MITL's projects. There are other cases where MITL projects provided something of a springboard for Pavlos and his students to explore new research areas. Over the years, Hamilton has proven to be an excellent "laboratory" for a great deal of research from Pavlos, his students and other collaborators and some of the insights that have emerged will be captured below.

Like Pavlos' career, this chapter will cover a lot of ground and this observation is also true of MITL's body of work. Since the chapter is project- and research focused, we take the opportunity now to acknowledge Pavlos' leadership of MITL's activities in outreach and bringing people together. Of particular note, MITL hosted a half-dozen TRANSLOG conferences under his leadership beginning in 2009. These events brought academic, government, and private sector stakeholders together to exchange ideas and seek real-world practical solutions for pressing problems, especially related to the freight sector.

The basic organization of the chapter is as follows. Initially, an overview is given of the spatial setting for the studies covered. In some cases, the relevant setting is the Greater Toronto and Hamilton Area (GTHA) and in other cases, the focus is on the immediate City of Hamilton. For the majority of the freight research covered, it is necessary to assess an area larger than Hamilton which is in keeping with the spatial scope of freight movements in the region and the dynamics by which freight modes are chosen. Many of the non-freight themes can be comfortably analyzed at the Hamilton level or are purely Hamilton based. After describing the setting, the core of the chapter is a series of sections organized around some of the main themes apparent in MITL's research. Associated with each of the main themes are overview discussions of two or more projects, often associated with MITL reports that were released. As distinct from peer-reviewed academic articles, Pavlos considered MITL's open-access reports to be of great importance for promoting the Institute as a center for applied research. Upon completion of the brief thematic reviews, a concluding section summarizes the lessons that have been learned from this exercise and from the guiding hand of Pavlos.

2 Hamilton as a Setting for the Research of Pavlos and MITL

The general setting of Hamilton (See Fig. 1) is as part of the GTHA, which is Canada's largest metropolitan agglomeration. However, Hamilton has always been thought of as a city distinct from Toronto, and in fact they are classified into separate census metropolitan areas. The fact that Hamilton has its own old and historic urban core is part of what is making the city attractive as a place to live for a new generation of young professionals. As of the 2016 Census, the Toronto metro had a population of 5.9 million while Hamilton's was approximately 750,000 (including the adjacent City of Burlington).

Figure 1 outlines some important rail, road, air, and marine infrastructure and gives a sense of how Hamilton fits into the overall region. With regard to air, the dominant airport is Pearson International which is located in western Toronto. This is an important hub for both the movement of people and air freight. Hamilton International Airport is located to the south of Hamilton and is relatively much more oriented to air freight (mostly domestic) than Pearson International. The most important rail freight infrastructure in the region, showing in red in Fig. 1, is located within a few kilometers of Pearson international. There are two intermodal yards and a major rail classification yard that are operated by Canada's two major Class 1 rail operators (Canadian National and Canadian Pacific). As well as having extensive national networks that converge on Toronto, both firms have extensive operations in the United States. A concentration of rail infrastructure also shows in Hamilton's industrial area some of which is linked to the Port of Hamilton.

Highways are a critical lifeblood for the region. The most important of these is Highway 401 which in Fig. 1 runs from bottom to top through Kitchener-Waterloo and the heart of Toronto. This major road links Ontario to the US border at Michigan and to the Province of Quebec. Significantly, Highway 401 does not pass through

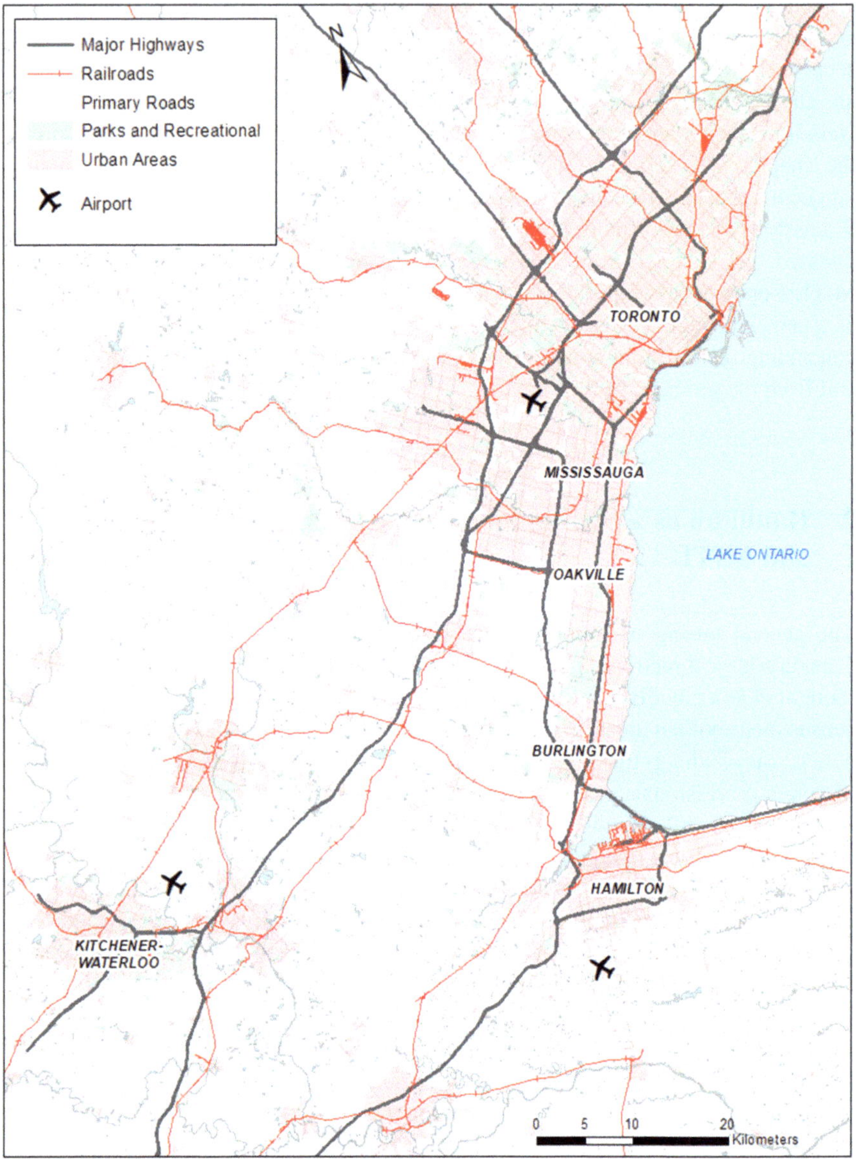

Fig. 1 The general setting of Hamilton within the region

Hamilton but is directly served by two major highways: the Queen Elizabeth Way (QEW) which follows the shoreline of Lake Ontario and links to the US border at New York State and to Toronto in the other direction; and secondly by Highway 403 which merges with Highway 401 in the west and with the QEW to the east. The system of major highways in the immediate vicinity of Hamilton achieves something of a "ring road" effect with the completion of the Lincoln Alexander Expressway in 1997 and the Red Hill Valley Parkway (RHVP) in 2007. These highways are city-run while other highways mentioned are run by the Province of Ontario. Another noteworthy piece of highway infrastructure is the Burlington Skyway: twinned bridges allowing the QEW to cross the western tip of Lake Ontario.

Focusing more specifically on Hamilton, one of its important distinguishing features is the Niagara Escarpment which bisects the city into an upper half and a lower half, separated by approximately 100 m of height. The upper half is affectionately referred to as "The Mountain" by residents which is an overstatement. The dotted outline of the escarpment is shown in Fig. 2. The effective separation into two cities has posed challenges for properly linking the road network. The completion of the RHVP, for example, through an environmentally sensitive area, caused considerable controversy. Years in advance of the opening the RHVP, Pavlos and his team had been running traffic scenarios for the addition of the new expressway, seeking to understand the impacts on traffic and emissions.

In socio-economic terms, the suburban and outlying areas of the Hamilton metropolitan area, including Burlington are the most prosperous. Some of the earliest suburbs, which are in the eastern lower city close to the heavily industrialized areas, are more challenged in terms of socio-economic and health outcomes. Figure 2 reveals that much of the city is fairly low density in character. While there are pockets of high densities in the city core, much of the city is in the range between 2500 and 7500 residents per square kilometer. Among other aspects, these types of densities have implications for public transit and particularly its light rail form as will be discussed below.

3 Freight-Oriented Research

The original impetus for the formation of MITL came from the outside and from stakeholders who were quite focused on freight, logistics, and goods movement. From each participant organization came a 5-year commitment to provide "seed" funding for MITL to carry out work on projects that were agreed upon by the MITL advisory board. Original partners who populated the advisory board included: City of Hamilton, Hamilton Port Authority, Hamilton International Airport, and the CareGo Group—a trucking and logistics firm based in Hamilton. As such, the first few years of MITL's existence were much more focused on freight than had been the case with Pavlos' research; notwithstanding the exceptions of Kanaroglou and Buliung (2008) which studied the emissions associated with commercial vehicle movements and Woudsma et al. (2008) which sought to explain patterns of logistics land use. If anything, research at Pavlos' Center for Spatial Analysis had been more focused on the movement of firms than the movement of goods—Maoh and Kanaroglou (2007) had studied the issue of firm migration while Maoh and Kanaroglou (2006) assessed clustering of firms and its relationship to urban form.

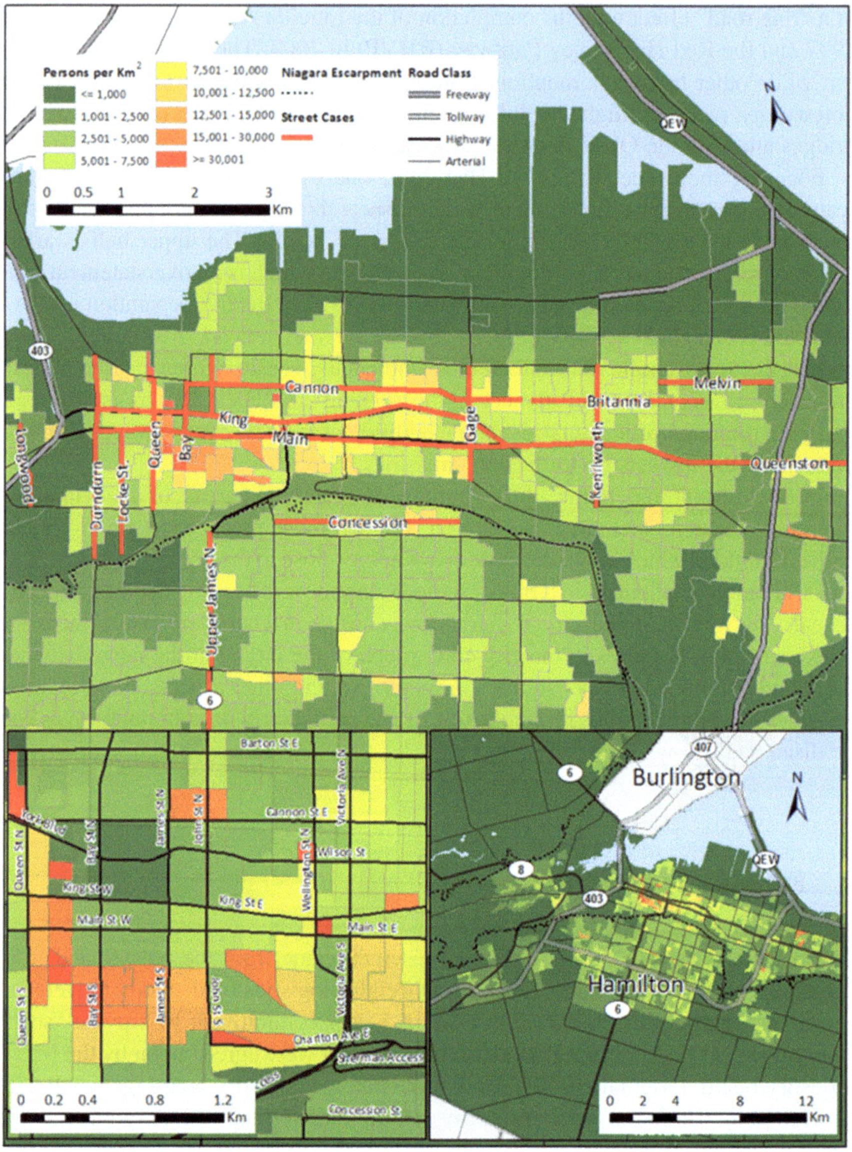

Fig. 2 Hamilton's local setting and residential densities

Up until MITL's first major project, considerable expertise had been built up, in particular, in the field of integrated urban modelling including the estimation of vehicular emissions. Scott et al. (1997), for example, used the team's first system (IMULATE) to assess how greater commuting efficiency in Hamilton (smaller average trip lengths) would reduce emissions. A later version of IMULATE, in

Behan et al. (2008), was used to assess the strategy of "smart growth" in Hamilton. Prominent in the discussion was the proposition that increasing population densities within an urban core can reduce overall traffic congestion and emissions.

3.1 A Sustainable Strategy for Developing Hamilton as a Gateway

Most of MITL's freight-oriented research included Hamilton, or was relevant to Hamilton, but was not directly focused on Hamilton. An exception in this regard was MITL's first major research project which was launched in April 2009. The work was presented by Pavlos himself at a large event at Hamilton's waterfront center. The report was entitled: "A Sustainable Strategy for developing Hamilton as a Gateway" and its content was certainly influenced by many aspects of the research of Pavlos and his teams. The integrated urban model IMULATE proved prominent in the analysis.

While the report emphasized Hamilton's potential to develop further as a goods movement hub, the research was not prototypically freight oriented. Concepts of sustainability were highly prominent and a great deal of attention was paid to issues such as urban form and the emissions impacts of more intensive development of the industry and freight sectors. Significant effort was also dedicated to measures taken by other gateways or goods movement hubs and whether similar approaches could work in Hamilton. Key modelling results of the MITL report were also captured in Ferguson et al. (2012a). The report emphasized the use of scenarios and assessed their impacts as opposed to making bold predictions about levels of increased economic activity.

Figure 3 shows an illustration of scenario outcomes from the study in terms of NO_x. IMULATE was run for four different scenarios at three different points in time. This included baseline assumptions about the growth of population and employment in Hamilton in the case of the business-as-usual scenario and the addition of about 60,000 new jobs in the case of "gateway" scenarios. The gateway scenarios differed on how the growth was implemented. A sprawled scenario assumed considerable dispersion of residential growth whereas a compact scenario assumed the opposite and a high degree of intensification of the urban core. Even in 2008–2009, prospects for light rail were very much under discussion and, accordingly, a scenario was devised as a combination of a compact city and an implemented light rail system which would help to get more people out of their cars. Results showed that even while absorbing a large increase in jobs, the compact-LRT scenario would not ultimately generate a great deal more in emissions than the business-as-usual scenario in which focused gateway development did not take place.

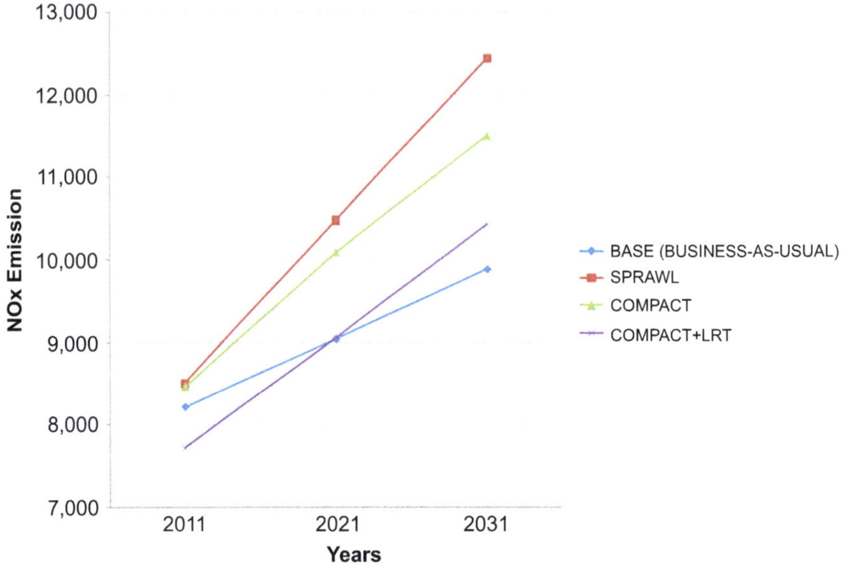

Fig. 3 Daily AM peak NO*x* emissions (in kg) in Hamilton Census Metropolitan Area under varied scenarios

3.2 Estimating Urban Commercial Vehicle Movements in the Greater Toronto Hamilton Area

This was MITL's second significant freight-oriented project (delivered in July 2010) and it was carried out for Metrolinx, a provincial transportation agency which had first been formed in 2006 to help solve major transportation problems in the GTHA. Metrolinx developed a major transportation plan called "The Big Move" and aspects of this plan were focused on goods movement. To help improve the understanding of commercial vehicle movements in the study region, MITL generated estimated origin-destination matrices of commercial vehicle movements for light, medium, and heavy commercial vehicles. This was a project in which Pavlos was particularly interested because it involved a heavy dose of discrete choice modelling, which alongside microsimulation, was one of his interests and specialities. Since early in his career, Pavlos had collaborated with some of the pioneers in choice modelling (Ben-Akiva et al. 1986). As well as generating results for the entire study region, this project generated outputs specifically relevant to small transport analysis zones in Hamilton.

Three types of urban commercial vehicle movements were estimated: tour-based, fleet allocator, and internal-external. Tour-based movements are those associated with a commercial vehicle leaving an establishment and carrying out a series of stops to deliver/pick-up goods or provide services. Fleet-allocator movements are involved with systematically covering a territory within an urban area and are associated with functions such as mail delivery, courier services, garbage, and recycling pick-up, and others. Internal-external movements include trips that start in the GTHA and terminate elsewhere, trips which start elsewhere and terminate within the GTHA, and trips which start and finish elsewhere but pass through the GTHA in the process.

To characterize the tour-based movements in particular, a microsimulation framework was applied and this was based on transferring statistical parameters from the work of Hunt and Stefan (2007) in Calgary, Alberta. The tour generation piece of this framework was driven by the firm-level InfoCanada data set which contained information on approximately 185,000 establishments located in the GTHA. Data from the Peel Commercial Vehicle Movement Survey of 2006–2007 carried out by Dr. Matthew Roorda and his team at the University of Toronto was also indispensable for this analysis. The results of this research were captured also in a peer-reviewed article (Ferguson et al. 2012b).

Several years after this study was completed, researchers in the region are still fundamentally relying on some of the same data sources for vehicle movements that were in place years ago. In Ontario, the two major sources are the trip-based Commercial Vehicle Survey, which is an intercept survey, led by the Ontario Ministry of Transportation and the Trucking Commodity Origin Destination Survey led by Statistics Canada. The latter samples carriers in the business of moving goods for other firms. Emerging more and more into the picture are data linked to GPS sources and telematics devices which promise to vastly improve the collective and understanding of commercial vehicle movements in this region.

3.3 An Exploration of the Freight Village Concept and Applicability to Ontario

In 2010, the Ontario Ministry of Transportation offered 5 years of research funding to support MITL's research on topics of interest to the Ministry. The leadership of Pavlos was instrumental in achieving this outcome. The first study, on freight villages, was proposed by the Ministry and it was released in October 2011. The study was a departure in the sense that there was nothing in the past work of Pavlos or his teams that addressed this topic. While offering insights for Hamilton, the study was not focused on Hamilton. The report was successful in generating attention and interest from the Ministry, and many freight stakeholders in the region and MITL received, and indeed continues to receive, inquiries/questions from other jurisdictions about the research.

A freight village was noted to have the following elements:

- A localized cluster of transport and logistics facilities which are co-located and coordinated for synergies

- Among the facilities is an intermodal terminal located near container storage, handling areas and warehouses linked to rail to reduce cargo handling costs and time and reduce the use of roads for containers
- Access to shared facilities, equipment, and services (e.g., customs services, truck cleaning, post office, conference, and training facilities and other services and amenities)
- Centralized management and ownership structure for long-term planning, investment, governance, environmental management, and other issues

Of the four points, the latter two are most powerful in differentiating freight villages from other logistics centers. The concept of "shared access" is fundamental and indeed a freight village acts as an incubator for small and medium size firms. A freight village can raise the likelihood that smaller firms utilize intermodal transport. The list of potential shared services is potentially very large and can range from those associated strictly with the movement of goods to those associated with serving either the employee base or a nearby residential population. For example, daycare and restaurants can be part of the mix in certain circumstances.

The report made an important distinction between different "orders" of freight villages. A higher order of freight village is one that is referred to as "integrated" meaning that modal changes occur on-site as part of a suite of logistics processes. The "non-integrated" version will not accomplish modal changes on-site. While an intermodal terminal will be nearby, it is not well integrated with the rest of the development. The non-integrated case is likely to come about when a development progresses incrementally rather than having been master planned.

One significant contribution from the research, which also is presented in Higgins et al. (2012), was based on the observation that too many terms were being used interchangeably in the freight literature to describe various types of logistics centers. As such, a standardized hierarchy of logistics centers was developed (Fig. 4), using very specific definitions of terminology. It was concluded that the freight village concept ranked very highly on the hierarchy and that there could be a possibility to develop such a complex in Ontario. Such discussions are continuing in the region to this day.

3.4 Seaway Under-Utilization: Are Regulations to Blame?

The Great Lakes-St. Lawrence Seaway is an engineering marvel of its time. It allows ocean-going and domestic vessels to travel deep into the North American interior. A set of locks on the St. Lawrence River allow movements between Montreal and Lake Ontario; and, via the Welland Canal, to bypass Niagara Falls for movements between Lake Ontario and Lake Erie. The Seaway does not operate near full capacity and accordingly the purpose of this MITL report was to assess the extent to which the regulatory environment might explain this under-utilization.

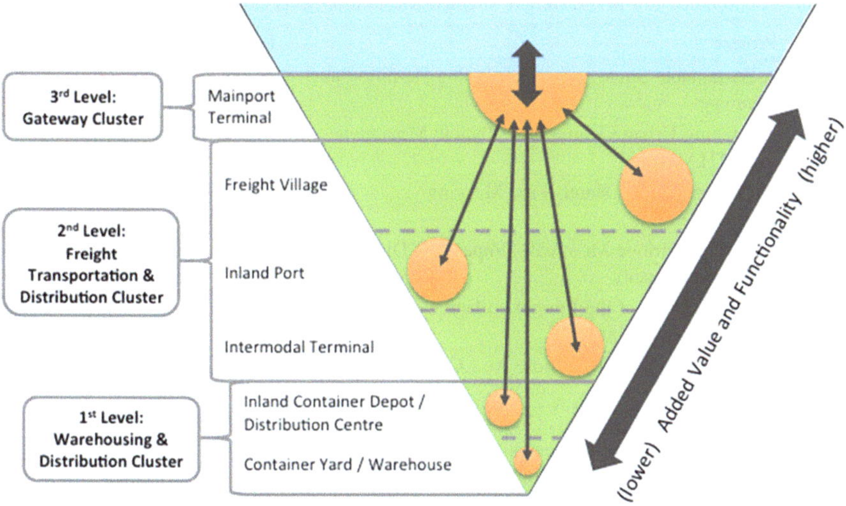

Fig. 4 Standardized logistics center hierarchy

The study found a series of factors to explain the phenomenon and regulatory factors were not at the top of the list. The most important aspects that reduce demand for the Seaway, especially for higher value goods, are the relative dominance of rail and trucking along the Montreal-Toronto corridor and the fact that the Seaway, unlike the Port of Montreal, is closed for the winter. Most of the really important logistics infrastructure for the region is located close to suburban highways (see Fig. 1) as opposed to marine ports more oriented to central cities. For the Port of Hamilton, and the Seaway as a whole, there is a dependence on bulk commodities. Important commodities for Hamilton include iron ore, which supports the local steel industry, and agricultural products, which have increased in prominence. The time of Seaway closure during the winter at the Port is essential for the berthing of vessels as opposed to cargo movements and this is something that meaningfully reduces the potential for marine to achieve larger mode shares in Ontario.

3.5 Other MITL Freight Projects

Over the years, freight-oriented themes have made up a significant share of MITL's projects and a comprehensive list is given in Table 1. Many of these have not been discussed in this chapter but most reports are available for download at mitl.mcmaster.ca. Pavlos was involved with each of these projects as a sounding board for the research team and also attended many meetings in support of the research either in-house or off-site.

Table 1 A chronology of MITL freight-oriented research projects

Year	Project	Partner/client
2009	A Sustainable Strategy for Developing Hamilton as a Gateway	Various Partners
2010	Estimating Urban Commercial Vehicle Movements in the GTHA	Metrolinx
2010	An Assessment of Hands Free Mooring	St. Lawrence Seaway Management Corporation
2010	Champlain Bridge Montreal—Impacts of Disruptions to Bridge Capacity	Federal Bridge Corporation
2011	An Exploration of the Freight Village Concept and its Applicability to Ontario	Ontario Ministry of Transportation
2011	Delivery Route Optimization: An LCBO Case Study	Liquor Control Board of Ontario (LCBO)
2011	Movements of Dangerous Goods Across the Credit Valley Conservation Watershed	Credit Valley Conservation
2011	Green Technology and Trucking	Transcare Logistics Corporation
2012	Hamilton Truck Route Study	City of Hamilton
2012	Seaway Under-Utilization: Are Regulations to Blame?	Various Partners
2013	Maximizing the Potential of the Foreign Trade Zone Concept in Canada	Various Partners
2014	Truck Freight Generators and Attractors in the Province of Ontario	Ontario Ministry of Transportation
2015	An Evaluation of the Ontario Green Commercial Vehicle Program	Ontario Ministry of Transportation
2017	A Set of Strategic Freight Performance Measures for Ontario	Ontario Ministry of Transportation

4 Traffic Congestion, Multi-Modalism, and Urban Form

Pavlos had long been interested in how land uses and transportation form the foundation of urban spatial structure and he believed that land uses, especially sprawled ones, explained a good share of the traffic congestion problem. He thought that while congestion is also a problem in very dense areas, these same dense areas provide options for other modes of travel. In the sprawled but congested context, other modal options are much more difficult to apply. Early in his career, Pavlos worked in the private sector in Mississauga, which is a generally sprawled area immediately west of Toronto (see Fig. 1). Decades later, when driving to meetings in this vicinity, Pavlos would openly criticize the sprawled development he was seeing and the negative implications that he had seen unfold in the intervening years.

With his focus on the nature of travel in sprawled circumstances, it is not surprising that Pavlos authored several papers on the topic of "excess commuting," which examines the balance between the locations of a city's jobs and its residents as a measure of the efficiency of a city's urban form. This includes the use of IMULATE to investigate the issue in Hamilton (Scott et al. 1997) as was mentioned earlier, the difference between a required minimum theoretical commute and observed commuting flows in the Greater Toronto and Hamilton Area (Buliung and Kanaroglou 2002), and a multi-city comparison of several excess commuting benchmarks with urban form across Canada (Chowdhury et al. 2012). Finally, Pavlos' most recent contribution to this area came through a comprehensive review, critical comparison, and assessment of the policy applications of the recent methodological developments that had occurred in the excess commuting literature over the course of his career (Kanaroglou et al. 2015).

Other aspects of Pavlos' research focused on multi-modalism in people movement and mobility for vulnerable populations, which also ties into aspects of complete streets. This work includes: Newbold et al. (2005) which showed high levels of automobile reliance among older Canadians, a spatial analysis of the trip-making behavior of older cohorts in the Greater Toronto and Hamilton Area in Páez et al. (2007), and the development of the IMPACT integrated urban model, which utilized a geographic information system to simulate how demographic changes and an aging population affect transportation in Hamilton (Maoh et al. 2009).

4.1 Metropolitan Traffic Congestion in Canada

Pavlos' considerable experience in relevant areas was instrumental in securing a significant grant from the Social Sciences and Humanities Research Council of Canada in 2013 to study the topic of metropolitan freight congestion in Canada. Traffic congestion is a major issue within the GTHA with the worst problems occurring in Toronto and along its major highways. In the case of Hamilton, the Mountain in particular, experienced significant sprawl over time which has been problematic since highway links to the lower city have become more and more congested. There is also little doubt that the highway links that have been added have induced more development and indeed more traffic. Since 2005, sprawled development on the Hamilton Mountain and elsewhere in the metropolitan complex is being controlled by a greenbelt and also the provincial "Places to Grow" Act which stresses intensification in development. Population on the Mountain continues to increase, albeit at a lower rate than might have otherwise been the case.

One useful tool in the study of traffic congestion is data enabled by GPS and smartphones. MITL acquired Canada-wide traffic speed data from INRIX Corporation, headquartered in Seattle, when it became available. Based on such data, Fig. 5 is a 2014 heat map of average traffic speeds along the Highway 403 corridor. This is a very important route for Hamilton. The rows are associated with locations and columns are associated with time of day. The eastbound direction, associated with trips toward Toronto, forms the top half of the figure and westbound travel is covered by the bottom half. Results suggest that PM congestion is generally worse than AM congestion along this corridor. Westbound PM congestion through the Burlington parts of the Hamilton metropolitan area is particularly bad.

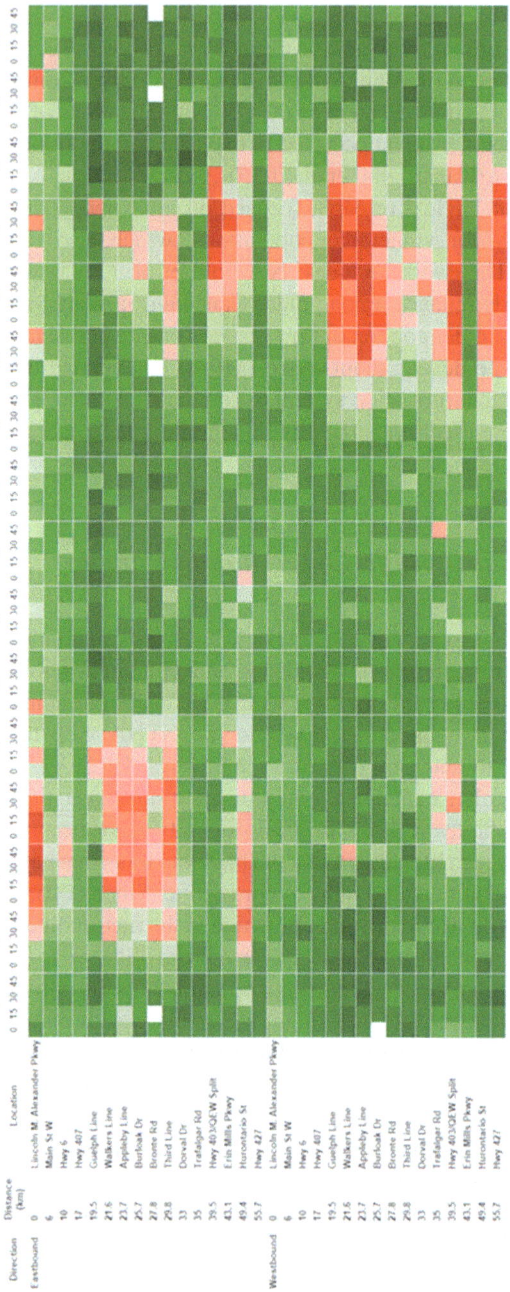

Fig. 5 Heat map of speeds on Highway 403 corridor

The pattern is reversed up on the Niagara escarpment where Highway 403 joins the Lincoln Alexander Parkway. Here, there is a serious eastbound AM bottleneck for traffic travelling in the direction of Toronto but this same location is much less congested during PM in the westbound direction. Part of the explanation is road capacity: only two lanes on Highway 403 descend the escarpment eastbound while three lanes are used to climb the escarpment westbound. The geography of the escarpment would appear to exclude the possibility to add more lanes in the eastbound direction. Meanwhile there remains considerable residential development taking place on the escarpment which is likely to fuel further deterioration in the eastbound flow of traffic.

Congestion in the Greater Toronto Hamilton Area is sufficiently severe that accidents and their aftermath can cause gridlock. When a dump truck, driven inexplicably with its bucket in a raised position, collided with one of the bridges that makes up the Burlington Skyway in 2014, the damage to the bridge caused its closure for several days along with one direction of travel for the Queen Elizabeth Way. Noting that events such as these put tremendous traffic congestion stress on an already burdened system, Pavlos was quoted in the local Hamilton Spectator as saying that the event had caused something of an "economic heart attack" for the region.

Although Pavlos' passing has shortened the funding window for this research, MITL's work on metropolitan traffic congestion has produced some useful outputs. Sweet et al. (2015) used INRIX data to carry out a spatial analysis of congestion intensity in Toronto. Recent papers have examined the health and equity impacts from traffic congestion, such as how gender differences and travel time affect the subjective well-being of commuters in Canada (Sweet and Kanaroglou 2016) and how an individual's satisfaction with their commute deteriorates as travel time and levels of congestion increase (Higgins et al. 2017b).

4.2 *Shaping Hamilton with Complete Streets*

With regard to complete streets, a quote from Pavlos echoes: "In Athens, every street is a complete street." While this may have been an exaggeration, the statement reflected his fundamental belief that concentrated development and high densities solved a lot of problems in big cities. The statement could certainly not be considered true for Hamilton as a whole. To look into the ways of raising the profile of complete streets in Hamilton, the study "Shaping Hamilton with Complete Streets" was launched in June 2015 and combined an extensive literature review with an analysis of some specific "street cases" that were selected by the City for potential treatment. Interestingly, one of the leading champions of complete streets for Hamilton was the Chamber of Commerce. Pavlos and the research team participated in several meetings with Chamber personnel and other stakeholders as the project developed and progressed.

Fig. 6 Main street in Hamilton: an arterial one-way expressway?

A particular sore point from the Chamber perspective was Hamilton's core city system of one-way streets that had been implemented in the mid-1950s when it was thought that maximizing vehicle throughput was the ideal objective. Figure 6 shows the one-way and five-lane flow of traffic travelling east on Main Street from Highway 403 toward the downtown. Many in Hamilton consider these streets to function more like highways than arterials and the fact that they act as a gateway for the City is distressing to some. Unfortunately, these streets could never be reversed overnight to two-way travel as they were in 1956 because expensive ramp and overpass infrastructure associated with Highway 403 was built in the 1960s that assumes the one-way configuration. Although the city's plans for light rail transit restore two-way travel on much of King Street, the plan works around the legacy of these highway junctions.

Despite fears about dominance of the automobile along the Main-King corridors that bisect the lower city, Fig. 7 shows that multi-modal trip making is nevertheless present. It captures trips that originate in a narrow catchment area defined by Main and King Streets sourced from the 2011 Transportation Tomorrow Survey. Walking trips, possibly associated with trips to and from school, are quite spiked in AM and PM periods. The automobile-as-passenger trip can be thought of as a significant "mode" in its own right based on the results. Automobile-as-driver trips predominate but not to the extent that they do in other areas of Hamilton. Cycling trips make up a small percentage of the totals observed in the survey.

With this background in mind, consider that a complete street is essentially a marketing term that bundles together an array of concepts. At a high level, a street

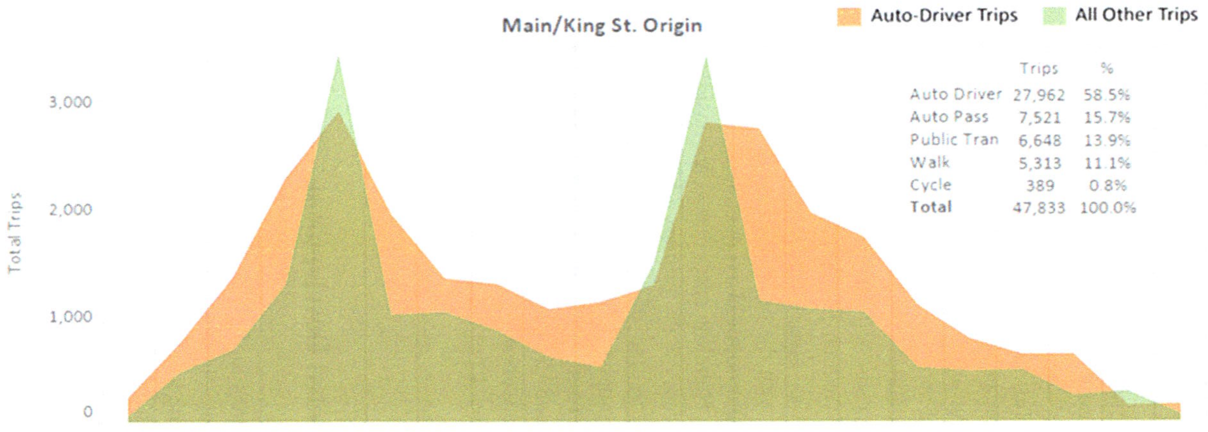

Mode	4 AM	5 AM	6 AM	7 AM	8 AM	9 AM	10 AM	11 AM	12 PM	1 PM	2 PM	3 PM	4 PM	5 PM	6 PM	7 PM	8 PM	9 PM	10 PM	11 PM	12 AM
Auto Driver	229	743	1,353	2,277	2,886	1,933	1,330	1,272	1,038	1,109	1,267	2,774	2,728	1,931	1,717	1,081	758	622	613	147	154
Auto Pass	37	183	244	428	1,040	441	466	335	134	209	303	724	627	524	646	327	345	261	130	117	
Public Tran	18	255	340	774	1,034	351	446	340	320	210	615	723	234	226	298	71	97	119	90	51	36
Walk		24	58	80	1,281	154	95	173	110	79	543	1,873	240	206	75	105	25	96		96	
Cycle			24		54	35			24		14	72	14	89					14	18	31
Total	284	1,205	2,019	3,559	6,295	2,914	2,337	2,120	1,626	1,607	2,742	6,166	3,843	2,976	2,736	1,584	1,225	1,098	847	429	221

Hourly % of Daily Trips for Given Mode 0.27% – 35.25%

Fig. 7 "Green" trips versus auto-driver trips for originations near the Main-King Corridor in Hamilton

that is complete is one that aims to make the immediate locality safer, more appealing, vibrant, and accessible for all users of the built environment, from children to the elderly. As opposed to promoting one strict type of built form, the term *complete streets* refers to a broad package of elements, and the degree of "completeness" of a street is determined not by the number or type of complete street elements in place but by the overall balance that is achieved between the users of an area and the activities that they engage in on the one hand, and the built form that supports these activities or, in some cases, suppresses these activities from taking place on the other.

Having said this, one of the main conclusions from the research was that complete streets are largely about managing the impact of the automobile to help other modes thrive. Research indicated that just about anything that reduces vehicle speeds is likely to improve safety. A vehicle speed increase from 30 to 50 km/h is associated with five times the level of pedestrian fatalities in the event of a pedestrian/vehicle collision. It is one thing to manage speeds and another to manage automobile dependence and the locational context in which people make their mode choices. Automobile ownership in Hamilton was something that Pavlos had studied (Potoglou and Kanaroglou 2008) along with the contexts in which Hamilton residents considered modes other than the automobile (Lavery et al. 2013).

Quite simply, some of the low-density Hamilton cases that were considered in the study are challenging complete streets cases if the primary objective is to reduce automobile dependence. A street case in one of Hamilton's upscale, outlying communities featured one of the best complete street implementations in the city but also had some of the highest shares of automobile travel. The implication is that while design helps, a lack of density and diversity in land uses can be hard to overcome if unfavorable for a complete streets implementation. In general, many of the more promising complete streets cases from the research were associated with the denser, lower city areas of Hamilton.

5 Light Rail and Transit-Oriented Development

One of the most prominent research themes for MITL and its work for Hamilton is on light rail transit and transit-oriented development. It has been a central issue in public discourse within Hamilton over the last several years. MITL's research outputs in this field for Hamilton and for the broader research community have been enabled by the collaboration between Pavlos, his last graduated doctoral student, and their colleagues at MITL.

5.1 *The North American Light Rail Experience: Insights for Hamilton*

With Hamilton actively planning its own light rail transit (LRT) line since the mid-2000s, MITL was engaged by the City of Hamilton to review experiences with LRT in other cities around North America. The report gathered important information and best practices related to ridership, transit-oriented development, and other issues and found that while more than 30 cities had built light rail, levels of success with the mode varied. In response, the team outlined key lessons for ensuring LRT in Hamilton achieves a high return on investment for the city, and these lessons have ultimately been incorporated into the city's revised east-west LRT plan that is scheduled to start construction in 2019.

With an estimated cost of $1 billion, planning for LRT is a complex undertaking, and the project has attracted a great deal of attention from planners, city councilors, the news media, and the general public in Hamilton. In this regard, outputs from the report were often utilized as a source of debate by various parties both for and against the project, and MITL quickly found itself at the center of some of the city's public discourse surrounding the topic. However, guided by Pavlos' determination to improve public discourse through fact-based academic research, the project team sought to bring clarity to the experiences of other cities and how they relate to Hamilton, making a number of presentations to the mayor, city councilors, and planning staff, as well as contributing to the local media over the past 5 years.

5.2 *Transit-Oriented Development and Value Planning*

Some of MITL's most recent research in the area of public transportation and transit-oriented development (TOD) examined land value capture as a policy and planning solution for financing rapid transit infrastructure in Southern Ontario. Land value capture is based on the idea of the "unearned increment"—if rapid transit infrastructure raises land values for private landowners around stations through greater accessibility and land use planning, there is a rationale for the public sector to engage in economic planning and recapture some of these positive externalities (Higgins et al. 2014; Higgins and Kanaroglou 2016).

From this, the team prepared two reports for policymakers in Ontario. The first is a consulting report (not shown in Table 2 below) completed in association with CPCS Transcom Limited for the Ministry of Transportation of Ontario that collected the available evidence to provide a rationale for engaging in value capture alongside the Province's new investments in public transit infrastructure. The second (see Table 2) was completed in 2016 through the Ministry of Municipal Affairs and Housing of Ontario's (MMAH) Places to Grow Implementation Fund. Using a latent class method for capturing the built environment context of different transit station areas, the project distills information on the accessibility and land use characteristics

Table 2 A chronology of assorted MITL research projects

Year	Project	Partner/client
2010	A Vision for Hamilton's Future	City of Hamilton
2011	Hamilton's Rapid Transit Future: The Role of Light Rail Transit	City of Hamilton
2012	The North American Light Rail Experience: Insight for Hamilton	City of Hamilton
2014	Estimating Vehicular Emissions for the Toronto and Hamilton Census Metropolitan Areas	Environment Canada
2015	Transportation Concepts for the Burlington Innovation District	City of Burlington
2015	Shaping Hamilton with Complete Streets	City of Hamilton
2015	Toronto Road System Performance—Stretching the State of Knowledge	City of Toronto
2016	The Development of a Transportation Demand Management Plan for McMaster University	McMaster Parking and Security Services
2016	Benchmarking, Planning, and Promoting Transit-Oriented Intensification in Rapid Transit Station Areas	Ministry of Municipal Affairs and Housing
2016	Children's Exposure to Criteria Air Pollution Due to Drop-off Programs at School	Ontario Ministry of the Environment and Climate Change
2016	The Safety Impacts of Red Light Cameras in the City of Hamilton	City of Hamilton
2016	Enhancing Transportation Corridors to Support Southern Ontario Innovation Ecosystems	Ontario Ministry of Economic Development, Employment and Infrastructure
2016	Connectivity Scenarios Along the Toronto-Waterloo Corridor	Ontario Ministry of Economic Development, Employment and Infrastructure
2017	Value Planning for New Highway Infrastructure in Ontario: Evidence and Extensions for Policy	Ontario Ministry of Transportation

of present and future rapid transit stations in the region, and uses this information to benchmark TOD. Contrasting these benchmarks with planning and policy allows municipalities and the MMAH to ensure supportive planning is implemented and achieve the goals of the Places to Grow Act.

An example of the spatial analysis framework used to measure transit-oriented land use and population and employment characteristics around rapid transit stations in Toronto from Higgins and Kanaroglou (2017) is presented in Fig. 8. Using data on real estate transactions, the team found that buyers of single-detached homes in Toronto were willing to pay a premium to live in more transit-oriented urban environments, including a premium of up to 20% and 23% for locations in *Urban* and *Suburban Neighbourhood* stations, respectively.

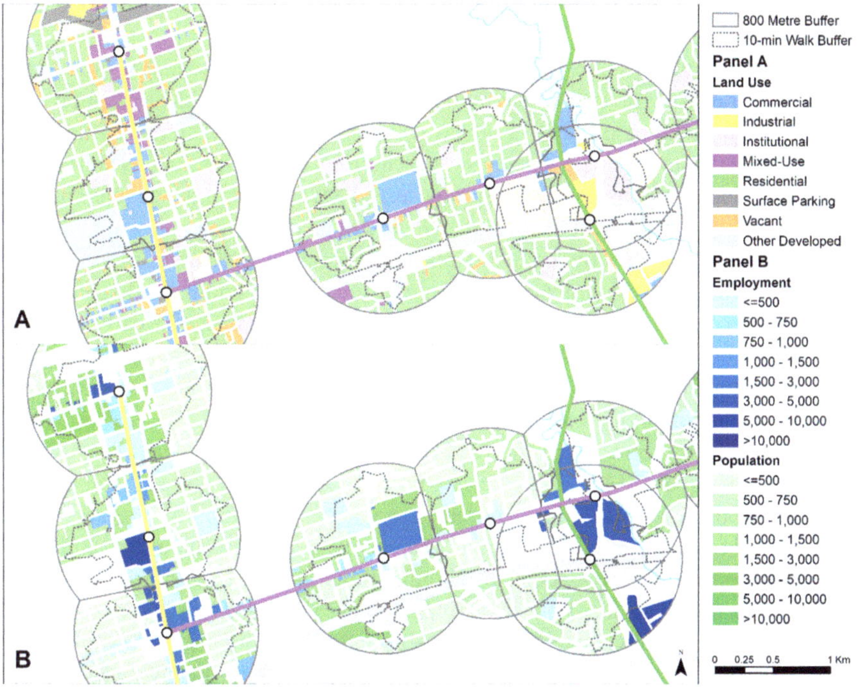

Fig. 8 Station area land use (Panel A) and population and employment (Panel B) in Toronto

6 Electric Mobility Research

In mid-2012, an opportunity was identified to pursue funding through Automotive Partnership Canada. Pavlos considered the success of our proposal and partnership for this initiative to be of the utmost priority. The basic form of the idea for the research had derived from a roadmap process led by Electric Mobility Canada which was the leading association of its type in Canada. Pavlos and the team ran with the basic concept and came up with a nine-module research proposal "*The Social Costs and Benefits of Electric Mobility in Canada*" with several industry partners and collaborators in place. The process to secure the funding took over a year and culminated with a site visit event at McMaster in August 2013 where Pavlos eloquently stated the case for our proposal. Pavlos and MITL were notified of the successful outcome in October 2013. Unfortunately, the second half of this 5-year project has been pursued without Pavlos as its leader.

Apart from generating useful results for the partnership and for electric vehicle stakeholders in Canada, the research has generated a number of useful academic outputs to this point. Mahmoud et al. (2016) carried out a review study of electric buses and a review of alternative powertrain options. Pavlos had strongly encouraged the pursuit of this study which has been well received. As the result of a national set of in-person and transcribed stakeholder interviews, the prospects for electric buses in Canada's public transit context are much better understood than before (Mohamed et al. 2017).

With respect to consumer adoption, Pavlos had not done a great deal of research in the general area of alternative fuel vehicles but one notable exception was Potoglou and Kanaroglou (2007) which conducted a stated preference survey of Hamilton residents to assess openness to acquiring alternative fuel vehicles. The experience gained in this consumer survey served MITL well in the development of a national survey instrument for Canada which was launched nationally in mid-2015 and gathered data from over 20,000 respondents via a consumer survey panel. Analysis of this survey is generating multiple outputs including Mohamed et al. (2016) which employed a structural equation approach to characterize early adopters and Higgins et al. (2017a) which assesses how preferences about vehicle body types (e.g., economy car versus pick-up truck) shape preferences for powertrains. Other publications are emerging on this topic.

Figure 9 shows some results derived from that survey for the Hamilton Census Metropolitan Area (CMA) that we wish Pavlos could have seen. The eastern boundary of the CMA is outlined in red. A synthetic population of households for all of Canada was developed to expand the survey in a representative manner and actual mapped results are associated with a national latent class choice model that was developed for Canada. Four latent classes were identified and two are depicted for the battery electric vehicle (BEV-oriented) and plug-in hybrid (PHEV-oriented) classes. For BEV transport, gasoline is not an option unlike for PHEVs where travel can be powered by electricity or gasoline. Clearly, the latter option is perceived in 2015 as offering a safety net.

The BEV-orientation is very much an urban phenomenon with little appetite for BEVs being detected in rural areas. Several factors are at play: range anxiety is felt most sharply in outlying areas, electric vehicle awareness is generally lower in rural areas, and rural areas have a larger share of the types of vehicles (e.g., pick-up trucks) where BEV options are essentially not available. The fact that rural areas are tending to have older and less educated populations is also relevant. In Hamilton, the BEV-orientation is strongest in close proximity of McMaster University and near the downtown core. The BEV mindset is notably weaker in the City of Burlington where perceived issues of range and long commutes into the Toronto region are presumably playing a role. The PHEV orientation is also stronger in urban areas but is much more suburban in character. This mindset is preferred in Burlington over BEV and is also seen as more palatable in rural areas.

7 Other MITL Research Projects

Table 2 below offers an overview of the large subset of MITL's work that has developed mostly outside the original freight-orientation of the Institute. The list is quite comprehensive and, for the sake of completeness, includes studies that were described earlier.

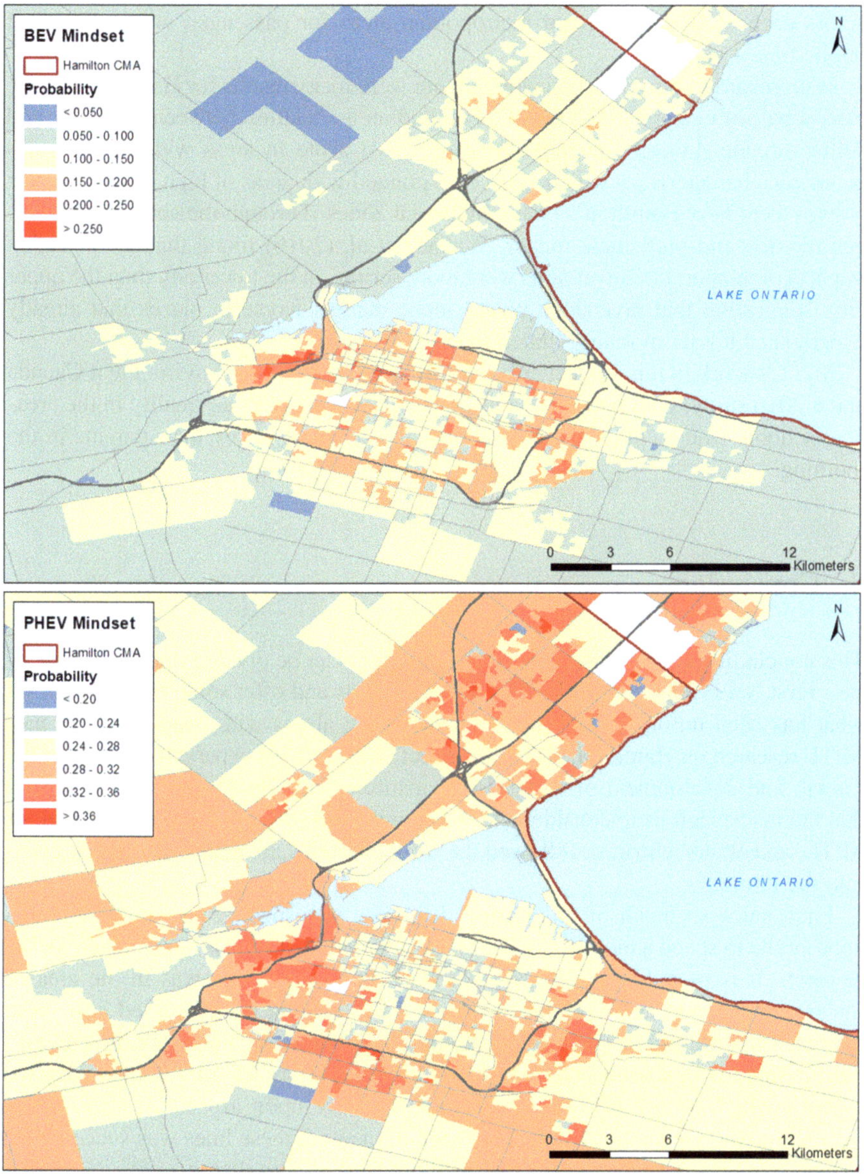

Fig. 9 Prevalence of BEV and PHEV mindsets in Hamilton Census Metropolitan Area

This chapter does not strongly emphasize vehicular emissions, environmental quality and linkages to health as the major aspects of Pavlos' research that they really were. But these aspects are also covered elsewhere in this volume. However, in the early years of MITL's existence, Pavlos led an effort through his Center for Spatial Analysis to estimate road segment-based emissions for a wide range of pollutants across each of Canada's major metropolitan areas for passenger, commercial and public transit vehicles.

With regard to his published works, there were many results for Hamilton and we offer a pair of examples. Jerrett (2004) found an association between mortality and sulfur dioxide. Impacts on mortality were most acute in areas with lower socio-economic characteristics which also corresponded with areas of high manufacturing employment near Hamilton's heavy industrial zones. Through the analysis of nitrogen dioxide and particulate matter, Wallace et al. (2010) found that the pollution impacts of temperature inversions were more serious in the lower city than the upper city, suggesting that inversions would make matters worse in places that already experienced higher mortality linked to emissions.

MITL's work in this subject area includes a 2014 study for Environment Canada and a 2016 study, both listed in Table 2. The latter considers air quality in the areas near schools where children are systematically dropped off by their parents in the morning.

8 Conclusions

This concluding section offers a discussion which can be divided into three categories. First, we briefly discuss Pavlos' research plans and offer some observations on what has been unfolding in Hamilton and how it aligns with conclusions of past MITL research on Hamilton. Next, we reflect on how these experiences relate to the growth and development of an academic institute. Finally, there are some lessons that Pavlos has left us to consider and which will hopefully help provide the fuel for MITL, and others who have followed the work of Pavlos, to make contributions well into the future.

First, since so much of this chapter has been retrospective in nature, it seems appropriate to spend a moment on the future direction that Pavlos had planned for his research. It is probably fair to say that his first research love was in the area of integrated transport and land use models and this chapter has discussed how these approaches were used in MITL research for Hamilton. In later years, he came to believe that the best avenue for progress was through the field of complex adaptive systems which he felt held the potential to improve upon aspects of agent-based microsimulation techniques. A research program along these lines was successfully funded but he was unable to proceed with the work due to illness.

With regard to Hamilton, MITL's research has coincided with an upswing in fortunes which of course is driven by much bigger forces than the Institute's research outputs. MITL's first major report was released at a time when Hamilton and much of the world were mired in a deep recession but nevertheless it painted an optimistic picture of what could happen if Hamilton emphasized a vital urban core over sprawled forms of development. Concepts that had been advocated in multiple MITL research reports have been aligned with provincial and municipal policies and are being implemented, most noticeably in the central city. At present, the city is experiencing a wave of new investment in its downtown core, where offices, commercial, and residential development and multiple generations of people have combined to increase vitality. Light rail transit, first written about by MITL in 2009, will break ground in Hamilton in 2019. While further decentralized development has occurred elsewhere in the city, it has been at lower rates and higher densities than in the past. But traffic congestion does appear to be gradually getting worse on the area's major highways.

On the freight side, it became clear through research that freight movements operate on a large spatial canvas. There is a tendency for strong agglomeration effects in the formation of logistics clusters and the one that is centered on Peel Region in Ontario is of an impressive scale. It is fueled by excellent access to a major market, extensive highway infrastructure, and strategic access to indispensable rail and air infrastructure. However, its success has generated problems such as heavy traffic congestion that persists throughout much of the day. There may be an opportunity for Hamilton to further capitalize on its competitive advantages and strengthen its position as a regional goods movement hub.

In developing an Ontario-based academic institute, and Pavlos knew this fact well: there is little time for complacency. MITL, as an example, operates without dedicated sources of funding and must earn its privilege to carry out research projects on an ongoing basis. In-kind forms of support from McMaster University are nevertheless very helpful. Achieving a significant scale of operation is a very big challenge indeed. Research outputs from MITL are not particularly standardized (no two projects are alike) so in many ways the Institute operates opposite to a business which would seek to generate profit through a repeatable formula. Partnerships are an important lifeblood and there are many that must be maintained. However, often the types of people who can do high-quality research are different from the people who can maintain partnerships and generate interest from prospective partners. Sustaining engagement with partners over a long period of time and over long duration projects is an ongoing challenge. The casts of characters can change at partner organizations and, in any case, people are always time constrained. Occasionally, a study is requested that lacks a proper or focused research question and decisions must be made whether or how to proceed. These are the types of issues and challenges that Pavlos and the team wrestled with over his tenure as Director and similar issues persist today.

That said, MITL has accomplished much over the past 10 years and although Pavlos' term as Director has concluded, his legacy lives on through the Institute and its team. As this retrospective has shown, MITL has been home to many achievements, securing a number of funded projects, making contributions to knowledge on a variety of topics in transportation and logistics, and working with planners, policymakers, and the general public to improve the environment, increase social equity, and promote economic prosperity in Hamilton and the GTHA. In the course of this, the Institute has also developed a number of highly qualified personnel who have found success in the public and private sectors and the academy.

In terms of what we have learned from Pavlos, the importance of having strong determination and resolve has to rank near the top of the list. His marvelous leadership and motivational skills act as an aspirational target for many. He reminded us of the importance of being humble, caring for others, enjoying the process of working hard and stopping once in a while for a good laugh. In his near decade at the head of MITL, it was clear that Pavlos hoped for Hamilton to become a sustainable city in the true sense of the word: socially, environmentally, and economically, and he did much in his last 10 years to help guide the city along this path. By the same token, he has offered much guidance for the future of MITL and for the futures of students and colleagues who had the pleasure to work closely with him. There is no better evidence of his impact than in the chapters that make up this volume.

References

Behan, K., Maoh, H., & Kanaroglou, P. (2008). Smart growth strategies, transportation and urban sprawl: Simulated futures for Hamilton, Ontario. *Canadian Geographer / Le Géographe canadien, 52*(3), 291–308. https://doi.org/10.1111/j.1541-0064.2008.00214.x.

Ben-Akiva, M., De Palma, A., & Kanaroglou, P. (1986). Dynamic model of peak period traffic congestion with elastic arrival rates. *Transportation Science, 20*(3), 164–181. https://doi.org/10.1287/trsc.20.3.164.

Buliung, R. N., & Kanaroglou, P. S. (2002). Commute minimization in the Greater Toronto Area: Applying a modified excess commute. *Journal of Transport Geography, 10*(3), 177–186. https://doi.org/10.1016/s0966-6923(02)00010-8.

Chowdhury, T. A., Scott, D. M., & Kanaroglou, P. S. (2012). Urban form and commuting efficiency: A comparative analysis across time and space. *Urban Studies, 50*(1), 191–207. https://doi.org/10.1177/0042098012452324.

Ferguson, M., Maoh, H., & Kanaroglou, P. (2012a). Simulating sustainable urban gateway development. *Transportation Research Record: Journal of the Transportation Research Board, 2269*, 135–144. https://doi.org/10.3141/2269-16.

Ferguson, M., Maoh, H., Ryan, J., Kanaroglou, P., & Rashidi, T. H. (2012b). Transferability and enhancement of a microsimulation model for estimating urban commercial vehicle movements. *Journal of Transport Geography, 24*, 358–369. https://doi.org/10.1016/j.jtrangeo.2012.04.013.

Higgins, C. D., & Kanaroglou, P. S. (2016). Infrastructure or attraction? Image-led planning and the intangible objectives of rapid transit projects. *Journal of Planning Literature, 31*(4), 452–462. https://doi.org/10.1177/0885412216667899.

Higgins, C., & Kanaroglou, P. (2017). Rapid transit, transit-oriented development, and the contextual sensitivity of land value uplift in Toronto. *Urban Studies,* 004209801771268. https://doi.org/10.1177/0042098017712680.

Higgins, C., Ferguson, M., & Kanaroglou, P. (2012). Varieties of logistics centers. *Transportation Research Record: Journal of the Transportation Research Board, 2288*, 9–18. https://doi.org/10.3141/2288-02.

Higgins, C., Ferguson, M., & Kanaroglou, P. (2014). Light rail and land use change: Rail transit's role in reshaping and revitalizing cities. *Journal of Public Transportation, 17*(2), 93–112. https://doi.org/10.5038/2375-0901.17.2.5.

Higgins, C. D., Mohamed, M., & Ferguson, M. R. (2017a). Size matters: How vehicle body type affects consumer preferences for electric vehicles. *Transportation Research Part A Policy & Practice, 100*, 182–201. https://doi.org/10.1016/j.tra.2017.04.014.

Higgins, C. D., Sweet, M. N., & Kanaroglou, P. S. (2017b). All minutes are not equal: Travel time and the effects of congestion on commute satisfaction in Canadian cities. *Transportation,* https://doi.org/10.1007/s11116-017-9766-2.

Hunt, J. D., & Stefan, K. J. (2007). Tour-based microsimulation of urban commercial movements. *Transportation Research Part B: Methodological, 41*(9), 981–1013. https://doi.org/10.1016/j.trb.2007.04.009.
Jerrett, M. (2004). Do socioeconomic characteristics modify the short term association between air pollution and mortality? Evidence from a zonal time series in Hamilton, Canada. *Journal of Epidemiology & Community Health, 58*(1), 31–40. https://doi.org/10.1136/jech.58.1.31.
Kanaroglou, P. S., & Buliung, R. N. (2008). Estimating the contribution of commercial vehicle movement to mobile emissions in urban areas. *Transportation Research Part E: Logistics and Transportation Review, 44*(2), 260–276. https://doi.org/10.1016/j.tre.2007.07.005.
Kanaroglou, P. S., Higgins, C. D., & Chowdhury, T. A. (2015). Excess commuting: A critical review and comparative analysis of concepts, indices, and policy implications. *Journal of Transport Geography, 44*, 13–23. https://doi.org/10.1016/j.jtrangeo.2015.02.009.
Lavery, T. A., Páez, A., & Kanaroglou, P. S. (2013). Driving out of choices: An investigation of transport modality in a university sample. *Transportation Research Part A: Policy and Practice, 57*, 37–46. https://doi.org/10.1016/j.tra.2013.09.010.
Mahmoud, M., Garnett, R., Ferguson, M., & Kanaroglou, P. (2016). Electric buses: A review of alternative powertrains. *Renewable and Sustainable Energy Reviews, 62*, 673–684. https://doi.org/10.1016/j.rser.2016.05.019.
Maoh, H., & Kanaroglou, P. (2006). Geographic clustering of firms and urban form: A multivariate analysis. *Journal of Geographical Systems, 9*(1), 29–52. https://doi.org/10.1007/s10109-006-0029-6.
Maoh, H., & Kanaroglou, P. (2007). Business establishment mobility behavior in urban areas: A microanalytical model for the City of Hamilton in Ontario, Canada. *Journal of Geographical Systems, 9*(3), 229–252. https://doi.org/10.1007/s10109-007-0043-3.
Maoh, H., Kanaroglou, P., Scott, D., Páez, A., & Newbold, B. (2009). IMPACT: An integrated GIS-based model for simulating the consequences of demographic changes and population ageing on transportation. *Computers, Environment and Urban Systems, 33*(3), 200–210. https://doi.org/10.1016/j.compenvurbsys.2008.10.004.
Mohamed, M., Higgins, C., Ferguson, M., & Kanaroglou, P. (2016). Identifying and characterizing potential electric vehicle adopters in Canada: A two-stage modelling approach. *Transport Policy, 52*, 100–112. https://doi.org/10.1016/j.tranpol.2016.07.006.
Mohamed, M., Ferguson, M., & Kanaroglou, P. (2017). What hinders the adoption of electric buses? Perspectives of transit providers. *Transportation Research D-Transport & Environment.*
Newbold, K. B., Scott, D. M., Spinney, J. E. L., Kanaroglou, P., & Páez, A. (2005). Travel behavior within Canada's older population: A cohort analysis. *Journal of Transport Geography, 13*(4), 340–351. https://doi.org/10.1016/j.jtrangeo.2004.07.007.
Páez, A., Scott, D., Potoglou, D., Kanaroglou, P., & Newbold, K. B. (2007). Elderly mobility: Demographic and spatial analysis of trip making in the Hamilton CMA, Canada. *Urban Studies, 44*(1), 123–146. https://doi.org/10.1080/00420980601023885.
Potoglou, D., & Kanaroglou, P. S. (2007). Household demand and willingness to pay for clean vehicles. *Transportation Research Part D-Transport and Environment, 12*(4), 264–274. https://doi.org/10.1016/j.trd.2007.03.001.
Potoglou, D., & Kanaroglou, P. S. (2008). Modelling car ownership in urban areas: A case study of Hamilton, Canada. *Journal of Transport Geography, 16*(1), 42–54. https://doi.org/10.1016/j.jtrangeo.2007.01.006.
Scott, D. M., Kanaroglou, P. S., & Anderson, W. P. (1997). Impacts of commuting efficiency on congestion and emissions: Case of the Hamilton CMA, Canada. *Transportation Research Part D: Transport and Environment, 2*(4), 245–257. https://doi.org/10.1016/s1361-9209(97)00015-1.
Sweet, M., & Kanaroglou, P. (2016). Gender differences: The role of travel and time use in subjective well-being. *Transportation Research Part F: Traffic Psychology and Behaviour, 40*, 23–34. https://doi.org/10.1016/j.trf.2016.03.006.

Sweet, M. N., Harrison, C. J., & Kanaroglou, P. S. (2015). Gridlock in the Greater Toronto Area: Its geography and intensity during key periods. *Applied Geography, 58*, 167–178. https://doi.org/10.1016/j.apgeog.2015.01.011.

Wallace, J., Corr, D., & Kanaroglou, P. (2010). Topographic and spatial impacts of temperature inversions on air quality using mobile air pollution surveys. *Science of the Total Environment, 408*(21), 5086–5096. https://doi.org/10.1016/j.scitotenv.2010.06.020.

Woudsma, C., Jensen, J. F., Kanaroglou, P., & Maoh, H. (2008). Logistics land use and the city: A spatial–temporal modeling approach. *Transportation Research Part E: Logistics and Transportation Review, 44*(2), 277–297. https://doi.org/10.1016/j.tre.2007.07.006.

An Ontological Framework for Spatial Socioeconomic Units

Athanasia Darra and Marinos Kavouras

Abstract Spatial units often serve as reference to socioeconomic phenomena constituting what is known as spatial socioeconomic units (SSEUs). SSEUs find a wide application in geography. This chapter presents the development of an ontological framework for the systematic definition of SSEUs on the basis of their properties, and particularly the spatial ones. Special attention is paid to the most common case of SSEUs, i.e., administrative units. SSEUs defined based on this framework can be effectively compared not only at the class level but also at the instance level. Furthermore, the restructuring of SSEUs and their unification into larger ones is greatly facilitated by the proposed property-based definition.

1 Introduction

Spatial units often serve as reference to socioeconomic phenomena, constituting what is known as spatial socioeconomic units (SSEUs). SSEUs find a wide application in geography. Two pivotal elements are combined to produce an infinite number of SSEU cases: geographic space as well as social and economic activities. This agrees with the definition of SSEUs provided by Frank et al. (2001:4):

> "*. . . socially created large geographic objects, used to subdivide space for administrative and economic purposes. Social processes, mostly political, subdivide geographic space in delimited areas, some as large as a continent, some as small as a parcel of land owned by a person.*"

A. Darra
Laboratory of Geography and Spatial Analysis, School of Rural and Surveying Engineering, National Technical University of Athens, Athens, Greece
e-mail: nancyd@mail.ntua.gr

M. Kavouras (✉)
Cartography Laboratory, School of Rural and Surveying Engineering, National Technical University of Athens, Athens, Greece
e-mail: mkav@mail.ntua.gr

H. Briassoulis et al. (eds.), *The Practice of Spatial Analysis*,
https://doi.org/10.1007/978-3-319-89806-3_3

SSEUs are represented at different scales. Most SSEU types constitute reference areas with crisp boundaries (e.g., a municipality) which are assigned a number of properties with their values. On the other hand, there are SSEUs the spatial extent of which is not predefined but is the result of property values (e.g., a service area). A slightly different SSEU type with a varying spatial extent is that of agglomerations (e.g., nomadic populations). The fact that SSEU concepts are defined quite variably and there are also differences in their instances, makes their comparison, evaluation, and association very difficult.

In order to understand and compare different typologies of SSEUs, it is important to examine the most important aspect of their definition; namely, their essential spatial but also aspatial properties/relations. Indicative spatial properties/relations include: location, shape (figure/compactness), structure (unified/fragmented), size, morphology-relief, adjacency, etc. More complex aspatial characteristics include: legal status, location stability, level in hierarchy, boundary type, life and history, motion, etc.

This chapter presents an ontological framework for representing SSEUs. Four ontological elements are used: concepts, properties/relations, axioms, and instances. SSEUs defined within this framework can be qualitatively compared on the basis of their properties. This is useful for a number of reasons:

- It reveals similarity and helps form clusters of similar SSEUs.
- It helps understand and deal with heterogeneity between different SSEUs.
- It facilitates the restructuring and unification of SSEUs.

The remaining of the chapter starts with an analysis of SSEUs and their behavior (Sect. 2). The identity of different SSEUs with special attention to administrative units is elaborated in Sect. 3. A list of key research issues in the representation of SEEUs is presented in Sect. 4. Out of these, spatial properties and boundaries are of great importance and are dealt with in Sect. 5. The ontological elements of SSEUs are presented in the ontological framework (Sect. 6). Finally, in the concluding paragraph (Sect. 7), the usage of the ontological approach in a number of tasks such as comparison, integration, spatialization, and space reorganization is advocated. Thereinafter, the abbreviations SSEU and SEU are used interchangeably, with SSEUs used mostly when the spatial aspect is more prominent.

2 Analysis of Spatial Socioeconomic Units

The term "spatial socioeconomic units" (SSEUs) was probably first introduced at the ESF GISDATA Specialist Meeting on "Formalizing and Representing Change of Spatial Socio-Economic Units in GIS," held in Nafplion Greece, on May 20–23, 1996, as well as in the volume that followed (Frank et al. 2001). At this meeting, an attempt was made towards a theoretical classification and identification of differences of SSEUs, mainly with respect to their spatiotemporal behavior.

The analysis of SSEUs leads to the following general observations:

- The establishment of SSEUs seems to meet a number of objectives that may not necessarily remain the same in the future. That is, the parameters that led to the configuration of a particular spatial situation may not fit the use of a SSEU for other purposes. For example, if the goal is to designate SSEUs exhibiting as much uniform infrastructure as possible (i.e., with substantial internal heterogeneity and little external heterogeneity), it probably does not serve to solve a problem that requires internally homogeneous (with respect to some parameters) areas. Hence, perspective (also known as context), whether it relates to initial creation or later use, plays an important role in the understanding of SSEUs.
- The spatial dimension is inextricably linked to the SSEUs. This is the case when the SSEUs are initially defined and their spatial extent is detailed. Once a SSEU is established, it affects all surrounding areas and thus creates a new "space."
- The multidimensionality of SSEUs makes it very difficult to develop a "generic" model that leads to their optimal definition. This is only possible for specific problems of a certain theme.
- The relatively indisputable elements that largely affect all kinds of analysis are those related to the spatial properties of SSEUs, including communication and accessibility.

The differences between the types of SSEUs, in relation to their use, are so many that it is extremely difficult, if not impossible, to be treated uniformly and effectively. This is because the terms used in the definition of SSEUs introduce new concepts such as "autonomy," "sufficiency," "standard of living," "services provided to citizens," "participation," "accessibility," and many others, which require further clarification. Thus, it is more realistic, but also practical, to include general features that apply to most SEU types and place emphasis on spatial SEUs, with particular interest in the administrative units.

In order to better understand the above, the rest of the chapter presents the main issues related to indicative examples of SSEUs and the key research questions that need to be addressed.

3 Identity and Examples of SSEUs

The primary characteristic of a SSEU is its socioeconomic dimension determined by the human activity in space. This includes the demographics of a region, social, cultural and historical data, economic data, property data, accessibility issues, living standards, infrastructures, participatory management, and others. Common SSEUs are census tracks, administrative units, design areas, zones, electoral wards, parishes, etc. The process of definition and spatial delimitation of SSEUs relates to the notion of *regionalisation*, that is, the need/trend to designate regions on the earth surface. The Lincoln Institute of Land Policy (LILP 2008) proposes the following combination: (a) "sense of place," (b) "problemshed," and (c) "people's interests." As a result, they use the following types: megaregions, ecological regions, hydrological regions, metropolitan regions, economic regions, administrative regions, service regions, and cultural regions.

As stated in the Introduction, it is quite common for a socioeconomic phenomenon to be studied within a predefined reference area, e.g., "average income in Luxembourg," "percentage of rural population in Sicily," or "social services in the municipality of Rhodes." On the other hand, a phenomenon can be studied without a predefined reference area, e.g., "urban fabric—city," "low income area," "epidemic outbreak points," and "hospital service area." In this second case, a spatial extent with its boundaries can be created a posteriori. Finally, in a more loose case of the latter, some SEUs can be defined without a direct/fixed relation to space. An example may be a "religious group," a "social group without permanent residence," or the "nomadic population of a country." In this case, where spatial properties are neither fixed nor always important or decisive for the identity of SEU, these SEUs can be considered as "non-spatial." Because these SEUs are defined on the basis of the "aggregates" of units having a common thematic property, another term used to connote them is "agglomerations." According to Smith (1999:267), "agglomerations are aggregates of entities that are dispersed through space on geographic scales."

A look at the above three cases, especially the first two, reveals that any difference between them may be occasional and one SSEU type may change to the other in the course of their lives. That is, the first case of units with a given spatial reference may have originated from the second or third case. And still, any subsequent merging of existing spatial SEUs into larger ones resembles case 2. Finally, at some point in time, agglomerations which are connected thematically but not spatially can undergo a transformation to become a spatially coherent unit, which again resembles SEU type two.

Below are specific examples of spatial and non-spatial SEUs.

3.1 Cities and Settlements

A very interesting case of SEUs, of great historical and geographical interest, is the Cities and Settlements. Their importance owes to a number of reasons. Obviously, the term "city" ("polis") is used here in its contemporary sense, as the Greek word "το άστυ" meaning "urban," and not as a civil society almost synonymous to "citizens," a sense that it had in antiquity (Hansen 2000).

There are various and certainly not uniform definitions of cities using a variety of criteria (Hubbard 2006). If we exclude the administrative importance they may have, for example, the capital of the administrative unit (state capital, county center, etc.), they can also be characterized by other criteria such as their population. Depending on countries, cultures, languages, and historical times, there has been a lot of ambiguity about the definition/separation between concepts such as city, town, settlement, locality, metropolis, village, suburban, pueblo, villa, ciudad, ancient city, city-state, colony, and divided city.

GaWC, a Research Network on Globalization and World Cities, uses a ranking similar to the "global city" (Sassen 2005), which has the main socioeconomic nature and refers to a city in which "the dynamics and processes that get territorialized are global" (p. 27). The world's top GaWC cities are classified as "global cities" and "world cities." The ranking of GaWC is based on specific criteria, which for 2008 were developed by evaluating 24 indicators in the following five areas: business activity, human capital, information exchange, cultural experience, and political engagement.

The concept of city as a basic SEU is very interesting and a broad research theme in itself. One of the main challenges is that there is no strictly agreed set of criteria for characterizing a city and evaluations vary according to cultural background, scientific assessments, and/or personal views. With the expansion of cities, the creation of suburbs and daily commuting, and the ensuing fuzziness of separating the urban area from its surroundings, the classical concept of city has weakened. This change, also known as the "urban sprawl," characterizing areas with low population density and high dependence on the car, has many socioeconomic implications. Recently, a shift has been developing in Europe towards redefining the concept of the city known as "new urbanism" or "urban renaissance" (Hall 2000). At the same time, other terms such as "intelligent city" and "digital city" are also being adopted.

In any case, the definition of these concepts, their essential qualities and, in particular, the spatial qualities of relations to be studied constitute an open field of geographic research which has not been adequately addressed yet. As a result, any analysis, comparison, and evaluation of different cities is ad hoc and not systematic.

3.2 Census Units

These units, internationally known as census tracts/areas/districts, are geographic areas designated for research or inventory. They usually have clear boundaries that at some level are linked to administrative boundaries. Although the latter is not mandatory, the fact that knowledge on administrative units is the basis of policy making, makes the linkage of census and administrative boundaries very practical.

3.3 Design Regions

These units are created for a specific purpose of studying a socioeconomic phenomenon. They are usually scientific research units designed to explore alternative scenarios, so their life span is limited, i.e., they are not permanent. In theory, many of the permanent SEUs have originated from design regions.

3.4 *Electoral Wards*

These units are created for the sole purpose of conducting political elections. Again, at some level of hierarchy, electoral districts are structured on the administrative divisions. Obviously, "objective" planning of electoral districts can often be circumvented by redistricting to serve political expediency in the election of representatives. This tactic is also known as *gerrymandering* (Monmonier 2001). Political expediency may go as far as to change the administrative subdivision.

Because electoral SEUs are based on different electoral systems, their comparison is an interesting subject. The definition of units also includes the setting of the International Electoral Standards (2002).

3.5 *Postcodes*

In many countries, areas designated by postal sectors (postal codes), beyond their classical utility for postal service, are now also used as reference unit for socioeconomic information. Non-postcode use of the postcodes includes: population censuses, marketing studies, health and education services, urban planning, emergency services, etc. (Raper et al. 1992). Dickson (2008:1) states: "*Postcodes are the country's greatest example of the Power of Information. I believe we would unlock significant power if we enshrined postcodes as our key national geography...*".

3.6 *Cadastral Units*

The SSEU type that is perhaps more directly connected to citizens is that related to cadastral units. These are characterized by large scale, limited size, and precise boundaries. They are also a key economic development element for both citizens and the administration. The existence of a multidimensional cadastral system (Dale and McLaughlin 1999) is a prerequisite of economic robustness and the criterion that differentiates developed from developing regions. Therefore, it is a key objective of agencies such as the World Bank. One of the main differences between cadastral units of different regions/countries is the way of defining and the priority assigned to the spatial description of their boundaries. These can be natural boundaries, demarcated boundaries with special survey monuments, described by coordinates of the perimeter points, etc. The significance and diversity of cadastral systems and the differentiation of their boundaries have been analyzed by Dale (1976).

3.7 Service Zones

A common category of SEUs is that created by the area covered by social services such as hospitals, schools, policing, ambulance, and fire-fighting. These areas do not always have clear boundaries, and they can also change according to socioeconomic situations (new facilities/technologies, budget fluctuations, demographics, commercial changes, etc.). A common example is school integration or change of area serviced (consolidation/redistricting). Such changes are becoming more common in recent years to reduce service costs and achieve economies of scale. For example, the change of service schools in the US MAINE State (MAINE.gov 2010) aimed at reducing hundreds of schools to less than 80 in order to reflect the drop in the school population. Service zones may have clear or fuzzy boundaries. A school area that does not allow students to move to other schools outside their own area has clear boundaries. A hospital, on the other hand, may serve people from different regions without being strict in their origin.

A special category of SEUs serving the proximity area of a church is that of parishes. Parishes are often referred to as small spatial administrative rural areas, sometimes as the community of the faithful within the reach of a specific church.

3.8 Administrative Units

These are spatial SEUs resulting from an administrative subdivision. Traditionally and principally, they refer to country subdivisions, but they also involve aggregations at a higher level (among countries). The terms used by different countries sometimes resemble, sometimes vary considerably. These are the most universal units in the sense that, despite their differences, they are found everywhere, they are widely accepted, and they are the basis for most inventories and analyses of geographical areas. They are also used as electoral constituencies but also in implementing any land policy. From the moment of their designation, administrative units "create new space," i.e., they create a framework where changes occur. At the same time, after their *de jure* definition, their societal acceptance of and adaptation to this subdivision creates an inertia and resistance to future changes. In the case of administrative reform, the usual case is merging upwards, i.e., aggregating small neighboring units to larger ones, an action that often causes negative reactions of local communities. As a result, administrative SEUs exhibit a notable spatiotemporal stability.

Typical administrative SEUs are: Country, Region, Prefecture, Metropolitan Municipality, Local Authorities (Municipality or Community), and Local (Communal or Community) division. In the case of SEUs that are superordinate to countries, such as unions or coalitions of countries (UN, EU, Commonwealth countries, defense alliances, etc.), there are significant differences in scale and detail.

Administrative SEUs are also important for attaching to them a plethora of descriptive features. FAO (Food and Agriculture Organization of the United Nations) uses the "Global Administrative Unit Layers (GAUL 2006)" system based on the UN Cartographic Unit co-ordination of co-operation between international organizations that maintain and update spatial information on administrative SEUs. GAUL, in addition to adopting known standards, uses specific procedures for updating SEUs, partitioning, naming, and identifying disputed areas including a geopolitical ontology (Kim et al. 2013), i.e., an ontology developed by FAO to describe, manage, and exchange data related to geopolitical entities.

In the United Kingdom, a typology of seven geographical levels (Administrative Units Typology—BRITAIN) (GBHGIS 2012) is used to record the evolution of administrative units from 1801 to 2001. The system is an extension of Alexandria Digital Library's Feature Type Thesaurus (ADL 2002) developed at the University of California in Santa Barbara.

Administrative SEUs with their boundaries are also subject to the visualization of Country Administration History such as Great Britain, Sweden, and Estonia, as presented in the development of an IST STREP (QVIZ 2008) research project. López-Pellicer et al. (2008) consider that proper documentation and comparison of administrative SEUs requires the use of appropriate ontological tools.

At the European level, mapping administrative SEUs has been dealt with in various projects. One of the first projects was SABE (Seamless Administrative Boundaries of Europe), a CERCO/MEGRIN product in co-operation with EUROSTAT. In SABE2004 (Fig. 1), 35 countries participated. This product was developed by Eurogeographics (a merger between CERCO and MEGRIN) to EuroBoundaryMap (EUROBOUNDARIES 2006), which is a single harmonized base of 1:100,000 scale geographic data including geometry, names, and coding of administrative and statistical units, coming from and constantly updated by the European Cartographic and Cadastral members of Eurogeographics.

The EBM v3.0 version includes 39 countries. In any case, the administrative units are also the basis for the INSPIRE Directive. In analyzing the administrative boundaries of countries including transnational boundaries, the EuroBoundaryMap Project (EUROBOUNDARIES 2006) shows different characteristics of boundaries with their legal features and disputed areas or controversial borders.

An extension of the SABE program was the ABDS (Administrative Boundary Data Services) project for the CEEC (Central and Eastern European Countries) that dealt with the preparation and partial implementation of online administrative data service. In addition to the common data model proposed, the program also envisaged a generalization process in order to harmonize data of different scales across countries. In addition to the CEEC countries (Bulgaria, Czech Republic, Estonia, Hungary, Latvia, Lithuania, Poland and Romania), two other EU countries participated—Finland and Greece (Fig. 2).

The description of the administrative units varies significantly from country to country. In advanced administration systems, detailed descriptions are used to significantly reduce the ambiguity that may result from generic descriptions or "common sense" approaches. The latter are useful, however, they can often create

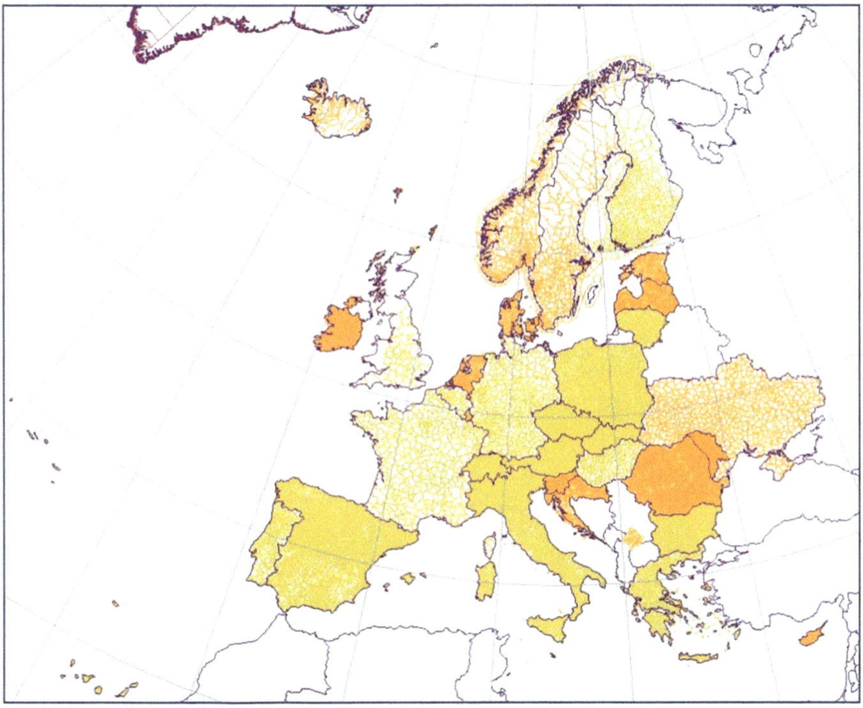

Fig. 1 Dataset SABE2004 (Seamless Administrative Boundaries of Europe) (SABE 2004)

problems because they are based on weak views of the type "it is known/obvious what is meant by the term ..." The Canadian Statistical Service uses a highly analytical but also complex system to describe its concepts. A typical example is the definition of census metropolitan area (CMA) and census agglomeration (CA) (STATCAN 2007). The disadvantage, on the other hand, is that highly specialized definitions, while describing in detail the content of a dataset, do not cover generic uses making it very difficult to align even slightly different categorizations.

4 Key Issues About SSEUs

In order to better represent and analyze spatial SEUs, it is necessary to identify a list of questions concerning their identity and behavior. Particularly important seem to be the following:

1. What is the main thematic feature that defines and identifies a SSEU?
2. What SSEUs are the most common and should be given priority?
3. How critical/important is the perspective/context in the diversity of SSEUs, and how can it be taken into account?

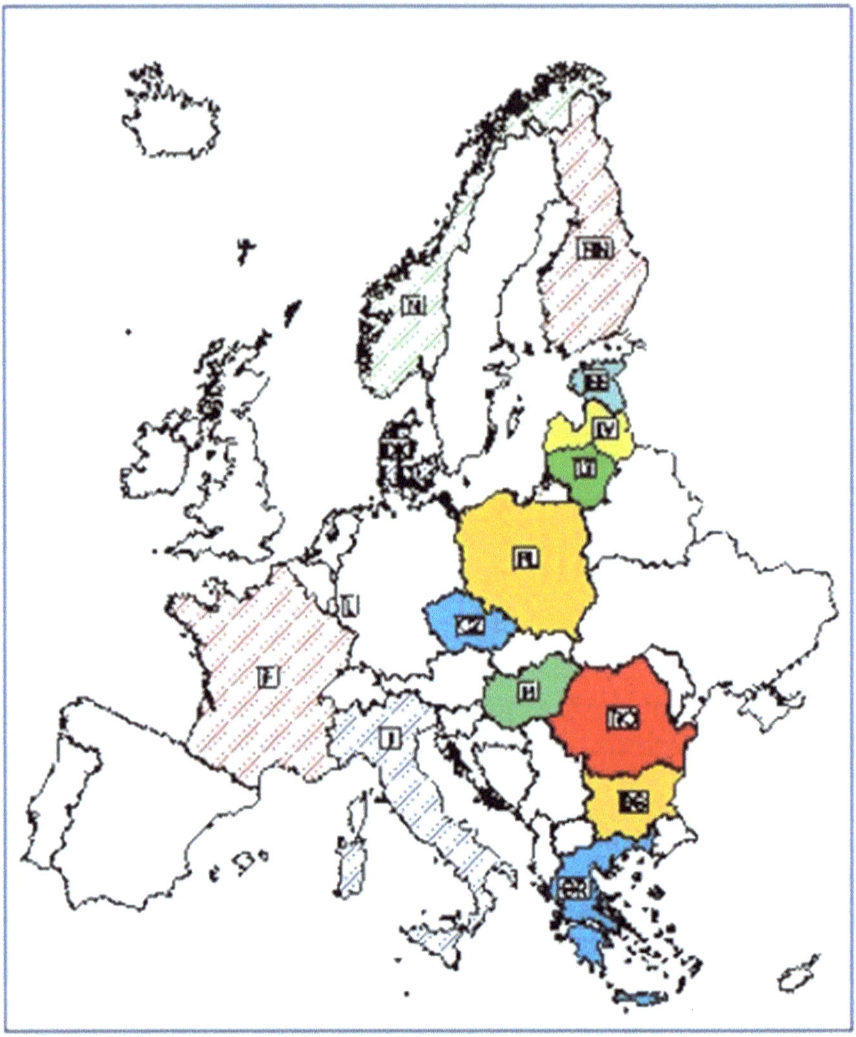

Fig. 2 Partner countries of project: ABDS (Administrative Boundary Data Services) for the CEEC (Central and Eastern European Countries)

4. How decisive for the identity of a SSEU are its spatial properties?
5. Are there some indisputably essential properties of SSEUs that directly or indirectly affect most of the other properties?
6. How can small SSEUs be combined/aggregated into larger ones?
7. If some properties change in a SSEUs' life, how does this affect the SSEU's identity? For what properties, and for what degree of change?
8. How can SSEU categories or instances be compared between different space organization systems? How should SSEUs be sufficiently defined so that they can be compared?

9. How can SSEUs and their qualities be visually and eloquently portrayed?
10. What is the role of vagueness/indeterminacy in the definition of a SSEU and its boundaries? Can, for example, a SSEU be certain while its boundaries are uncertain (and vice versa)?
11. The importance of scale and detail in observation, perception, and cognition. How "crisp" boundaries and "crisp" entities become vague and vice versa.
12. What are the strengths and weaknesses of IT systems and tools in addressing the representation, analysis, and management of SSEUs?

Each one of the above questions constitutes a full research topic. But the spatial aspects of SEUs and the special role of their boundaries seem to be quite important to understand and as such, they are further elaborated thereinafter.

5 Boundaries and Spatial Properties of SSEUs

As geographic concepts, SSEUs are characterized by their spatial properties to a significant extent. With their importance depending on the context, the following spatial properties may play a particular role, among others:

- location
- area
- compactness
- shape
- adjacency
- distance/proximity
- altitude
- slope
- aspect

Undoubtedly, one of the most important spatial concepts is the concept of *boundaries*. The lives of people, their claims, and often their disputes revolve around boundaries, spatial or not. As a consequence, SEUs as social constructs are also often based on boundaries and especially spatial boundaries (Couclelis 1996; Fleck 1996; Smith 1997; Smith and Varzi 1997). It is characteristic that border studies, especially national ones, involve many organizations and research units (IBRU 2018).

Two kinds of boundaries are distinguished: *bona-fide* (prominent/undeniable) and *fiat* (arbitrary/conditional) (Smith and Varzi 2000). They may be physically realized or be virtual/imaginable, and they also entail ambiguity (Burrough and Frank 1996). The nature of boundaries does not necessarily characterize the entity itself (Fig. 3). That is, there can be an entity with crisp recognizable identity, but some or all of its boundaries may possess ambiguity (e.g., mountain). The case of an entity with vague identity but clear boundaries is hard to document, and it is also a complex philosophical question (Weintraub 2004); unless, we consider something

		Entities	
		crisp	vague
Boundaries	crisp	street, land parcel	uninterpretable structure on a satellite image
	vague	city, mountain	cloud, pollution

Fig. 3 Vagueness of entities and boundaries

that we can clearly delimit but do not know what it is (e.g., a geometric structure appearing on a planet).

Ambiguity, an inherent feature of some boundaries should not be confused with the accuracy of their measurement. For example, if one was able to delineate a SSEU, its measurement would be very accurate. If however, there is any doubt as to what/where the boundary may be, then any surveying method, no matter how precise, will not reduce ambiguity.

In the case of three-dimensional boundaries, usually, but not always, the two dimensions (horizontal) precede the third (elevation). That is, the coordinates of the boundaries are determined, or their position is described, and the elevation/relief is simply an attribute. It may happen though, that a border definition is based on the relief (e.g., ridge); therefore, an elevation change shifts horizontally the boundary.

Another parameter that undoubtedly affects boundaries and their description is the fact that they vary in detail according to the observation scale. Especially in the case of changes in methodology and scale, minor to significant changes are expected.

A boundary change is something that may occur in the life of a SSEU. This change may be limited and/or insignificant, so it does not change the identity of the SSEU, but it may be extensive and/or important so that a new SSEU is created. However, even if the identity of the SSEUs is preserved, the spatial changes that take place must be recorded so that the life and the motion of the SEU can be monitored (Frank et al. 2001).

There are also weaknesses in software systems in representing boundaries. Some systems support multi-attribute boundaries, while others support multiple boundaries by replicating them in the database.

In the case of administrative units, boundaries are defined by the central administration, so they are mainly human creations (fiat boundaries). Regarding the usual case of cadaster and SSEUs, the boundary issue is crucial, as it affects properties and their owners. The preservation of the close attachment of people with their land property requires the preservation and security of boundaries. Critical questions here are:

- What is the prevailing element in the definition of a land property? Coordinates, natural boundaries, demarcation monuments, landmarks, or other?
- Does/should/can cadaster guarantee the land boundaries?
- What change of boundaries causes a property change?

Such problems are encountered everywhere in the world. For example, a series of many related borderline problems are described for Great Britain in "The Boundary Problems website" (Maynard 2016).

6 A Formal Ontology for SSEUs

In order to effectively represent SSEUs, it is necessary to standardize the knowledge that exists, or more precisely, is more relevant to them. The most common or popular approach to fully capturing the concepts of a scientific field or application in a standardized way is that of ontologies. The term ontology, especially when it is written "Ontology" (i.e., singular and capital "O") refers to the philosophical use of the term, the subject of which is the study of existence. In the practical adoption of the term "ontology" or "ontologies" (with lower "o") by information science, the focus is on documenting and specifying knowledge as understood in a particular domain. This is also the popular definition given by Gruber (1993:199): "An ontology is an explicit specification of a conceptualization."

Documenting the knowledge of a field through an ontology has many common characteristics with the use of terms and definitions in dictionaries and thesauri, with "typologies" and "categorizations," and with the "conceptual design" in databases. Whatever the commonalities, SSEUs constitute an important subset of geographical concepts that need to be described in a systematic way.

Common components of a formal ontology are the following (Fig. 4): concepts, properties, relations, constraints (axioms), and instances. Instances are not obligatory.

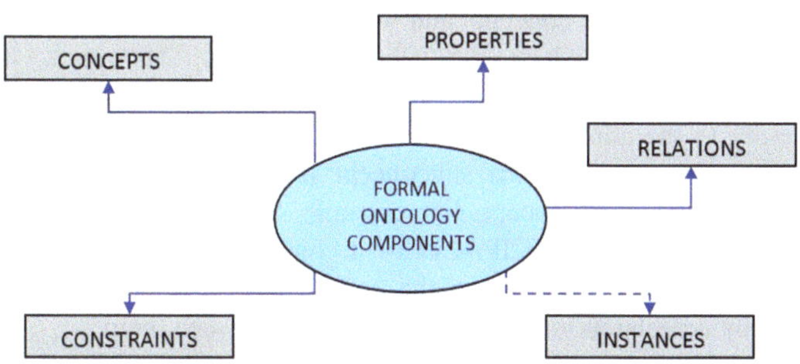

Fig. 4 Basic components of an ontology

Concepts are units of meaning on which categories of things are based. Often the notions of concept, category, and class are correlated and, if there is no other specific reason, they are used interchangeably. Every concept (e.g., city) is usually described by a term ("city") and a definition. The definition describes/delimits the concept using its main properties. For example, "A city is an urban area with a high population density and a particular administrative, legal or historical status." Concepts are associated by relations. The constraints that may exist, delimit concepts and properties. Finally, although it is not strictly necessary, since an ontology is based on conceptualization, an ontology may also contain instances (Athens, London, etc. as instances of the concept "city").

In the present case, it is necessary to describe the main concepts and relations describing spatial SEUs, especially those of an administrative nature. This issue is developed further below.

6.1 Concepts and Relations

There are numerous SSEU concepts (categories) with variable characteristics. Looking into SSEUs with administrative interest, some typical cases are:

- Country
- Region
- Prefecture
- Metropolitan Municipality
- Local Authorities (Municipality or Community)
- Local (Communal or Community) Division

Other common SSEU concepts are: cities, settlements, census sectors, design regions, electoral districts, postal sectors, tax sectors, cadastral parcels, parishes, etc.

Looking into the concepts often used, two important observations can be made. Firstly, the concepts expressed by the above terms, even the synonymous ones, differ significantly from country to country and cannot be easily associated. Secondly, the often used hierarchy between administrative SSEUs is not necessarily ontologically hierarchical. This is explained as follows. An ontology by its nature describes and explains the meaning of its concepts and what they refer to. Hence, it is an "IS-A" (subsumption) hierarchy where the subconcepts (e.g., car, track, bus) of a superconcept (e.g., vehicle) are its kinds and inherit all its properties. This obviously does not apply to the administrative SSEUs in Fig. 5, so it is not an ontological hierarchy. On the contrary, each subconcept is associated with a "PART-OF" relationship with the superconcept. For example, SEU "PREFECTURE" is composed of all the SEUs "MUNICIPALITY" it contains. Therefore, in the ontology discussed, the hierarchical relation of the units is not implemented by superclass-subclass, but by a declaration of direct relationship between them.

After defining the concepts, the definition of properties and relations follows. What matters here is to identify important properties and especially spatial ones,

Fig. 5 The administrative hierarchy of SSEUs

which are often encountered in the analysis and understanding of administrative units. For the ontological documentation of SSEUs, ontology editors such as Protégé (2016) can be used. Properties and relationships at the implementation level, that is, in an ontology editor environment, are recorded in the so-called slots.

The importance of properties depends on the questions to be asked and the application. So it is not possible to claim that a list of relationships-properties can be complete or a closed set. The following is a list of such properties:

Spatial properties/relations	Non-spatial properties/relations
1. Location	1. Legal status
2. Form (shape, coherence)	2. Stability of location
3. Structure (single/fragmented)	3. Super-SEU
4. Size (area)	4. Sub-SEU

(continued)

Spatial properties/relations	Non-spatial properties/relations
5. Morphology—Rrelief	5. Contact with sea
6. Neighborhood	6. Boundary type
	7. Life-temporal history
	8. Motion
	9. Internal self-sufficiency
	10. Distance from city poles/potential
	11. Infrastructures
	12. Standard of living
	13. Inland waters
	14. Maritime economic zone

Specifically, with regard to the main spatial properties of administrative SSEUs—metric and topological—these may include the following:

LOCATION (metrics)

Each SEU has a place in the space that is always identified relative to a reference system or other SEUs.

NEIGHBORHOOD (topology)

This property describes whether, for example, the SEU is isolated (e.g., an island) or is in the vicinity of other SEUs; how many are the neighbors of a SEU; if the SEU has contact with the sea.

STRUCTURE (Single/Fragmented) (topology)

This property describes whether the SEU consists of a single unit or is separated into more (non-adjacent) segments. A special case is island clusters. Here for each pair of two segments of a SEU, a measure of the fragmentation can be calculated as the minimum of the outlines divided by the centerline distance. The more distant the segments, the higher the value of the fragmentation.

SIZE (metrics)

1. Measurement of the size (area) of the SEU in an ordinal scale or in an interval scale.
2. Determination of the rank of the SEU with respect to the size of neighboring SEUs.

SHAPE (SPATIAL COHERENCE) (metrics)

This property measures the magnitude of the internal spatial coherence (SC) of the SEU. Larger SC implies that the distances within the SEU are minimized. For homogeneous and isotropic space, this can be measured by the ratio:

$SC = A/P$(area/perimeter).

SEUs with large SC are, for example, circular in shape. Longitudinal SEUs have small area with respect to their perimeter, that is, SC is small. If this expression is normalized, then a coherence measure can be expressed as follows:

$[SC] = (1/P) \sqrt{(4\pi A)}$

A SEU ideally shaped as a circle has $[SC] = 1$. A 10:1 shaped longitudinal SEU has $[SC] = 0.5$.

6.2 SSEU Boundaries

Several types of boundaries in relation to types of SSEU were presented and discussed above (Sect. 5). In the ontological formalism, in terms of their definition, geographical boundaries can be bona-fide or fiat. In terms of their implementation, they can be realized (natural or artificial) or virtual (conceivable). Regardless of the ambiguity of SEUs themselves, their boundaries may be crisp or vague. Although one would expect the boundaries of the SSEUs to be clear, there are many cases where this is not the case, so the ontology should depict it.

7 Conclusion

Establishing sound theoretical frameworks based on ontologies is essential for geographic analysis, planning strategies, and policies. This chapter introduced several aspects related to spatial socioeconomic units (SSEUs). The emphasis was placed on ontological issues, that is, the important elements that are necessary for the understanding of SSEUs, their formal definition, and the basis of their comparison. A very important and common case is that of administrative SSEUs.

An ontological framework provides a systematic base to compare different SSEUs at the concept level but also at the instance level; and furthermore to use visual means, such as spatialization (Kuhn and Blumenthal 1996; Skupin and Buttenfield 1997) to better understand/convey their characteristics. In this respect, Darra (2009) has presented an approach employing conceptual tools and structures such as similarity measures, MDS-Multidimensional Scaling (Kruskal and Wish 1978), SOM-Self Organizing Maps (Kohonen 1995), FCA-Formal concept analysis (Ganter and Wille 1999), and Cartograms. Furthermore, ontologically based definition with properties and values proves valuable when integration tasks, reorganization of space, etc. have to be performed.

References

ADL. (2002). *Alexandria digital library feature type thesaurus*. Santa Barbara, CA: University of California. Retrieved from http://legacy.alexandria.ucsb.edu/gazetteer/FeatureTypes/ver070302/index.htm.

Burrough, P. A., & Frank, A. U. (Eds.). (1996). *Geographic objects with indeterminate boundaries, European Science Foundation, GISdata Series, 2*. London and Bristol: CRC Press, Taylor and Francis.

Couclelis, H. (1996). Typology of geographic entities with Ill-defined boundaries. In P. A. Burrough & A. U. Frank (Eds.), *Geographic objects with indeterminate boundaries* (pp. 45–56). London and Bristol, PA: Taylor and Francis.

Dale, P. F. (1976). *Cadastral surveys within the commonwealth*. London: HMSO.

Dale, P. F., & McLaughlin, J. D. (1999). *Land administration*. Oxford: Oxford University Press.

Darra, A. (2009). A framework for the determination and comparison of socioeconomic units. Ph.D. thesis (in Greek), National Technical University of Athens.

Dickson, S. (2008). *The power of postcodes, Simon blogs about e-government, online news and politics*. Retrieved from http://puffbox.com/2008/06/23/the-power-of-postcodes/.

EUROBOUNDARIES. (2006). *State of the Art of National Boundaries, Analysis and Open Issues, EuroBoundaries Project—WP2 Scope and Strategy*. Retrieved from http://www.eurogeographics.org/eng/documents/EB_StateOfArt_QuestionnaireAnalysis_v14.pdf.

Fleck, M. M. (1996). The topology of boundaries. *Artificial Intelligence, 80*, 1–27.

Frank, A. U., Raper, J., & Cheylan, J.-P. (Eds.). (2001). *Life and motion of socio-economic units (GISDATA 8)*. Taylor and Francis: London.

Ganter, B., & Wille, R. (1999). *Formal concept analysis. Mathematical foundations*. Berlin: Springer.

GAUL. (2006). *The Global Administrative Unit Layers (GAUL)*. Retrieved from http://www.fao.org/GIEWS/ENGLISH/shortnews/GAUL1.pdf/.

GBHGIS. (2012). Administrative Units Typology, A Vision of Britain through Time, Great Britain Historical Geographical Information System, University of Portsmouth. Retrieved from http://www.visionofbritain.org.uk/expertsearch#tab04.

Gruber, T. R. (1993). A translation approach to portable ontologies. *Knowledge Acquisition, 5*(2), 199–220.

Hall, P. (2000). Urban renaissance/new urbanism. *Journal of the American Planning Association, 66*(4), 359–360.

Hansen, M. H. (Ed.). (2000). *A comparative study of thirty city-state cultures: An investigation* (Vol. 21). The Royal Danish Academy of Sciences and Letters, 636p.

Hubbard, P. (2006). *City, series: Key ideas in geography*. London: Routledge, Taylor & Francis Group.

IBRU. (2018). Boundary links, Centre for Borders Research, International Boundaries Research Unit, Durham University. Retrieved from https://www.dur.ac.uk/ibru/resources/links/index.php?cat=55.

International Electoral Standards. (2002). *Guidelines for reviewing the legal framework of elections Chapter 4*. Boundary Delimitation, Districting Or Defining Boundaries Of Electoral Units, International Institute for Democracy and Electoral Assistance, Publications Office, SE 103 34 Stockholm, Sweden. Retrieved from http://aceproject.org/ero-en/topics/election-integrity/UNPAN016077.pdf/at_download/file

Kim, S., Iglesias-Sucasas, M., & Viollier, V. (2013). The FAO geopolitical ontology: A reference for country-based information. *Journal of Agricultural & Food Information, 14*(1), 50–65.

Kohonen, T. (1995). *Self-organizing maps. Series in information sciences* (Vol. 30, 2nd ed.). Heidelberg: Springer.

Kruskal, J. B., & Wish M. (1978). *Multidimensional Scaling*. Sage University Paper series on Quantitative Applications in the Social Sciences, number 07–011. Sage Publications, Newbury Park, CA.

Kuhn, W., & Blumenthal, B. (1996). Spatialization: Spatial metaphors for user interfaces. In A. U. Frank (Ed.), *Geoinfo series, 8*. Vienna, Austria: Department of Geoinformation, TU Vienna.

LILP. (2008). Defining the Region, Lincoln Institute of Land Policy. The Lincoln Institute of Land Policy.

López-Pellicer, F. J., Florczyk, A. J., Lacasta, J., Zarazaga-Soria, F. J., & Muro-Medrano, P. R. (2008). Administrative units, an ontological perspective. In I.-Y. Song et al. (Eds.), *ER Workshops 2008, LNCS 5232* (pp. 354–363). Berlin/Heidelberg: Springer.

MAINE.gov. (2010). School Administrative Reorganization, Maine Department of Education. Retrieved from http://www.maine.gov/education/reorg/reorgnarrative120410.pdf.

Maynard, J. (2016). The Boundary Problems web site. Retrieved from http://www.boundary-problems.co.uk/boundary-problems/frontpage.html.

Monmonier, M. (2001). *Bushmanders and Bullwinkles: How politicians manipulate electronic maps and census data to win elections*. Chicago and London: University of Chicago Press.

Protégé. (2016). A free, open-source ontology editor and framework for building intelligent systems. Protégé resource supported by grant LM007885 from the United States National Library of Medicine. Retrieved from http://protege.stanford.edu/

QVIZ. (2008). Administrative Unit Ontology Report and Schema D3.2, Deliverable IST STREP Project Number: 032518: Query and context based visualization of time-spatial cultural dynamics, Umeå University.

Raper, J., Rhind, D., & Shepherd, J. (1992). *Postal codes: The new geography*. Harlow: Longman.

SABE. (2004). SABE—Seamless Administrative Boundaries of Europe. Retrieved from http://www.cartographic.com/TEMP/Samples/Euro%20Global%20Map///SABE2004.PDF.

Sassen, S. (2005). The global city: Introducing a concept. *Brown Journal of World Affairs, XI*(2), 27–43. http://www.saskiasassen.com/pdfs/publications/the-global-city-brown.pdf.

Skupin, A., & Buttenfield, B. P. (1997). *Spatial metaphors for display of information spaces*. In: Proceedings, AUTO-CARTO 13, Seattle, WA, Apr. 7–10. pp. 116–125.

Smith, B. (1997). Boundaries: An essay in mereotopology. In L. E. Hahn (Ed.), *The Philosophy of Roderick M. Chisholm* (pp. 533–561). Chicago/La Salle, IL: Open Court.

Smith, B. (1999). *Agglomerations. COSIT 1999, Lecture Notes in Computer Science* (Vol. 1661, pp. 267–282). Berlin/Heidelberg: Springer.

Smith, B., & Varzi, A. (1997). The formal ontology of boundaries. *The Electronic Journal of Analytic Philosophy, 5.*

Smith, B., & Varzi, A. (2000). Fiat and bona fide boundaries. *Philosophy and Phenomenological Research, 60*(2), 401–420.

STATCAN. (2007). Census metropolitan area (CMA) and census agglomeration (CA), Statistics Canada. Retrieved from http://www12.statcan.ca/census-recensement/2006/ref/dict/geo009-eng.cfm.

Weintraub, R. (2004). On sharp boundaries for vague terms. *Synthese, 138*(2), 233–245.

Spatial Microsimulation and Agent-Based Modelling

Dimitris Ballas, Tom Broomhead, and Phil Mike Jones

Abstract This chapter critically reviews the state-of-the-art in spatial microsimulation and agent-based modelling approaches with an emphasis on efforts to combine them in order to address applied geography problems. Spatial microsimulation typically involves the merging of census and social survey data to simulate a population of individuals within households (for different geographical units) whose characteristics are as close to the real population as it is possible to estimate (and for small areas for which this information is not available from published sources). Microsimulation is closely linked conceptually to another type of individual-level modelling: agent-based models (ABM). ABM are normally associated with the behaviour of multiple agents in a social or economic system. This chapter offers an overview of the state-of-the-art of both modelling approaches as well as a discussion of attempts to combine them, with an articulation of a relevant research agenda.

1 Introduction

> The era of a computational urban or regional geography based on the behaviour of individual households or firms not only provides a stimulating and exciting prospect for the years ahead but it is one which is now realistic and achievable.
>
> (Clarke 1996: 202)

D. Ballas (✉)
Department of Economic Geography, Faculty of Spatial Sciences, University of Groningen, Groningen, The Netherlands
e-mail: d.ballas@rug.nl

T. Broomhead
Department of Automatic Control and System Engineering, University of Sheffield, Sheffield, England

P. M. Jones
School of Law, Bartolomé House, University of Sheffield, Sheffield, England

H. Briassoulis et al. (eds.), *The Practice of Spatial Analysis*,
https://doi.org/10.1007/978-3-319-89806-3_4

Over the past 20 years, micro-level simulation methods, such as spatial microsimulation and agent-based modelling, have been increasingly established as tools for applied regional, urban and local analysis, and they are also a research area where the work of Pavlos Kanaroglou with colleagues has made an impact (see, for example, Maoh and Kanaroglou, 2005, 2006, 2012; Ferguson et al. 2012; Ryan et al. 2009; Svinterikou and Kanaroglou 2006). This chapter critically reviews the state-of-the-art in spatial microsimulation and agent-based modelling approaches with an emphasis on efforts to combine them in order to address applied geography problems.

Spatial microsimulation typically involves the merging of census and social survey data to simulate a population of individuals within households (for different geographical units) whose characteristics are as close to the real population as it is possible to estimate, and for small areas for which this information is not available from published sources (Ballas et al. 2007a, b).

Microsimulation is closely linked conceptually to another type of individual-level modelling: agent-based models (ABM). ABM can be associated with the behaviour of multiple agents in a socio-economic system. These agents are capable of interacting constantly with each other and with the environment they live or move within. Their actions are driven by certain rules. Although this methodology sounds similar to microsimulation (where agents could be the individuals within households), it has long been argued (see, for example, Davidsson 2000) that ABM may offer a better framework for including behavioural rules into the actions of agents (including an element of random behaviour) and for allowing interactions between agents.

There are a number of good illustrations in a geographical setting (Batty and Densham 1996; Heppenstall et al. 2005, 2006, 2007; Malleson et al. 2010; Wu et al. 2008), and there is a research agenda to link these two complementary approaches more effectively. Microsimulation could be used to give the agents in ABM their initial characteristics and locations while ABM could then provide the capacity to model individual adaptive behaviours and emergence of new behaviours (also see Boman and Holm 2004). This chapter considers these possibilities, offers an overview of the state-of-the-art and articulates a research agenda.

2 Spatial Microsimulation

> Simulation is a critical concept in the future development of modelling because it provides a way of handling complexity that cannot be handled analytically. Microsimulation is a valuable example of a technique that may have increasing prominence in future research.
> (Wilson 2000:98)

Spatial microsimulation has been becoming increasingly established as a key quantitative method in human geography, building on a long successful history of non-geographical models (mainly developed by economists; for example, see Hancock and Sutherland 1992; Harding 1996; Mitton et al. 2000). Microsimulation has been used in the social sciences since at least the 1950s (Orcutt 1957, Orcutt et al., 1961) and, therefore, it has a long history of development and application: '... it can be argued that microsimulation modelling methodologies have long become accepted tools in the evaluation of economic and social policy' (Ballas et al. 2012: 3). Geographers and regional scientists have long been involved in adding a geographical dimension to this work and, thus, developing *spatial* microsimulation methods. These are underpinned by techniques used to estimate data about individuals when this data is not readily available. Individual units can be people, organisations, businesses or any other discrete entity. Most of the applications in geography to date have been focusing on people and typically involved the creation or synthesis of small area population microdata. This can be achieved by combining different small area census cross-tabulations or by merging survey data such as census and other geographical area data to simulate a population of individuals within households (for different geographical units), whose characteristics are as close to the real population as it is possible to estimate. In other words, the models simulate *virtual populations* in given geographical areas, so that the characteristics of these populations are as close as possible to their 'real-world' counterparts. The simulation outputs include a wide range of policy relevant variables such as pre-tax and disposable income, tenure status, household type, socio-economic group, consumption patterns, car ownership and so forth. The outputs can then be used to address questions such as the following (Ballas et al. 2007a, b):

- How does the lifestyle and quality of life of individuals and households vary across different regions, cities or neighbourhoods?
- What are the interdependencies of household characteristics with geographical factors such as the presence of hospitals, community centres or schools in an area?
- To perform *what-if* scenario analysis: i.e. answer questions such as 'what would happen to personal accessibilities if the patterns of service provision change?'
- What would be the geographical impact of national social policies on personal incomes and how effective would they be compared with alternative area-based policies?

As discussed above, the first examples of using spatial microsimulation techniques can be traced back to the 1960s, but they gradually became more common during the 1970s and 1980s as computers became more powerful and accessible. It can be argued that the conceptual framework was first set by Wilson and Pownall (1976) who presented a theoretical foundation of microsimulation models and suggested a new spatial modelling framework for representing the urban system based on the micro-level interdependence of household and individual characteristics. This seminal work was the basis for further groundbreaking work in the 1980s which involved the development of the first applied spatial microsimulation models (Clarke et al. 1984; Clarke and Wilson 1985; Birkin and Clarke 1988) based on an approach often described as synthetic reconstruction (see below). These in turn have been the basis for further development of increasingly sophisticated models making use of high quality data, software and hardware and involving for the first time the use of hill-climbing, simulated annealing and genetic algorithms (Williamson et al. 1998), iterative proportional fitting deterministic reweighting (Ballas et al. 2005a, b) and further applications and refinements of these approaches (Farrell et al. 2013; Tanton and Edwards 2013; Lovelace and Ballas 2013).

As also noted above, the key procedure behind the various spatial microsimulation methods is the combining of census data and survey data, by reweighting individual-level microdata (usually the product of comprehensive social surveys at the national level) so that they match census population totals. The 'small area level' at which the analysis is conducted varies depending on data availability and common variables in both datasets. There are numerous types of spatial microsimulation models. One distinction is between static and dynamic models, with the former based on 'a single snapshot', or cross-sectional view of a population at one point in time (Ballas et al. 2005a: 8). Alternatively, dynamic models are used to 'age the attributes of each micro unit', producing datasets that project the characteristics of these micro-units into the future. There are a number of different types of static microsimulation models, including:

- Synthetic probabilistic reconstruction models—these involve random sampling to select records from microdata sources
- Probabilistic reweighting models—these typically involve the reweighting of existing national microdata to fit geographical areas, using random sampling and optimisation techniques
- Deterministic reweighting models—these typically reweight microdata to fit small area characteristics, but without the use of random sampling, so results are consistent.

The first efforts to build synthetic reconstruction models include the work of Birkin and Clarke (1988 and 1989) and Williamson (1992). These methods were (and still are) particularly suitable and appropriate when there are no good quality population microdata. The method typically involves the use of iterative proportional fitting (also known as raking) methods combined with Monte Carlo sampling to synthesize small area microdata by combining different small area cross-tabulations (typically census tables). In particular iterative proportional fitting can be used to estimate joint probability distributions of particular characteristics (e.g. being in a particular age-group, socio-economic group and tenure status) on the basis of small area cross-tabulations. Monte Carlo can then be used to sample from probability distributions, which are then applied to individual characteristics (Birkin and Clarke 1989).

A key advantage of spatial microsimulation models is that they address the lack of spatially disaggregated data in surveys, due to the creation of the new datasets at either the individual or household level for small areas. Secondly, surveys may underestimate the presence of certain groups within a population therefore reweighting survey data at a national level would not include such groups in the data, whereas the matching techniques used in spatial microsimulation makes it possible to include such groups (Morrissey et al. 2008).

There are also a number of dynamic microsimulation approaches:

- Implicitly dynamic models—which use small area projections and then apply static microsimulation methodologies to create microdata statistically (e.g. see Ballas et al. 2005a)
- Probabilistic dynamic models—which use probabilities to project individuals into the future (e.g. see Ballas et al. 2005b)

A key advantage of the 'bottom-up' approach of spatial microsimulation is that outputs of such models can be used to study several social issues, such as social inequalities, at a variety of spatial scales. In the health domain, for example, Mitchell et al. (2002) have demonstrated the difficulty in using census data to study multiple factors associated with patterns of mortality, while also exhibiting the potential of microsimulation, specifically IPF, to solve such problems.

There have been a number of platforms and computer programming languages employed to build spatial microsimulation models ranging from FORTRAN programs in the early days to Java, C++ and more recently the use of R (Campbell 2011) with open source code available (Lovelace and Ballas 2013; Lovelace and Dumont 2016; Jones et al. 2017).

Over the past 20 years, there has been a rapidly growing number of applied spatial microsimulation models and studies in a wide range of themes building on the very successful and inspiring earlier work that was briefly discussed above. The application areas include the analysis of local labour market policies (Ballas and Clarke 2000), small area income and poverty (Ballas 2004; Panori et al. 2017), social and economic policy analysis (Ballas and Clarke 2000; Ballas et al. 2007a, b; Campbell and Ballas 2013), retail analysis (Nakaya et al. 2007), health inequalities (Jones 2017; Campbell 2011; Campbell and Ballas 2016; Edwards and Clarke 2009), education inequalities (Kavroudakis et al. 2013), industrial location and firm failure (Maoh and Kanaroglou 2012), residential mobility (Ryan et al. 2009), demography and population dynamics (Ballas et al. 2005a, b, 2006; Voas and Williamson 2000), water demand estimation (Williamson 2001), health inequalities (Tomintz et al. 2008; Procter et al. 2008; Edwards and Clarke 2009; Edwards et al. 2010; Campbell 2011; Campbell and Ballas 2016), (Kavroudakis and Ballas 2011) and transport analysis (Ferguson et al. 2012; Lovelace et al. 2014; Miller this volume).

This section provided a brief historical overview of microsimulation and spatial microsimulation (for more detailed overviews, see Ballas and Clarke 2009; Birkin and Clarke 2011; Ballas et al. 2013). It has long been argued (Williamson 1999; Ballas and Clarke 2009) that microsimulation is closely linked to another type of individual-level modelling: agent-based models (ABM). ABM can be associated with the behaviour of multiple agents in a socio-economic system, interacting constantly with each other and with the environment they live or move within. The following section provides an overview of ABM methods and discusses their key features and advantages.

3 Agent-Based Modelling

Agent-based models have been defined as 'computer representations of systems consisting of a collection of discrete microentities interacting and changing over discrete time steps that give rise to macrosystems' (Auchincloss and Diez Roux 2008: 3). Macal and North (2010) argue that the macrosystems emerging from these models, represented as patterns, structures and behaviours, are 'not programmed into the models, but arise through the agent interactions' (p. 151). Further, they state that the focus on modelling heterogeneous agents across populations, and 'the emergence of self-organisation' are key and distinguishing features that lead to agent-based models comparing favourably to other simulation methods, including system dynamics and discrete event simulation.

Agent-based models can be traced to the cellular automata models of the 1970s, including Gardner's (1970) Game of Life and Schelling's (1971) famous attempt to model human and societal behaviour, one of the first studies of its kind. Agent-based models gradually increased in complexity over time. Sugarscape, a model created by Epstein and Axtell (1996), is a good example of this in the social sciences, with agents free to move cell to cell and cells containing spatially distributed resources that agents could acquire from their environment. Agent-based models have evolved in line with computational capacity and are now used across a wide range of academic subjects including molecular modelling, biology, ecology, epidemic and pandemic modelling, computational sciences, economics, market analysis and numerous other 'real-world' systems including traffic, air traffic control, military exercises and physical infrastructure including electric power and energy markets (Macal and North 2010).

Similar to spatial microsimulation, agent-based models take a 'bottom-up' approach when investigating behaviours and characteristics at the individual level. The sum of these characteristics and behaviours and their interactions over time represent the system-level model (Teweldemedhin et al. 2004). This is opposed to top-down approaches which analyse global characteristics and system-level interactions. Systems are divided into smaller parts, but generally ignore individual-level characteristics, exploring these at a system level instead (Teweldemedhin et al. 2004). Bottom-up approaches are therefore more suited to analyses of individual interactions in small geographical areas. Features of such analyses are often simplified due to the difficulty in modelling human processes exactly. While there is a danger of oversimplification, this adheres to the 'KISS' principle ('keep it simple, stupid!') introduced by Robert Axelrod (1997), which emphasises the importance of simplicity in design so as not to make the model or its output too confusing to interpret.

While there is no agreed upon definition of what an agent is, Crooks and Heppenstall (2012) define three principles which each agent must adhere to:

1. Autonomy—agents should be 'governed without the influence of centralised control' (p. 87), able to absorb and exchange information with other agents they interact with. This, in turn, informs agent decision-making.

2. Heterogeneity—'Agents permit the development of autonomous individuals' (p. 87). Agents have unique individual attributes and any groups of agents that exist are created from amalgamations of these autonomous individuals.
3. Activity—'Agents are active because they exert independent influence in a simulation' (p. 87). Agents should therefore be: pro-active and goal directed, reactive and perceptive, interactive and communicative, mobile, adaptive and capable of learning and, finally, have 'bounded rationality'—it is assumed that the choices of agents are rational and that by bounding this rationality agents can 'make inductive, discrete, and adaptive choices that move them towards achieving goals' (p. 87).

Further, agent-based models contain rules that directly impact on agent behaviour and interactions. These rules are typically based on theory, expert knowledge or data analysis and can be applied across groups or individually. Rules are usually based around 'what-if' scenarios, while the behaviour of agents can also be specified when interacting with other agents and their environment. These interactions are key to the running of the models. Macal and North (2010) point out that 'the two primary issues of modelling agent interactions are specifying who is, or could be, connected to who, and the mechanisms of the dynamics of the interactions' (p. 154). Further, Torrens and McDaniel (2013) argue that 'agents' dynamics in simulation are not scripted; rather, they are processed or computed from a model that determines their behaviour given agent characteristics (states) and algorithms (rules) that feed on agents 'endogenous attributes' (p. 23).

Environments in agent-based models can be equally important. These are demarcated areas in which agents operate, or a 'miniature laboratory where the attributes and behaviours of agents, and the environment in which they are housed, can be altered and the repercussions observed' (Crooks and Heppenstall 2012: 90). The locations of agents in these environments may or may not be relevant, depending on the type of study being conducted. Agent-based models are extremely useful for observing events that only become obvious when the process has already started and is progressing (such as segregation), meaning that taking records of the event in real time becomes impossible (Batty et al. 2004). The method allows for the reconstruction of such processes and environments which can be followed from start to finish.

Examples of applied agent-based modelling work include the research of Malleson et al. (2013), who incorporated GIS into their agent-based model of crime hotspots in East and South East-Leeds (UK), and demonstrated that while crime generally decreased in regenerated areas, a small number of houses suffered increased burglary rates as a result of the regeneration. Batty et al. (2003) have also incorporated GIS into an ABM and showed how such approaches can evaluate specific events in urban environments. This research demonstrated how different routing methods for the Notting Hill Carnival led to reduced crowd densities and higher perceived public safety. Crooks (2008) has also emphasised the importance of both geography and geometry in such models by re-creating Schelling's (1971) segregation model with the inclusion of such features. This demonstrated that geometry can act as a barrier to segregation. Burke et al. (2006) have stated that the inclusion of physical space, individual heterogeneity and local interactions allows ABMs to produce different spatiotemporal dynamics when considering outcomes such as epidemic dynamics. The authors used the method to demonstrate that vaccinations, isolation of infected individuals and contact tracing can all help limit the spread of smallpox. There are also many examples of agent-based models of travel activity with an excellent overview of the state-of-the-art and research agenda provided by Miller (2018, this volume).

In addition, non-geographically explicit models also have great value as Auchincloss et al. (2011) demonstrate. Their model, based on behavioural economics, found that price was the key driver for food preferences even when low income families desired healthier food.

Some agent-based models have been concerned with the prediction and evaluation of certain future events and scenarios, particularly those related to the spread of diseases (Potter et al. 2012; Merler et al. 2013), and, in some historical evaluations, they have had access to past data to help parameterise models (O'Neil and Sattenspiel 2010).

A number of tools and software platforms enable the building of agent-based models. An increasingly popular platform is the multi-agent programming environment NetLogo (Wilensky 1999). NetLogo is object oriented, meaning that a system can be modelled as a set of objects which can be controlled and manipulated in a certain way depending on the purpose of the model or the system. This software has been described as being 'low threshold, no ceiling' (Tisue and Wilensky 2004), a central principle carried forward from modelling conducted in the original Logo language (The Logo Foundation 2016). As well as being relatively accessible to those without a history of writing scripts in programming languages, NetLogo also has the advantage of being able to read in a number of external file types. One such example includes the GIS extension, a function for reading in geographical data in both vector and raster format, which allows for models to be designed with specific landscapes, features, and contexts in mind. An example of the use of both types of GIS data in a NetLogo simulation is given in the work of Dawson et al. (2011) who designed a model investigating coastal flood management in the Welsh seaside resort of Towyn.

A key advantage of agent-based models is the ability to track agent characteristics of interest as simulations progress and interactions occur (Gorman et al. 2006). Through this it may be possible to gain better knowledge of the processes occurring at the small area level. Agent-based models can help research move beyond descriptive analysis as well as to test theoretical hypotheses which may offer better insights into a problem at hand (Johnson and Groff 2014; Cerda et al. 2014).

Nevertheless, what agent-based models lack when compared to microsimulation models is the richness in real-world information about the units of analysis (agents). In particular, agent-based models are often based on virtual units that 'behave' and 'interact' on the basis of a specified set of assumptions and theories but without the benefit of the use of 'real' data, such as the social surveys and census data typically used in the spatial microsimulation models reviewed in the previous section. It is, therefore, interesting to explore the possibilities of combining spatial microsimulation with agent-based modelling approaches. This would typically involve the replacement of microsimulated units driven by transition probabilities, with adaptive rule-based agents. The next section discusses this possibility with a brief review of the relatively limited number of studies to date that have attempted to carry out analyses based on such a combination.

4 Spatial Microsimulation and Agent-based Models

Amongst the few studies that attempted to combine spatial microsimulation and agent-based models is the work of Merler et al. (2013) who developed a stochastic microsimulation model combined with an agent-based model in their study of a flu outbreak in the Netherlands. The work of Wu et al. (2008) represents another rare example demonstrating the combination of these bottom-up approaches when investigating student migration patterns in Leeds. This hybrid model was shown to be very accurate in predicting the geographical spread of students within the city, compared to a microsimulation model which failed to capture the spatial clustering patterns of students. It is noted that microsimulation is driven more by statistics and probabilities, whereas agent-based models can rely on their rule-based (built-in intelligence) nature. However, the tried and tested techniques associated with microsimulation provide 'important statistical mechanisms that ensure the similarity of what it predicts and what is actually observed in the gathered data' (p. 446), which can aid in guiding patterns of population evolution.

Wu and Birkin (2012) suggest that the combination of the two modelling approaches allows large-scale data to be processed using list processing power, as well as to identify the consequences of behaviours and policies at a macro scale through predefined transition rates, while also modelling interactions and behaviours of individuals. Their mortality model demonstrated the importance of personal histories, with previous places of residence influencing health regardless of certain individual's current residences, showing that ABMs can 'complement MSM [microsimulation] by retrieving personal histories with great ease' (p. 356). Given the potential shown by these models it is surprising that this approach has not been used more often.

Cajka et al. (2010) further demonstrated the benefit of combining microsimulation data with agent-based models, commenting that 'although these models [ABMs] can simulate the realistic propagation of epidemics, they require input data about the social networks that are part of the agents' day-to-day activities' (p. 1). Although not modelling the spread of diseases, the paper demonstrates how personal attribute data agents from a microsimulation dataset can be used 'to create school, workplace, and public transit interactions and then code this information into the data' (p. 1), in attempting to build realistic social networks into interactive disease modelling.

The analysis of travel activity is another area where there has been considerable applied work aimed at combining spatial microsimulation and agent-based models and where there is great potential for further research. Miller (2018, this volume) offers an excellent overview of the state-of-the-art and presents the key issues and challenges as well as a research agenda.

Overall there is great potential for further applications of agent-based models for both longitudinal and historical analyses in a wide range of fields. The incorporation of interactions gives the method a different focus to spatial microsimulation modelling, while also demonstrating incredible adaptability in their set up. This includes connecting with other model types, incorporating other datasets, allowing them to 'borrow' the strength of these. The ability to base models on theory (for example, Cerda et al. 2014) also offers the exciting prospect of simulating conceptually valid and relevant studies. This theoretical grounding will be important for work to avoid being simply designated as 'blue skies research'. Despite all this, research combining multiple models within an urban environment backed by a specific theory is rare. Many studies have used hypothetical data in their analyses, so the opportunity to impute a representative population into a model using spatial microsimulation has huge potential for applied geographical analysis.

5 Concluding Comments

> Governments need to predict the outcomes of their actions and produce forecasts at the local level.
>
> (Openshaw 1995: 60)

It has long been anticipated that GIS-based simulation with the use of powerful computing and software will provide an enabling environment for comprehensive analysis of the impacts of government policy at different levels (including small area level). Significant progress in this direction has materialised with the rapidly expanding use of GIS in the Social Sciences (Ballas et al. 2017). It can be argued that, to some extent, this is due to the use of spatial microsimulation models and, to a lesser extent, of agent-based models. A key argument of this chapter is that the combination of spatial microsimulation and agent-based models has huge potential for the analysis of the short- and long-term impacts of social, regional, and urban policies upon different geographical areas and population segments. Furthermore, there is great potential to combine the two approaches to model all sub-systems that make up a local and regional socio-economic system, making the most of data on local labour market linkages, retail outlets, schools, hospitals, health services, etc. In particular, it is increasingly possible to build agent-based and spatial microsimulation models linking all datasets on the sub-systems that make up the local economy. For instance, data sources for groceries could be explored and the relevant database could be built, with the prospect of linking it to the household micro-database. Likewise, similar databases can be built for all existing hospitals, schools and major employers. An example of a set of specific aims and objectives that could be formulated in order to combine the approaches reviewed in this chapter and perform comprehensive spatial microsimulation and agent-based modelling can be summarised as follows:

- Develop a geographic synthetic micro-database of households and individuals using data from social surveys and small area data
- Estimate a geographic database of retail outlets. This could include actual or estimated floorspace and trading intensity, estimated attractiveness of the retailer, etc.
- Construct a database of amenities including schools, hospitals and health centres but also entertainment and recreation outlets (e.g. museums, theatres, cinemas but also spaces to socialise, bars, restaurants,)
- Build a database of transport networks and estimated trips (such as travel to work, travel to shop,)
- Link all these databases in order to explore the geography of quality of life in cities and regions—for example, link households to retail outlets using shopping flow data to see how well served different households are in different neighbourhoods for basic service provision.
- Use the spatial microsimulation outputs as a basis for agent-based models, representing the types of people that live in particular neighbourhoods, as well as the features and characteristics they would encounter in everyday life.
- Convert all micro-units to agents and model suitable interactions between them, including social interactions between agents, as well as family structures, and demographic turnover over the course of the model.
- Perform static *what-if* scenario analysis: i.e. answer questions such as 'what if the patterns of service provision change?'
- Perform dynamic analysis into the future, including interactions between sub-systems that make up urban and regional systems and micro-units (defined as agents).

This list is by no means exhaustive. It just gives a flavour of the huge potential that there is for applying agent-based spatial microsimulation in a wide range of inter-disciplinary thematic contexts.

References

Auchincloss, A. H., & Diez Roux, A. V. (2008). A new tool for epidemiology: The usefulness of dynamic-agent models in understanding place effects on health. *American Journal of Epidemiology, 168*(1), 1–8.

Auchincloss, A. H., Riolo, R. L., Brown, D. G., Cook, J., & Diez Roux, A. V. (2011). An agent-based model of income inequalities in diet in the context of residential segregation. *American Journal of Preventative Medicine, 40*(3), 303–311.

Axelrod, R. (1997). *The complexity of cooperation: Agent-based models of competition and collaboration*. Princeton: Princeton University Press.

Ballas, D. (2004). Simulating trends in poverty and income inequality on the basis of 1991 and 2001 census data: A tale of two cities. *Area, 36*, 146–163.

Ballas, D., & Clarke, G. P. (2000). GIS and microsimulation for local labour market policy analysis. *Computers, Environment and Urban Systems, 24*, 305–330.
Ballas, D., Clarke, G. P., Dorling, D., Eyre, H., Rossiter, D., & Thomas, B. (2005a). SimBritain: A spatial microsimulation approach to population dynamics. *Population, Space and Place, 11*, 13–34.
Ballas, D., Clarke, G. P., & Wiemers, E. (2005b). Building a dynamic spatial microsimulation model for Ireland. *Population, Space and Place, 11*, 157–172.
Ballas, D., Clarke, G. P., & Wiemers, E. (2006). Spatial microsimulation for rural policy analysis in Ireland: The implications of CAP reforms for the national spatial strategy. *Journal of Rural Studies, 22*, 367–378.
Ballas, D., Kingston, R., Stillwell, J., & Jin, J. (2007a). Building a spatial microsimulation-based planning support system for local policy making. *Environment and Planning A, 39*(10), 2482–2499.
Ballas, D., Clarke, G. P., Dorling, D., & Rossiter, D. (2007b). Using SimBritain to model the geographical impact of national government policies. *Geographical Analysis, 39*(1), 44–77.
Ballas, D., & Clarke, G. P. (2009). Spatial microsimulation. In A. S. Fotheringham & P. A. Rogerson (Eds.), *Handbook of spatial analysis* (pp. 277–298). Thousand Oaks: Sage.
Ballas, D., Clarke, G. P., Hynes, S., Morrissey, K., & O'Donoghue, C. (2012). Introduction. In C. O'Donoghue, D. Ballas, G. P. Clarke, S. Hynes, & K. Morrissey (Eds.), *Spatial microsimulation for rural policy analysis* (pp. 1–10). New York: Springer.
Ballas, D., Clarke, G. P., Hynes, S., Morrissey, K., & O'Donoghue, C. (2013). A review of microsimulation for policy analysis. In C. O'Donoghue, D. Ballas, G. P. Clarke, S. Hynes, & K. Morrissey (Eds.), *Spatial microsimulation for rural policy analysis* (pp. 35–54). Berlin: Springer.
Ballas, D., Clarke, G., Franklin, R. S., & Newing, A. (2017). *GIS and the social sciences: Theory and applications*. London: Routledge.
Batty, M., & Densham, P. (1996). Decision support, GIS and urban planning. *Systemma Terra, 1*, 72–76.
Batty, M., Desyllas, J., & Duxbury, E. (2003). Safety in numbers? Modelling crowds and designing control for the Notting Hill Carnival. *Urban Studies, 40*(8), 1573–1590.
Batty, M., Barros, J., Alves, S. Jr. (2004). *Cities: Continuity, transformation, and emergence.* CASA Working Paper—Paper 72. Retrieved from http://discovery.ucl.ac.uk/214/1/paper72.pdf
Birkin, M., & Clarke, M. (1988). SYNTHESIS—A synthetic spatial information system for urban and regional analysis: Methods and examples. *Environment and Planning A, 20*, 1645–1671.
Birkin, M., & Clarke, M. (1989). The generation of individual and household incomes at the small area level using synthesis. *Regional Studies, 23*, 535–548.
Birkin, M., & Clarke, M. (2011). Spatial microsimulation models: A review and a glimpse into the future. In J. Stillwell, M. Clarke, & J. Stillwell (Eds.), *Population dynamics and projection methods. Understanding population trends and processes* (Vol. 4, pp. 193–208). Netherlands: Springer.
Boman, M., & Holm, E. (2004). Multi-Agent systems, time geography, and microsimulations. In M. Olsson & G. Sjöstedt (Eds.), *Systems approaches and their application* (pp. 95–118). New York: Springer.
Burke, D. S., Epstein, J. M., Cummings, D. A. T., Parker, J. I., Cline, K. C., Singa, R. M., & Chakravarty, S. (2006). Individual-based computational modeling of smallpox epidemic control strategies. *Academic Emergency Medicine, 13*(11), 1142–1149.
Cajka, J. C., Cooley, P. C., & Wheaton, W. D. (2010). Attribute assignment to a synthetic population in support of agent-based disease modeling. *Methods Report RTI Press, 19*(1009), 1–14.
Campbell, M. H. (2011). *Exploring the social and spatial inequalities of ill-health in Scotland: A spatial microsimulation approach.* PhD thesis, University of Sheffield. Retrieved from http://etheses.whiterose.ac.uk/1942/

Campbell, M. H., & Ballas, D. (2013). A spatial micro simulation approach to economic policy analysis in Scotland. *Regional Science Policy and Practice, 5*(3), 263–288.
Campbell, M., & Ballas, D. (2016). SimAlba: A spatial microsimulation approach to the analysis of health inequalities. *Frontiers in Public Health, 4*(230).
Cerda, M., Tracy, M., Ahern, J., & Galea, S. (2014). Addressing population health and health inequalities: The role of fundamental causes. *American Journal of Public Health, 104*(4), 609–619.
Clarke, G. P. (Ed.). (1996). *Microsimulation for urban and regional policy analysis*. London: Pion.
Clarke, G. P., Clarke, C., Birkin, M., Rees, P. H., & Wilson, A. G. (1984). *A strategic planning simulation model of a district health service system: The in-patient component and results. Number Bd. 385-389* (p. 1984). Leeds: School of Geography, University of Leeds.
Clarke, M., & Wilson, A. (1985). The dynamics of urban spatial structure: The progress of a research programme. *Transactions of the Institute of British Geographers, 10*(4), 427–451.
Crooks, A. T. (2008). *Constructing and implementing an agent-based model of residential segregation through vector GIS*. UCL Working Paper Series, Paper 133. Retrieved from http://discovery.ucl.ac.uk/15185/1/15185.pdf
Crooks, A. T., & Heppenstall, A. J. (2012). Introduction to agent-based modelling. In A. J. Heppenstall, A. T. Crooks, L. M. See, & M. Batty (Eds.), *Agent-based models of geographical systems*. London: Springer.
Davidsson, P. (2000). Multi agent based simulation: Beyond social simulation. In S. Moss & P. Davidsson (Eds.), *Multi agent based simulations*. Berlin: Springer.
Dawson, R. J., Peppe, R., & Wang, M. (2011). An agent-based model for risk-based flood incident management. *Natural Hazards, 59*(1), 167–189.
Edwards, K. L., & Clarke, G. P. (2009). The design and validation of a spatial microsimulation model of obesogenic environments for children in Leeds, UK: SimObesity. *Social Science and Medicine, 69*(7), 1127–1134.
Edwards, K. L., Clarke, G. P., Ransley, J. K., et al. (2010). The neighbourhood matters: Studying exposures relevant to childhood obesity and the policy implications in Leeds, UK. *Journal of Epidemiology and Community Health, 64*(3). Retrieved from http://jech.bmj.com/content/64/3/194
Epstein, J. M., & Axtell, R. L. (1996). *Growing artificial societies: Social science from the bottom up*. Cambridge: The MIT Press.
Farrell, N., Morrissey, K., & O'Donoghue, C. (2013). Creating a spatial microsimulation model of the Irish Local Economy. In Tanton & Edwards (Eds.), *Spatial microsimulation: A reference guide for users* (pp. 105–125). New York: Springer.
Ferguson, M., Maoh, H., Ryan, J., Kanaroglou, P., & Rashidi, T. H. (2012). Transferability and enhancement of a microsimulation model for estimating urban commercial vehicle movements. *Journal of Transport Geography, 24*, 358–369.
Gardner, M. (1970). The fantastic combinations of John Conway's new solitaire game "Life". *Scientific American, 223*, 120–123.
Gorman, D. M., Mezic, J., Mezic, I., & Gruenewald, P. J. (2006). Agent-based modelling of drinking behaviour: A preliminary model and potential applications to theory and practice. *American Journal of Public Health, 96*(11), 2055–2060.
Hancock, R., & Sutherland, H. (Eds.). (1992). *Microsimulation models for public policy analysis: New frontiers*. London: Suntory-Toyota International Centre for Economics and Related Disciplines – LSE.
Harding, A. (Eds.) (1996). *Microsimulation and public policy, contributions to economic analysis 232*, Amsterdam, North Holland.
Heppenstall, A. J., Evans, A. J., & Birkin, M. H. (2005). A hybrid multi-agent/spatial interaction model system for petrol price setting. *Transactions in GIS, 9*(1), 35–51.
Heppenstall, A. J., Evans, A. J., & Birkin, M. H. (2006). Application of multi-agent systems to modelling a dynamic, locally interacting retail market. *JASSS, 9*(3).
Heppenstall, A. J., Evans, A. J., & Birkin, M. H. (2007). Genetic algorithm optimisation of a multi-agent system for simulating a retail market. *Environment and Planning B, 34*, 1051–1070.

Johnson, S. D., & Groff, E. R. (2014). Strengthening theoretical testing in criminology using agent-based modeling. *Journal of Research in Crime and Delinquency, 51*(4), 509–525.
Jones, P. (2017). *A spatial microsimulation approach to the analysis of health inequalities and health resilience*. Unpublished PhD thesis, University of Sheffield.
Jones, P. M., Lovelace, R., & Dumont, M. (2017). *rakeR: Easy spatial microsimulation (raking) in R*. https://doi.org/10.5281/zenodo.821506. Retrieved from https://cran.r-project.org/package=rakeR
Kavroudakis, D., & Ballas, D. (2011). Agent-based modelling for labour force analysis, *Annual meeting of European Association of Geographers (Eurogeo)*, Athens, Greece, 2-5 June 2011.
Kavroudakis, D., Ballas, D., & Birkin, M. (2013). A spatial microsimulation approach to the analysis of social and spatial inequalities in higher education attainment. *Applied Spatial Analysis and Policy, 3*, 1–23.
Lovelace, R., & Dumont, M. (2016). *Spatial microsimulation in R*. Boca Raton, FL: CRC Press/Taylor and Francis Group.
Lovelace, R, Ballas, D, Watson, M (2014), A spatial microsimulation approach for the analysis of commuter patterns: from individual to regional levels, Journal of Transport Geography, vol. 34, 282–296.
Lovelace, R., Birkin, M., Ballas, D., & van Leeuwen, E. (2015). Evaluating the performance of iterative proportional fitting for spatial microsimulation: New tests for an established technique. *Journal of Artificial Societies and Social Simulation, 18*(2), 21.
Lovelace, R., & Ballas, D. (2013). 'Truncate, replicate, sample': A method for creating integer weights for spatial microsimulation. *Computers, Environment and Urban Systems, 41*, 1–11.
Macal, C. M., & North, M. J. (2010). Tutorial on agent-based modelling and simulation. *Journal of Simulation, 4*(3), 151–162.
Malleson, N., Evans, A., Heppenstall, A., & See, L. (2010). *Evaluating an agent-based model of burglary*. University of Leeds Working Paper. Retrieved from http://www.geog.leeds.ac.uk/fileadmin/downloads/school/research/wpapers/10_1.pdf
Malleson, N., Heppenstall, A., See, L., & Evans, A. (2013). Using an agent-based crime simulation to predict the effects of urban regeneration on individual household burglary risk. *Environment and Planning B: Planning and Design, 40*(3), 405–426.
Maoh, H., & Kanaroglou, P. (2012). Modelling firm failure: Towards building a firmographic microsimulation model. In F. Pagliara, M. de Bok, D. Simmonds, & A. Wilson (Eds.), *Employment location in cities and regions* (pp. 243–261). Berlin: Springer.
Merler, S., Ajelli, M., Fumanelli, L., & Vespignani, A. (2013). Containing the accidental laboratory escape of potential pandemic influenza viruses. *BMC Medicine, 11*, 252.
Miller, H. (2018). Agent-based activity/travel microsimulation: What's next? In H. Briassoulis, D. Kavroudakis, & N. Soulakellis (Eds.), *The Practice of Spatial Analysis: Essays in memory of Professor Pavlos Kanaroglou*. New York: Springer.
Mitchell, R., Dorling, D., & Shaw, M. (2002). Population production and modelling mortality—an application of geographic information systems in health inequalities research. *Health and Place, 8*, 15–24.
Mitton, L., Sutherland, H., & Weeks, M. (Eds.). (2000). *Microsimulation modelling for policy analysis: Challenges and innovations*. Cambridge: Cambridge University Press.
Morrissey, K., Clarke, G., Ballas, D., Hynes, S., & O'Donoghue, C. (2008). Examining access to GP services in rural Ireland using microsimulation analysis. *Area, 40*(3), 354–364.
Nakaya, T., Fotheringham, A. S., Hanaoka, K., Clarke, G., Ballas, D., & Yano, K. (2007). Combining microsimulation and spatial interaction models for retail location analysis. *Journal of Geographical Systems, 9*, 345–369.
O'Neil, C. A., & Sattenspiel, L. (2010). Agent-based modeling of the spread of the 1918-1919 flu in three Canadian fur trading communities. *American Journal of Human Biology, 22*(6), 757–767.
Openshaw, S. (1995). Human systems modelling as a new grand challenge area in science. *Environment and Planning A, 27*, 159–164.

Orcutt, G. H. (1957). A new type of socio-economic system. *The Review of Economics and Statistics, 39*, 116–123.
Orcutt, G. H., Greenberger, M., Korbel, J., & Rivlin, A. (1961). *Microanalysis of socioeconomic systems: A simulation study*. New York: Harper and Row.
Panori, A., Ballas, D., & Psycharis, Y. (2017). SimAthens: A spatial microsimulation approach to the estimation and analysis of small-area income distributions and poverty rates in Athens, Greece. *Computers, Environment and Urban Systems, 63*, 15–25.
Potter, M. A., Brown, S. T., Cooley, P. C., Sweeney, P. M., Hershey, P. B., Gleason, S. M., Lee, B. Y., Keane, C. R., Grefenstette, J., & Burke, D. (2012). School closure as an influenza mitigation strategy: How variations in legal authority and plan criteria can alter the impact. *BMC Public Health, 12*, 977.
Procter, K. L., Clarke, G. P., Ransley, J. K., et al. (2008). Micro-level analysis of childhood obesity, diet, physical activity, residential and social capital variables: Where are the obesogenic environments in Leeds? *Area, 40*(3), 323–340.
Ryan, J., Maoh, H., & Kanaroglou, P. S. (2009). *Population synthesis for microsimulating urban residential mobility, Transportation Research Board 89th Annual Meeting*.
Schelling, T. C. (1971). Dynamic models of segregation. *Journal of Mathematical Sociology, 1*, 143–186.
Svinterikou, M., & Kanaroglou, P. (2006). *A microsimulation approach to the modelling of urban population and housing markets within an object-oriented framework, ERSA conference papers, European Regional Science Association*.
Tanton, R., & Edwards, K. L. (2013). *Spatial microsimualtion: A reference guide for users*. London: Springer.
Teweldemedhin, E., Marwala, T., & Mueller, C. (2004). Agent-based modelling: A case study in HIV epidemic. In *Fourth International Conference on Hybrid Intelligent Systems* (pp. 154–159).
The Logo Foundation. (2016). Retrieved from http://el.media.mit.edu/logo-foundation/
Tisue, S., & Wilensky, U. (2004). *NetLogo: Design and implementation of a multi-agent modelling environment*. Center for Connected Learning and Computer-Based Modeling Northwestern University, Evanston, IL. Retrieved from https://ccl.northwestern.edu/papers/2013/netlogo-agent2004c.pdf.
Tomintz, M. N., Clarke, G. P., & Rigby, J. E. (2008). The geography of smoking in Leeds: Estimating individual smoking rates and the implications for the location of stop smoking services. *Area, 40*(3), 341–353.
Torrens, P. M., & McDaniel, A. W. (2013). Modeling geographic behaviour in riotous crowds. *Annals of the Association of America Geographers, 103*(1), 20–46.
Voas, D. W., & Williamson, P. (2000). An evaluation of the combinatorial optimisation approach to the creation of synthetic microdata. *International Journal of Population Geography, 6*, 349–366.
Wilensky, U. (1999). *NetLogo*. Center for Connected Learning and Computer-Based Modeling. Northwestern University, Evanston, IL. Retrieved from https://ccl.northwestern.edu/netlogo/.
Williamson, P. (1992). *Community care policies for the elderly: A microsimulation approach*. Unpublished PhD Thesis, School of Geography, University of Leeds, Leeds.
Williamson, P. (1999). *Microsimulation: An idea whose time has come?*. Paper Presented at the 39th European Regional Science Association Congress, University College Dublin, Dublin, Ireland, 23–27 August 1999.
Williamson, P. (2001). Modelling alternative domestic water demand scenarios. In G. Clarke & M. Madden (Eds.), *Regional science in business* (pp. 243–268). Berlin: Springer.
Williamson, P., Birkin, M., & Rees, P. (1998). The estimation of population microdata by using data from small area statistics and samples of anonymised records. *Environment and Planning A, 30*, 785–816.
Wilson, A. G. (2000). *Complex spatial systems: The modelling foundations of urban and regional analysis*. London: Prentice Hall.

Wilson, A., & Pownall, C. E. (1976). A new representation of the urban system for modelling and for the study of micro-level interdependence. *Area, 8*, 246–254.

Wu, B. M., & Birkin, M. H. (2012). Agent-based extensions to a spatial microsimulation model of demographic change. In A. J. Heppenstall, A. T. Crooks, L. M. See, & M. Batty (Eds.), *Agent-based models of geographical systems*. London: Springer.

Wu, B., Birkin, M., & Rees, P. (2008). A spatial microsimulation model with student agents. *Computers Environment and Urban Systems, 32*, 440–453.

Spatial Patterns of Accessibility Between Islands of the North and South Aegean Regions and Attica

Dimitris Kavroudakis, Polydoropoulos Ioakeim, Kyriakidis Phaedon, and Kizos Thanassis

Abstract Accessibility is an important determinant of socio-economic development at all spatial levels because it critically affects the operations and success of economic activities. Measures of accessibility are frequently used to support development policy-making and spatial planning. The spatial sciences provide several quantitative approaches to assess accessibility and identify over- and under-served areas. This chapter contributes to the broader issue of quantifying accessibility by analyzing the spatial patterns of accessibility in the case of a very fragmented geographical space; the islands of the North and South Aegean Sea Regions, Greece. Following a preliminary spatial analysis of: transportation route frequency, passengers/population, two graph theory metrics and mean ticket cost, clustering methods, enhanced by introducing local indices of spatial association, are employed to determine possible groupings of islands and delineate spatial clusters based on their accessibility. "Hot spots" of accessibility in the study area are identified, suggesting that accessibility is more spatially clustered than would be expected under a random model.

1 Introduction

Accessibility reflects the ease of access (in units of time) between two points in the space. It constitutes a measure of spatial separation between human activities that are spread over geographical space. Island regions, such as the Aegean Sea in Greece, form an interesting case study for analyzing accessibility measures. This is a

D. Kavroudakis (✉) · P. Ioakeim · K. Thanassis
Department of Geography, University of the Aegean, Mytilini, Lesvos, Greece
e-mail: dimitrisk@aegean.gr; geom12018@geo.aegean.gr; akizos@aegean.gr

K. Phaedon
Faculty of Engineering and Technology, Department of Civil Engineering and Geomatics, Cyprus University of Technology, Limassol, Cyprus
e-mail: phaedon.kyriakidis@cut.ac.cy

H. Briassoulis et al. (eds.), *The Practice of Spatial Analysis*,
https://doi.org/10.1007/978-3-319-89806-3_5

very fragmented geographical space, comprising islands of multiple size and socio-economic dynamics. The economies of the islands are organized in a system of economic poles that are particularly interesting to study as population composition and administrative structures vary between the islands (Kizos 2007; Spilanis and Kizos 2015; Zervakis et al. 2000). The accessibility of the Aegean islands exhibits significant seasonality between summer and winter months that makes them even more vulnerable as transport significantly affects their economic activities (Karampela et al. 2015). Seasonal fluctuations have a significant impact on economic activities in most islands as tourism is an important economic sector. Tourism activities importantly contribute a 24% to the GDP of Greece, making this sector one of the most important for the Greek economy. The importance of tourism for the Greek economy as well as the particularities of the Aegean islands make the study and analysis of their economic activities even more necessary (Karampela et al. 2014, 2015; Kizos 2007; Kizos et al. 2005; Shih 2006; Spilanis et al. 2005, 2012, 2013a, b; Spilanis and Kizos 2015; Weidenfeld et al. 2011).

Farrington (2007, p. 320) describes a "new narrative of accessibility," which echoes the concept of Mosley's accessibility as "the degree to which something can be accessed" (Moseley 1979, p. 56). Accessibility is a general concept, whose definition depends on the scientific field and the objectives of analysis, and one which everyone understands differently based on expertise and experience (Scott 2000). The contemporary concept of accessibility naturally emerged in the context of land use and transport analysis and modeling (Farrington 2007), as well as in urban or rural analysis, and gained ground in the 1990s, also assisted by the development of Geographic Information Systems (GIS); see Kwan et al. (2003). A number of quantitative indices have been developed for calculating and interpreting accessibility, such as indicators of infrastructure, travel time, travel costs, and day-to-day accessibility (Schürmann and Talaat 2002). An interesting work regarding islands' accessibility is that of Ankre and Nilsson (2016), which focuses on the accessibility of the Faroe Islands by various transportation means (sea ferries, airplanes). Another notable work is that of Makkonen et al. (2013), which focuses on the identification of spatial discrepancies between population patterns and transport options in the peripheral archipelago of Finland.

The quantification of accessibility is important because it can contribute towards geographically informed economic policy-making. The European Union has funded a number of accessibility-related research projects, indicating the importance of accessibility in policy-making. As islands constitute unique and near-border regions in Europe, research on island accessibility was part of the "*Island Development-European Islands and Cohesion Policy*" (EUROISLANDS) under the European Spatial Planning Observation Network (ESPON) Action Plan 2007–2013. There are 362 European islands with more than 50 inhabitants and 228 with less than 50 inhabitants. In the context of this project, islands were classified into 15 large islands with more than 50,000 inhabitants, 44 islands with population of 5000–50,000 inhabitants, 303 small islands of 50–5000, and 228 very small islands of less than 50 inhabitants. The aim of the EUROISLANDS research was "a focused analysis to deliver a benchmarking work, policy proposals and orientations towards the sustainable development of European islands within the framework of the Single Market, ensuring equal conditions and opportunities with other non-disadvantaged areas." The study examined 31 island regions and nine islands. The status of islands and their attractiveness were described by means of various variables. Results revealed that one of the basic factors affecting the attractiveness of businesses and islands was accessibility that refers to the real time needed to reach an island relative to the corresponding time needed to cover the same terrestrial distance (Spilanis et al. 2012). In ESPON reports, islands are listed as areas with specific features that need to be studied in order to "provide a comparative understanding of territorial potential and challenges from the European perspective and to ensure that other areas with similar characteristics can benefit from this analysis." In the context of EUROISLANDS research, the ESPON multimodal accessibility index (EU 2010) was used to calculate the accessibility of islands. This approach has been identified as having significant shortcomings and inaccuracies for islands which include:

- Only island regions NUTS II and III were included in the calculations rather than individual islands, which conceals the problems the smaller islands are facing
- Potential air accessibility overstates the existence of a local airport in NUTS III regions while ignoring proximity to an international airport. Generally, the values of the total index depend on 90% of air accessibility.

Particular weight was given to daily accessibility, which ignores the reality of most islands, for which factors such as frequency, quality, and price are also equally important. Despite these weaknesses and the fact that a particular indicator overestimates the accessibility of islands, all the islands which were investigated had lower values compared to the overall average (Spilanis et al. 2012). In addition, the transfer to/from an island still divides on national borders, which may prevent islands from fully participating in a single market. Therefore, islands are disadvantaged in terms of accessibility to the mainland for choice of means of transport, travel time, and cost. The situation is getting worse for small islands. It was concluded that poor accessibility adversely affects residents and visitors (Spilanis et al. 2012).

In terms of quantifying accessibility, Spilanis et al. (2012) proposed a measure based on total travel time, and applied it to three different Greek islands in the Aegean Sea. The accessibility of inhabitants of these islands to selected services was also examined compared to that of settlements in mainland Greece with a similar population and distance from the capital city of Athens. Their findings showed clearly the disadvantages faced by islanders, especially for small islands, where access to selected services may require access to four consecutive destinations, with virtual distances 4–6 times greater than "real distances." As mentioned in Spilanis et al. (2012), this indicator has some significant advantages and disadvantages. The most important advantage is that it takes into account the frequency of connections and can be used to reveal seasonal differences in accessibility for the same island or group of islands. Another important advantage is that it is flexible and can be calculated separately for island groups with some assumptions. It can also be calculated for different ports of the same island. The weekly basis used to calculate connections and frequencies reflects the reality of maritime transport, which is planned on this basis. Its drawbacks include the fact that the quality and capacity of ferry boats are not included in the index and this may be of great importance. In addition, no other means of transport, such as aircraft or high-speed ferries are included. Lastly, travel costs are not included in the index due to different prices among seasons and island shipping companies.

This chapter focuses on assessing the accessibility of 50 islands of the Northern and Southern Aegean Regions, while coastal (other than high-speed) and airline connections are also included. According to Karampela et al. (2014), geographical distance is not determining the accessibility of islands and new geographies emerge based more on the availability of transport mode choices. Furthermore, Karampela et al. (2015) suggest that a number of different archipelagic clusters emerge within the area largely determined by existing transport routes and tourism development trajectories.

2 Spatial Analysis

The work undertaken in this chapter falls within the realm of spatial analysis and spatial statistics, fields where Pavlos Kanaroglou was particularly prolific. The broad field of spatial analysis includes quantitative techniques that study the topological, geometric, or geographic properties of entities and phenomena that evolve in space and time (Haining 2003). Spatial statistics have progressed significantly over the last 20 years and offer probabilistic modeling techniques to analyze three kinds of spatial problems: analysis of point locations (point patterns), analysis of processes that give rise to variables with continuous spatial distribution (geostatistics), and analysis of phenomena whose data appear in spatial objects, such as polygons corresponding to administrative regions (spatial econometrics). These analytical techniques have been developed in different scientific fields. Standard point pattern analysis was cultivated within quantitative geography, while geostatistics flourished in the geosciences. In 1958, Ackerman (1958) wrote that Geography "*is the science of spatial distribution*" and believed that an object of geographic research should be "*to understand the evolution of the constituents of space as it is influenced by natural, cultural and linguistic developments*" (p. 22). In general, the idea of spatial analysis is relatively simple but its definition is particularly difficult. Bailey (1994: 41) defined it as "*an overall possibility to manage and transform the spatial elements into different forms, giving them a different meaning.*" This definition coincides with what has been called process from data to information, where the terms "geographic data" and "geographic information" are clearly differentiated (O'Sullivan and Unwin 2010). As a result, spatial analysis is a set of "*quantitative procedures and techniques applied to spatial analytical work*" (Johnston 1986: 33). According to Haining (1994), the aims of spatial analysis include:

- Description of events in the field, which mainly includes the description of spatial models
- Systematic exploration of spatial patterns and spatial relationships with a view to better understanding spatial processes that are responsible for spatial patterns and relationships.
- To increase the ability to predict and control events occurring in a geographical area.
- To use these techniques and methods for informed spatial decision-making.

According to Haining (2003), the geographical area and the measures that represent it refer to four types of spatial distributions:

- Point distributions, where each measurement refers to a particular point in space (point entities).
- Line distributions, where each measurement is represented by a line (straight, polyline, or any other form) and, therefore, refers to the linear entities.
- Discontinuous surface distributions, where each measurement refers to a particular surface representing polygonal entities.
- Continuous surface distributions, where each measurement relates to all points of a surface.

In the case of discontinuous surface distributions, interest does not focus on the location of point events and the spatial patterns they create, but on understanding the spatial distribution of the characteristics of these events. Spatial analysis aims at cases where the feature under consideration is not considered to be constantly changing in the site, but it presents values only at one point for each of the total spatial units covering the entire study area. The aim is to examine the pattern or relations of values of characteristics observed for a set of spatial units of the study area. In the present work, the island regions of the North and the South Aegean were examined with polygon spatial units of the islands forming a discontinuous spatial distribution. The characteristics under consideration include variables related to the accessibility of the islands to mainland Greece and specifically to Attica. The methods used in this work include:

- Summary statistics and cartographic displays
- Clustering: encompasses the process of organizing a dataset in groups in such a way that observations within a group are more similar to each other than they are observations belonging to a different group.
- Spatial autocorrelation: quantifies spatial association between attribute values measured at different spatial units.

Clustering methods aim to assign objects into groups based on similarity. They are used for exploratory analysis to find patterns in data. Input data for clustering could be either numerical, categorical, or Boolean. A typical clustering approach must include a method for computing similarity or distance between objects in feature space. Distance is the most commonly used method of computing similarity for numerical datasets. Lower values of distance measures indicate more similar objects. Distance metrics can use Euclidean distance, Manhattan distance or any other similar distance calculation approach (Amato et al. 1998). A good clustering result is usually a set of clusters with high intra-class similarity and low inter-class similarity. Clustering methods include partitioning methods, hierarchical methods, model-based methods, and density-based methods. Partitioning methods construct various partitions and then evaluate them by some criterion. Hierarchical methods create a hierarchical decomposition of the set of objects using some criterion. Model-based methods hypothesize a model for each cluster and find best fit of models to data. Finally, density-based methods are guided by connectivity and density functions. Partitioning methods, which are used in this chapter, construct a partition of a dataset D of n objects into a set of k clusters. Given k, the algorithm identifies a partition of k clusters that optimizes the selected partitioning criterion.

Spatial autocorrelation can be defined as the average similarity between values of a variable derived from geographical distribution of regions within which these values occur. It quantifies the similarity of data within an area, under the assumption that the values of the characteristics a spatial phenomenon are self-related (Cliff and Ord 1973, 1981), the level of interdependence between variables, and the intensity of interdependence. It also evaluates whether the values are interrelated, and if so whether there is a spatial pattern in that relationship.

Goodchild et al. (1992) states that spatial autocorrelation examines the extent to which the values of a variable at a position are similar to the values of the same variable that are geographically close to it. Therefore, spatial autocorrelation is an evaluation of the spatial structure of a variable with respect to spatial position of its values. The existence of spatial autocorrelation essentially indicates the increased likelihood of observing similar values for less geographically distant regions (Anselin 2001). Spatial autocorrelation can be either positive or negative. Positive spatial autocorrelation corresponds to the case where similar values occur at smaller distances, while negative spatial autocorrelation corresponds to the case where dissimilar values occur at greater distances. When spatial autocorrelation is not statistically significant, the spatial distribution patterns appear as random.

To compute the level of spatial autocorrelation for a variable, similarity values must be computed between all pairs of observations. However, not all observation pairs are considered, only those that are deemed neighbors. The measurement of the correlation between neighboring observations in a pattern is done using spatial autocorrelation indices. There are many indicators of spatial autocorrelation (Cliff and Ord 1970, 1973). In this work, the local version of Moran's I index (Haining 2003) is adopted to evaluate the nature and extent of spatial autocorrelation for accessibility-related variables between the Aegean Sea islands and Attica.

3 Materials and Methods

The study area includes two island regions of Greece, the North Aegean Region and the South Aegean Region (Map 1).

The North Aegean Region includes nine inhabited islands belonging to five municipalities (Lesvos, Chios, Lemnos, Samos, and Ikaria). Its total area reaches 3836 km^2. Five islands are relatively large in terms of area (Lesvos, Chios, Samos, Lemnos, Ikaria) and four are small (Agios Efstratios, Inousses, Psara, Fournoi). The Region also includes many smaller uninhabited islands and many rocky islands. Its population is 199,231 inhabitants (Greek Statistical Agency 2011). The insular character of the North Aegean Region and the geographical location of its islands determine the relative position of the Region at country and the EU level. Its extent and position create unfavorable effects regarding its integration into existing communication and transportation networks. For example, islands of the North Aegean Region have limited transportation time slots, seasonality of ferry routes, and relatively increased transportation costs. The airports of the North Aegean Region are located in the five major islands, the most recent airport being in Ikaria.

The South Aegean Region includes the island clusters of Cyclades and Dodecanese, located at the southeastern edge of Greece. This Region has a total area of 5286 km^2 and covers 4% of the total area of the country. It consists of 79 islands, of which 48 are inhabited and 178 are islets. Its population is 308,610 inhabitants (Greek Statistical Agency 2011). The regional district of Dodecanese has an area of

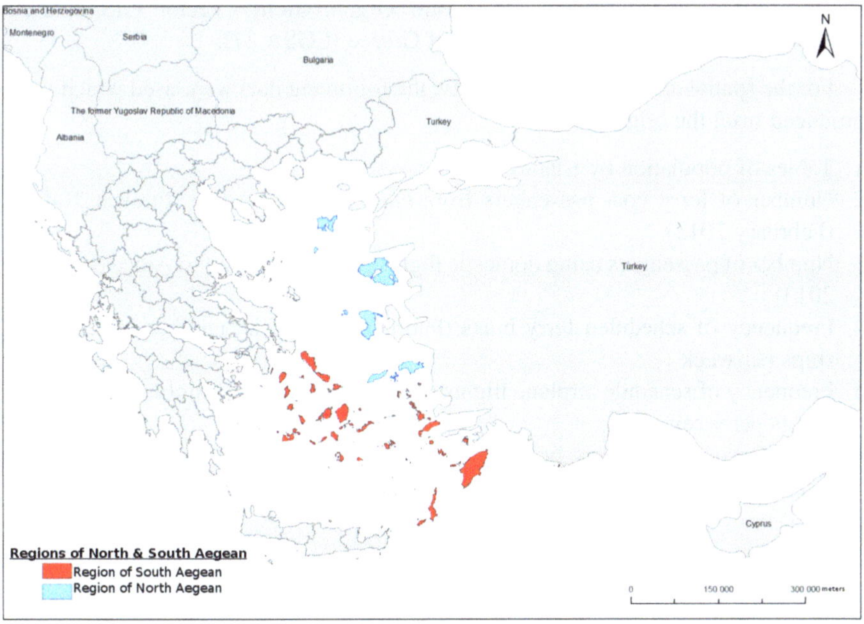

Map 1 Study are of the Aegean Sea in Eastern Mediterranean Sea

2714 km^2 and a population of 190,770 inhabitants. Its main islands are: Rodos, Karpathos, Kos, Astypalea, Kalymnos, Kasos, Tilos, Symi, Leros, Nisyros, Patmos, Chalki, Lipsi, Megisti or Kastelorizo, and Agathonisi. The prefecture of Cyclades has an area of 2572 km^2 and a population of 117,840 inhabitants. It consists of the following islands: Naxos, Andros, Paros, Tinos, Milos, Kea, Amorgos, Ios, Kythnos, Mykonos, Syros, Thira, Serifos, Sifnos, Sikinos, Anafi, Kimolos, Antiparos, Folegandros Iraklia, Schinoussa, and Koufonisia. There are other islands and islets such as Delos, Makronisos, Gyaros, and Despotiko. Of the 57 inhabited islands of the North and South Aegean Regions, seven small islands were not included in our work because of their small size.

Secondary data from the following authorities were obtained:

- Hellenic Statistical Authority
  - Census Population Year, 2011
  - Passengers per port of islands of the North and South Aegean per quarter and year 2011
- Civil Aviation Authority (Domestic passenger flights per airport and month of 2011)
- Greek Travel Pages
  - Time duration and frequency of schedule for ships and planes (February 2013)
  - Connectivity and number of intermediate stops between islands of the North and South Aegean (February 2013)
- OKXE (Greek Mapping and Cadastre Organization)—Vector GIS datasets depicting the level of the prefectures of Greece (EGSA 87).

For the spatial analysis of accessibility, the following data were used which were produced from the original data:

1. Tables of population by island
2. Number of ferry boat passengers from each island to Attica (Piraeus, Rafina) (February 2013)
3. Number of passengers using domestic flights from each island to Attica (February 2013)
4. Frequency of scheduled ferry boats (February 2013) and number of scheduled ships per week.
5. Frequency of schedule airplane flights (February 2013) and number of scheduled flights per week.
6. Real time of trip for ferry boats (in hours)
7. Real time of trip for flights (in hours)

8. Waiting time in hours for ferry boats. The waiting time was calculated for islands that do not have direct ferry connection to Attica and is the time from arrival to port until departure of the first ship to Attica.
9. Mean fare cost (ferry) from Attica (Piraeus, Rafina) to each of the islands.

In the present case, accessibility is being considered to concern the island's permanent residents. For this purpose, the data used refer to February, when travels are mainly made by permanent residents of the islands, while the passenger data refer to those arriving from the islands to Attica and not to those who depart. The variables used were weighted by the number of passengers for each transport mode in order to obtain a weighted measure for comparison purposes. The financial cost of travel was not directly taken into account as it is formed based on supply/demand basis which varies considerably between months.

The final variables, which were used after processing the original data sources are the following: Weighted total journey time (tt_st), Weighted frequency of scheduled weekly connections (fr_st), Passengers over population (p_p), Connectivity of the islands (con), Network standardized betweenness centrality (cen). tt_st: The total weighted journey time calculated from the real time of voyage by ship and airplane including the waiting time, alongside the travel times which were weighted by the number of passengers of ships and airplanes. These are defined as follows:

1. fr_st: The weighted route frequency calculated from the frequency of both sea and air routes weighted by the passenger numbers of the two transport modes
2. p_p: Passengers/population calculated from the number of passenger of ferries and airplanes (based on February) divided by the population of the islands, which reflects the number of journeys made by residents per year. It is used to assess the relationship between permanent residents of each island and passengers.
3. con: Connectivity of the islands calculated from the total number of transportation connections for each island in the study area
4. cen: Network standardized betweenness centrality which is a measure of the centrality of a network node based on the number of routes passing by each island standardized by the total number of routes in the study area.
5. mcost: Mean ticket cost with ferry from Attica (Piraeus, Rafina) to each of the islands.

The variables of total weighted time (tt_st) and weighted route frequency (fr_st) were based on Spilanis et al. (2005) and Spilanis et al. (2012). The weighted total time (tt_st) was calculated according to the following equation (Eq. 1). This is by summing the travel time trip by ferry boat (rt_ship) and the waiting time (wt) and then weighted by the number of passengers on ship (pas_ship) and airplane (pas_air) according to the following formula:

$$\mathrm{tt\_st} = \frac{(\mathrm{rt\_ship} + \mathrm{wt}) * \mathrm{pas\_ship}) + (\mathrm{rt\_air} * \mathrm{pas\_air})}{\mathrm{pas\_ship} + \mathrm{pas\_air}} \quad (1)$$

where

rt_ship: real time for trip ship
wt: waiting time
rt_air: flight time by airplane
pas_ship: number of passengers (ferry boat)
pas_air: number of passengers (airplane)

The weighted route frequency (fr_st) was calculated according to the following formula (Eq. 2). The calculation includes the weekly frequency of ferry boats and the weekly frequency of scheduled flights weighted by the number of passengers of both transport modes.

$$\text{fr\_st} = \text{fr\_ship} + \left(\text{fr\_air} * \frac{\text{pas\_air}}{\text{pas\_ship}}\right) \tag{2}$$

where

fr_ship: number of scheduled weekly ferry boats
fr_air: number of scheduled flights
pas_ship: number of ferry boat passengers
pas_air: number of airplane passengers

The third variable (Eq. 3) refers to the ratio of passengers/population (p_p) that expresses the mean number of trips per resident of an island and is the sum of the number of ferry boat passengers (pas_ship) and the number of airplane passengers (pas_air) divided by the population of the island (pop).

$$\text{p\_p} = \frac{\text{pas\_ship}_{\text{feb}} + \text{pas\_air}_{\text{feb}}}{\text{pop}} \tag{3}$$

where

$\text{pas\_ship}_{\text{feb}}$: number of ferry boat passengers. As the Hellenic Statistical Authority data are quarterly, the number of passengers in February was calculated based on the first quarter of the year
$\text{pas\_air}_{\text{feb}}$: number of airplane passengers
pop: population from the 2011 population census

The third variable is the connectivity between islands. This calculation involved the extensive analysis of published ferry schedules for the year 2011 in order to identify the number of stops of all ferry routes in the Aegean Sea. This revealed a number for each island which depicts the minimum number of stops required to travel by ferry boat from mainland Greece (Attica) to any island in the Aegean Sea. This variable may reveal which are the less connected islands in the area and those islands which are not so accessible from mainland Greece.

It was assumed that the transportation network between the islands by ferry boats could also be represented by a directed graph in the form of nodes (islands) and edges (transportation connections). This enabled the quantification of the relative centrality of each node (island) in terms of transportation in the Aegean Sea. A node with a higher centrality measure is a node with high involvement in many relations regardless of outbound/inbound direction or passengers. There are a number of centrality measures to apply. The standardized betweenness centrality (cen) measure per island was used. As can be seen in Eq. 4, for each island *I*, the standardized betweenness centrality is the sum of the proportions, for all pairs of islands *j* and *k*, in which island *i* is involved in a pair's ferry boat trip (from *j* to *k* or from *k* to *j*) as an intermediate stop.

$$\mathrm{BC}(i) = \sum_{\substack{j,k \in V \\ j \neq k \neq i}} \frac{\sigma_{jk}(i)}{\sigma_{jk}} \tag{4}$$

An island with relatively high betweenness centrality occupies a "between" position in the transportation network connecting many pairs of other islands in the network. As an intermediate point in the shortest path connecting two other islands, a "between island" might influence the flow of passengers. This measure is standardized by the total number of all intermediate ferry boat trips in the network of the Aegean Sea. The mean fare cost by ferry from Attica (Piraeus, Rafina) to each of the islands (mcost) has been calculated based on data from travel agencies. The 'mcost' variable is the weekly average fare cost of all ferry routes connected to an island.

The R statistical programming language (CRAN 2017) was used to make all necessary calculations and produce the above measures for the 50 islands of the North and South Aegean Regions. We then utilized a number of summary statistics (graphical and quantitative) for the effective presentation of data trends through summary measures, grouping, and depiction. Two types of descriptive measures were used. The measures of central tendency determine the distinctive positions into the data range for each island (average, median, etc.). The measures of variability express the fluctuation of values of a variable around the measures of central tendency such as the range, the interquartile width, the variance, and the standard deviation. Histograms of the relative frequencies of the variables, box-plot, calculation median, lower and upper quartile, standard deviation, and minimum and maximum value were used. The variables were also depicted on thematic maps.

Partitioning Around Medoids (PAM) is a clustering method similar to the popular *k*-means methods initially described in Kaufman and Rousseeuw (2009). This clustering algorithm is based on the extended search for *k* representative observations (medoids) among the observations of the dataset which should represent the structure of the data. Then, *k* clusters are constructed by assigning each observation to the nearest medoid. The goal of this method is to find *k* representative objects which minimize the sum of the dissimilarities of the observations to their closest representative object. This method searches for a good initial set of observations (Build phase) and then it finds a local minimum for the objective function, that is, a solution such that there is no single switch of an observation with a medoid that will decrease the objective (the SWAP phase) (Reynolds et al. 2006). Compared to the *k*-means approach, this approach is more robust because it minimizes a sum of dissimilarities instead of a sum of squared Euclidean distances.

The results of spatial analysis can form the basis for the development of transport policies and other relevant actions to improve the accessibility of the North and South Aegean Region islands. For this reason, visual data presentations with maps, diagrams, and selected exploratory methods of spatial statistics for analyzing geographic diversification of accessibility of the Northern and Southern Aegean islands were used.

4 Analysis and Results

Initially, an exploratory data analysis on each variable was conducted in order to evaluate the general tendency of its values and identify the characteristics of each distribution. The variable "weighted total journey time" expresses the travel time required by island residents to reach Attica either by ferry boat or by airplane. It includes the actual journey time by ferry boat and airplane as well as the waiting time, when it there is no direct ferry connection to Attica. It is weighted by the estimated number of passengers. Fig. 1 depicts the histogram and the box-plot of the distribution of values for all 50 islands in the study area, and Table 1 shows the descriptive statistics of the variable distribution. The results indicate that medium and large islands have relatively small weighted journey times, even if they are located at relatively longer distance from Attica. These islands include: Lesvos, Chios, Leros, Kalymnos, and Kos that have weighted journey time between 2 and 5 h. This is an indication of relatively increased number of available transport options. These islands have fully working airport facilities. Islands such as Ikaria, Samos, Astipalaia, Rodos, and Karpathos also have airplane connection which reduces the weighted total journey time to a certain extent, but they still appear in the range of 5–8 h of weighted total journey time. According to the results, islands closer to Attica irrespective of their size have relatively smaller values of weighted total journey time. The smallest values appear for the island of Kea which is the only island where the weighted total journey time is less than an hour (0.98) and follows Andros, Kythnos, Paros, Syros, and Serifos with less than 5 h, while Mykonos has exactly 5 h of weighted total journey time. Higher values appear in small islands located at medium and relatively long distance from Attica with no airport and direct ferry boat connection such as Arki, Lipsi, Symi, Halki, Tilos, Oinousses, Kasos, and Megisti.

The histogram of the weighted total journey time shows that the distribution of the variables resembles the normal distribution with the hollow slightly to the right, meaning that there are slightly smaller values and fewer large values which is an indication of positive asymmetry. Fifty percent of the values are between 6 and 11.31 h. The median is 8.16 h and the average is 9.27 h which confirms that there are

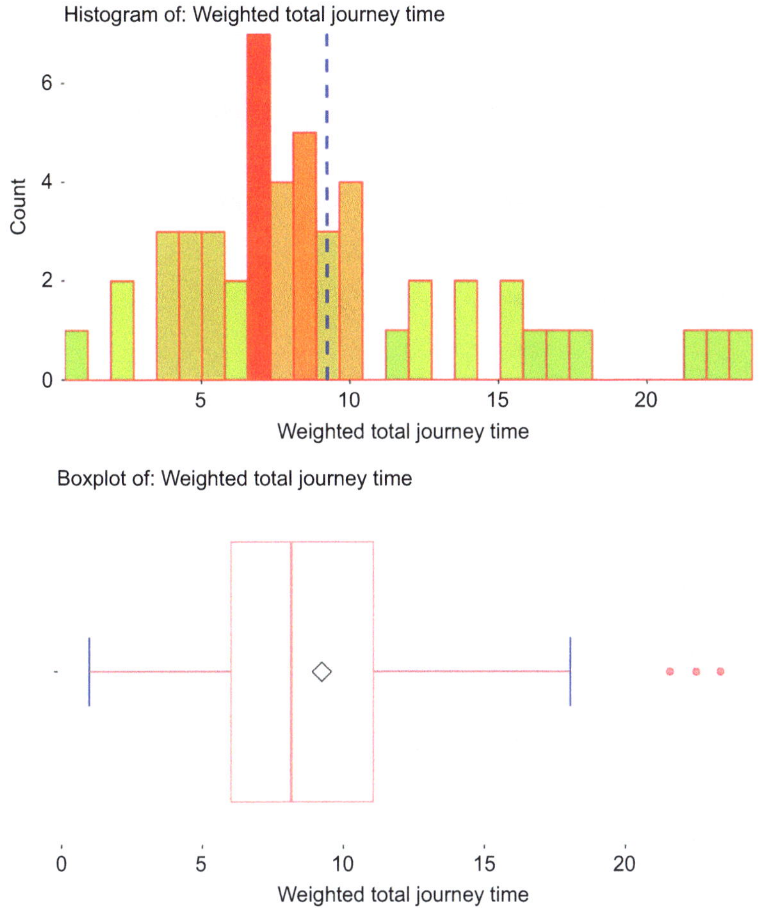

Fig. 1 Histogram and box-plot of distribution of "weighted journey time" variable for the transportation network of the Aegean Sea. Somehow symmetric distribution

Table 1 Descriptive statistics of "weighted total journey time" (in hours)

Average	Median	Lower quartile	Upper quartile	Min	Max	Standard deviation
9.27	8.16	6	11.31	0.98	23.37	5.11

smaller values. Seventy five percent of the values are less than 11.31 h. According to the box-plot (Fig. 1), the mean of the distribution is affected by some extreme values. These values are small islands at long distances that do not have an air connection; more specifically, the islands of Halki, Pserimos, and Oinousses. Pserimos and Oinousses do not have a direct connection to Attica, and their high values of weighted total journey time are also due to the overall waiting time between transportation connections (Map 2).

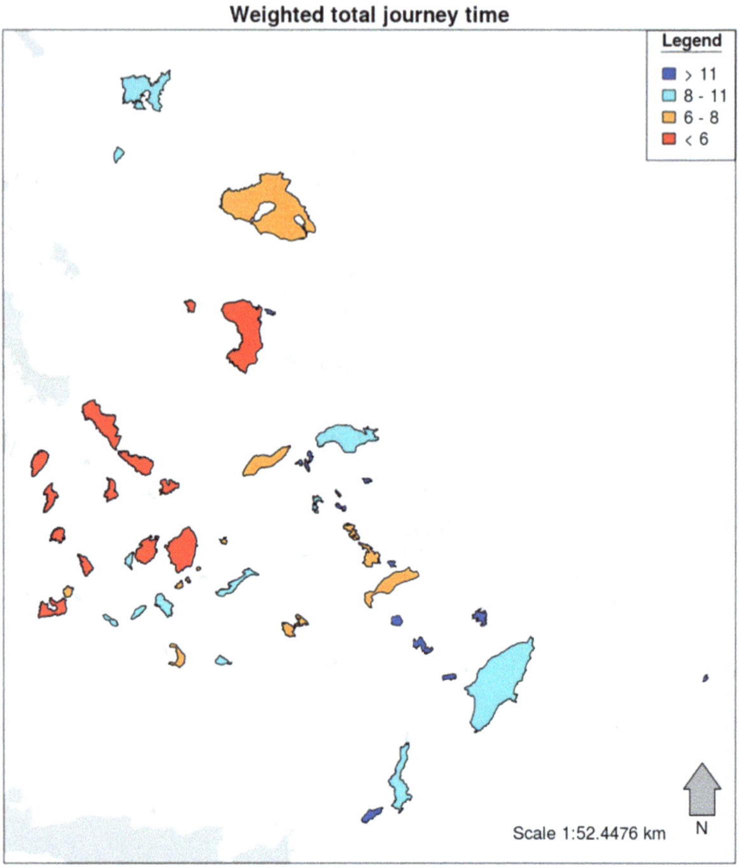

Map 2 Spatial distribution of "weighted total journey time" measure for the transportation network of the Aegean Sea. Smaller islands close to the border, show relatively increased weighted total journey time

The results of "weighted route frequency" indicate that Rodos and Lesvos have the higher values. These are two islands with fully working airport facilities and relatively higher population than other islands in the study area. Mykonos is the third in the row even though its permanent winter population is smaller than that of Lesvos and Rodos. Kos and Andros have also relatively high values of weighted frequency of scheduled weekly connections with values greater than 16 weekly connections. At the lower level with up to two weekly connections, there are the islands of Kasos, Tilos, Symi, Nisyros, Patmos, Halki, Agathonisi via Patmos, Megisti, Arki via Patmos, and Donousa. These smaller islands are located further away from Attica. Of course, there are some islands that even though they are located closer to Attica, they have low values of weighted frequency of scheduled weekly connections such as Folegandros and Kimolos. The histogram of "weighted frequency of scheduled

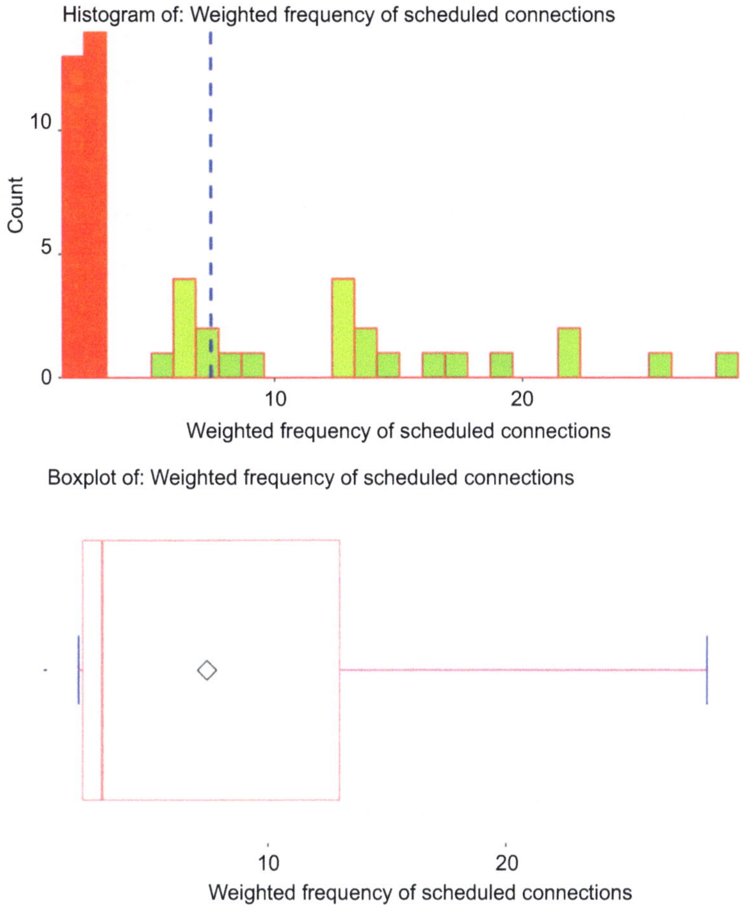

Fig. 2 Histogram and box-plot of distribution of "weighted route frequency" variable for the transportation network of the Aegean Sea. Right-skewed distribution with many small values

weekly connections" (Fig. 2) indicates that the distribution is located with the hollow to the right, which indicates the existence of relatively smaller values, i.e., positive asymmetry. Map 3 shows the spatial distribution of the variable, indicating that larger islands in the eastern part of the study area as well as popular islands of Cyclades possess the higher values. According to the descriptive statistics (Table 2), there is a substantial difference between median (three weekly connections) and mean (7.45 weekly connections) which confirms that there are smaller values. On quarter of the values are up to two (weekly connections) and half of the values are between two and 13.1 weekly connections. Three quarters of the values are below 13.1 while the higher value is found for the island of Rodos (28.449).

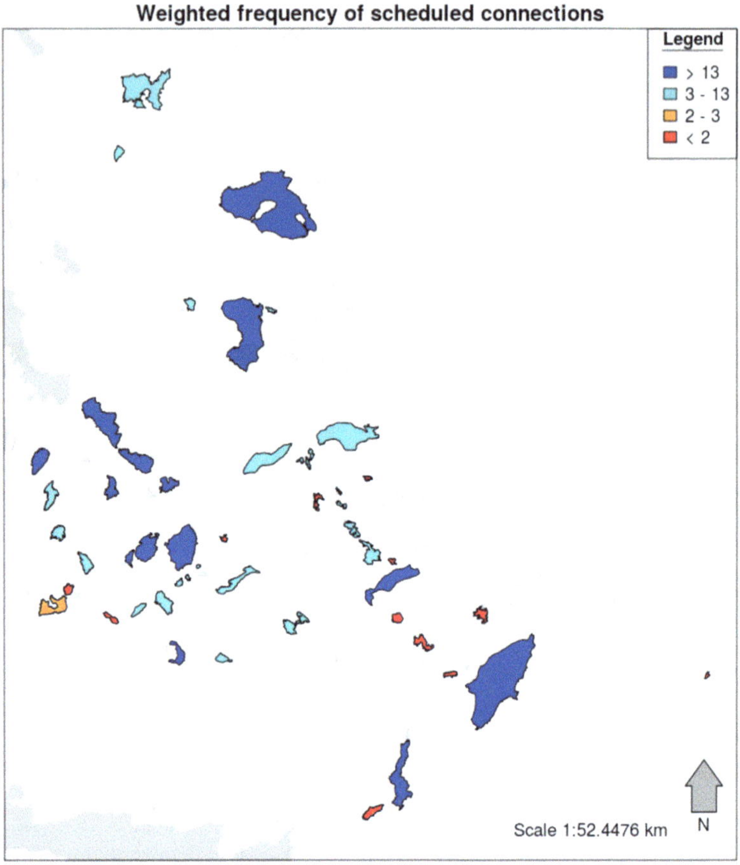

Map 3 Spatial distribution of "weighted route frequency" measure for the transportation network of the Aegean Sea. Major islands show increased weighted frequency of scheduled weekly connections

Table 2 Descriptive statistics of "weighted route frequency" variable

Average	Median	Lower quartile	Upper quartile	Min	Max	Standard deviation
7.45	3	2	13.1	2	28.84	7.02

According to the results, the values of the "weighted route frequency" is small for most islands with only a few greater values mainly due to the existence of airports in some islands, some with large population such as Rodos and Lesvos. A very interesting finding is the existence of a "gap" between values 9 and 13; namely, there is no island that has weighted frequency schedule from 8.8 up to 12.9 weekly connections. The results of this variable make clear that supply of weekly transportation connections follows the demand which is based on population size.

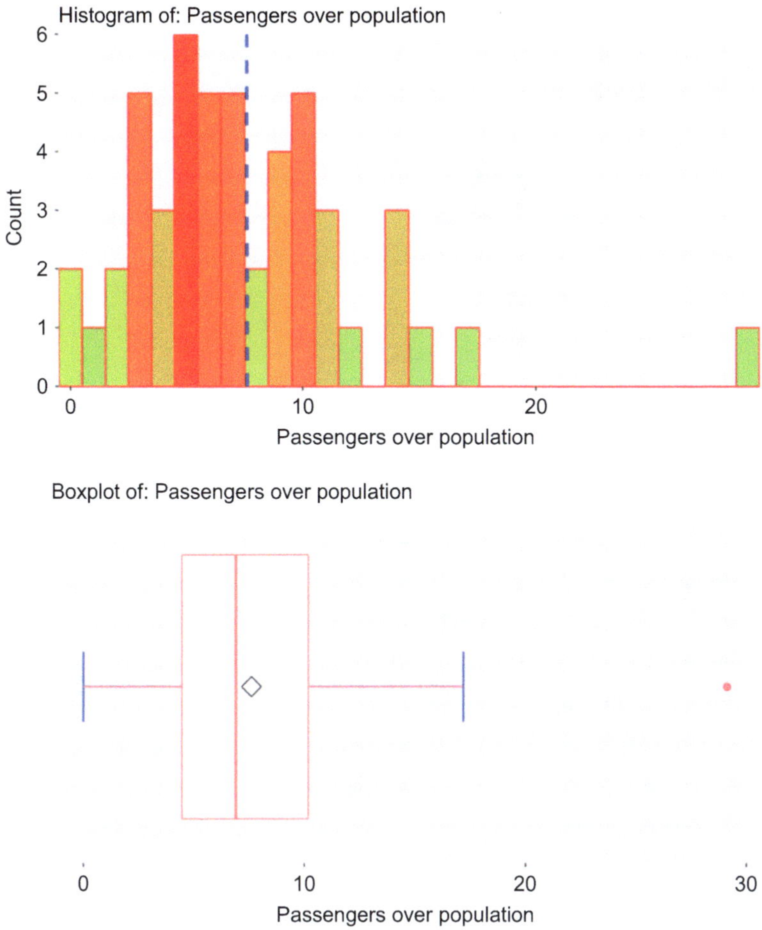

Fig. 3 Histogram and box-plot of distribution of "passengers over population" variable for the transportation network of the Aegean Sea. Somehow symmetric distribution

The results of the relationship between passengers and population of the islands can be seen in Fig. 3 and Map 4. The mean value of the variable "passenger to population ratio" is 7.6. According to these results, the islands of Pserimos, Antiparos, Kalimnos, Kasos, Lesvos (2.6), Lipsi, Rodos (3.2), and Samos (3.2) have somehow a smaller ratio, while the islands of Tinos, Iraklia, Kythnos, and Kea have greater values. Half of the islands show values between 4.4 and 10.1 while three quarters of the islands have values less than 10.1. This variable shows that there is a relationship between the total number of passengers and the total population of an island, under the assumption that more passengers are expected to be observed in islands with greater population. An interesting result from the analysis of this variable is that some larger islands, such as Lesvos, Rodos, Samos, and Chios,

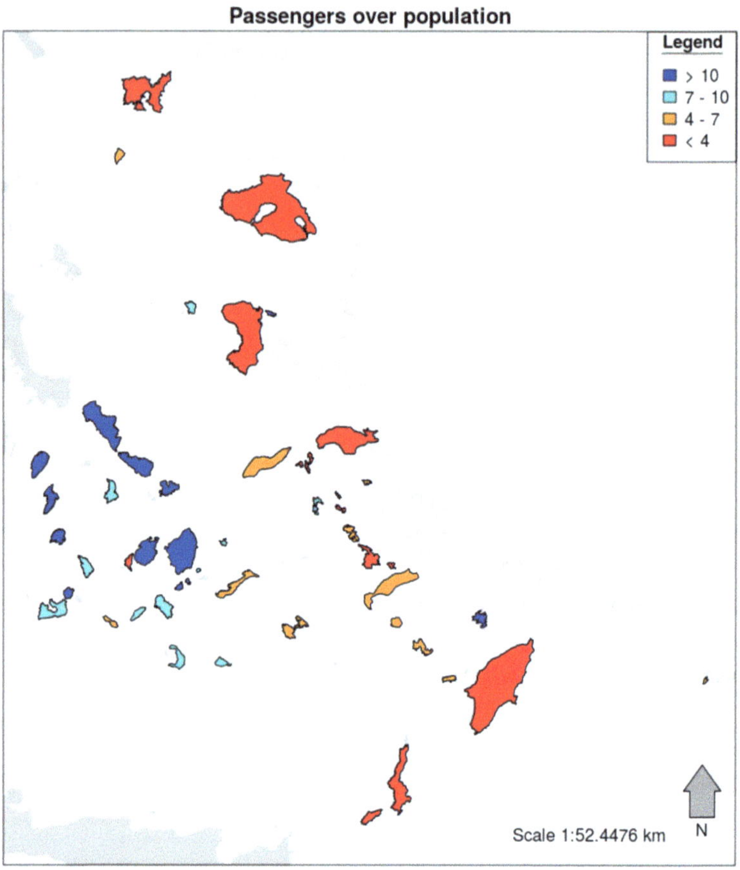

Map 4 Spatial distribution of "passengers over population" measure for the transportation network of the Aegean Sea. Major islands close to the border show relatively increased percentage of passengers over population

Table 3 Descriptive statistics of "passengers over population" variable

Mean	Median	Lower quartile	Upper quartile	Min	Max	Standard deviation
7.6	6.9	4.4	10.1	0	29.1	5

appear in the lower part of the distribution indicating that there is a relatively weak relationship between passengers and total population on each island. This may occur because of the existence of several facilities in these two islands (administrative facilities, health services, commercial businesses, and university facilities) (Table 3).

Figure 4 and Map 5 depict the results of the analysis of the variable "connectivity with other islands." The following Table (Table 4) shows the descriptive statistics for this variable. The total connectivity of each island shows how many islands are

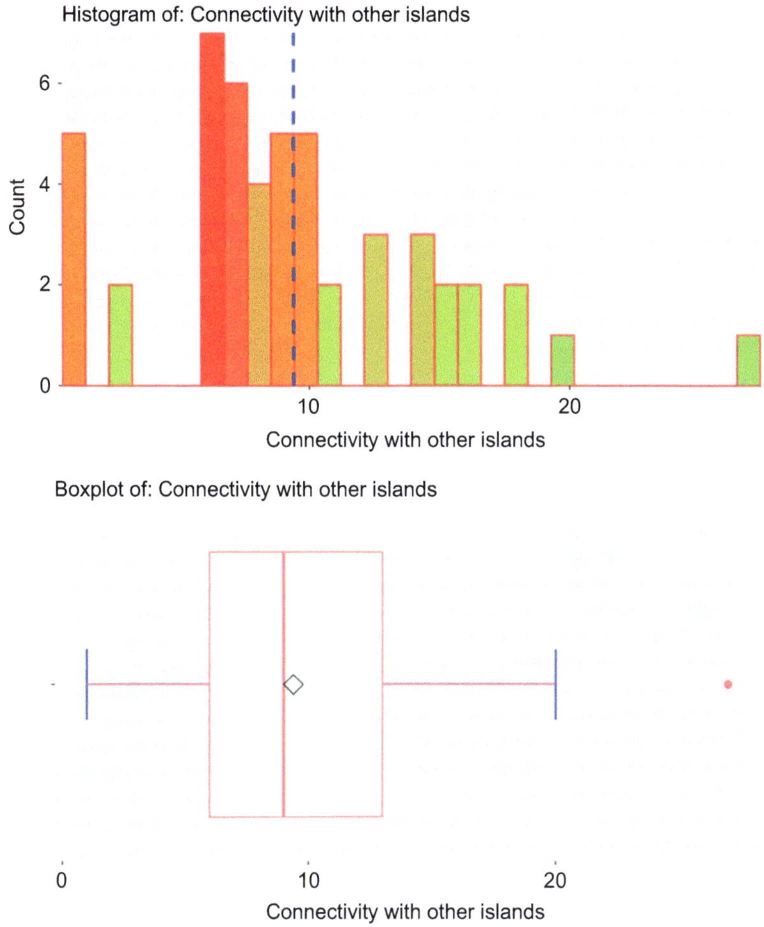

Fig. 4 Histogram and box-plot of distribution of "number of transportation connections" variable for the transportation network of the Aegean Sea. Somehow symmetric distribution

connected to other islands in the study area. Lower values indicate islands with fewer opportunities of commuting to other islands in the Aegean Sea while greater values indicate islands with more opportunities of commuting to other islands. Half of the islands show values between 6 and 13 islands. The mean value is 9.4 islands while the greater value appears for the island of Syros (the capital of the Cyclades). Islands with relatively small values include: Agios Efstratios, Psara, Oinousses, Pserimos, while the islands with greater values are Syros, Paros, Milos, Rodos, Santorini, and Naxos. The islands of the North Aegean Region have low connectivity options except for Samos, while the islands of the South Aegean Region appear to have generally higher connectivity. Higher values of connectivity, more than 20, can be found in the islands of Syros (27 connections) and Paros (20 connections). According to the histogram of the distribution, there is a hollow to the right with

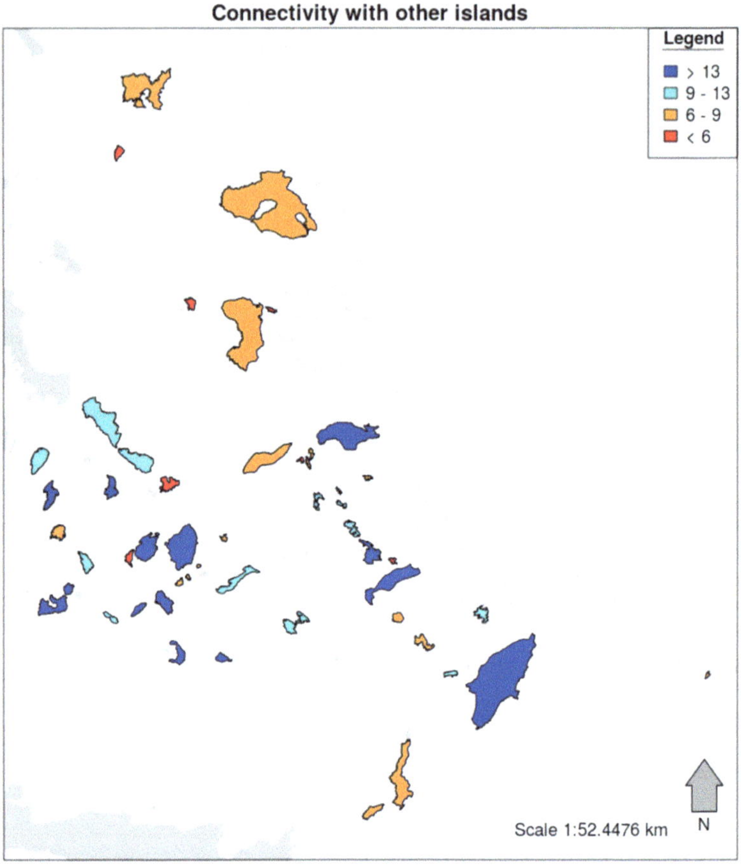

Map 5 Spatial distribution of "number of transportation connections" measure for the transportation network of the Aegean Sea. Islands in the central and south of the Aegean Sea, show relatively increased number of transportation connections

Table 4 Descriptive statistics of "connectivity with other islands" variable

Mean	Median	Lower quartile	Upper quartile	Min	Max	Standard deviation
9.4	9	6	13	1	27	5.3

more small and medium values and fewer high values, while Syros is located at the right end which is categorized as an extreme value. Map 5 depicts the connectivity of all islands and the number of stops required to connect two islands. Even though a dense communication network between the islands seems to exist, the frequency of schedules between islands is not taken into account since two of the islands appear to be connected although there is only one scheduled weekly connection. Values of 1 indicate that there is a direct connection between two islands, while values of

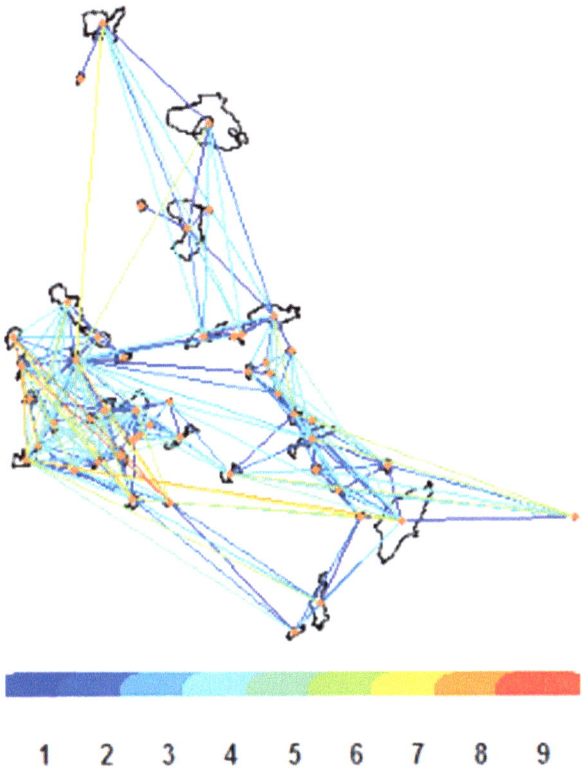

Map 6 Spatial Network visualization of transportation stops between islands in the Aegean Sea. Islands in the Cyclades group, show increased connectivity and some of them have very high number of transportation stops

2 indicate that an intermediate stop is required to connect two islands. Values of 9 indicate that eight intermediate stops are needed for an intra-Aegean travel. For example, Lesvos (north-east) is connected with six islands in total, it has direct connection with Limnos and Chios and indirect connections with Samos (one intermediate stop), Ikaria (two intermediate stops), Fournoi (three intermediate stops), and Syros (five intermediate stops) (Map 6).

The results of network analysis showed that the majority of islands have relatively small standardized betweenness centrality. Half of the islands appear between 0.2 and 8.6%. This is that relatively few islands act as an intermediate stop in the transportation network. On the other hand, there are some islands which appear to have frequent intermediate stops in the overall transportation network such as Andros and Paros. Figure 5 shows the histogram and the box-plot of the distribution of the weighted network betweenness centrality measure for the islands of the Aegean Sea. Map 7 depicts the spatial distribution of this measure. Islands in the eastern border region show relatively increased betweenness centrality as they are served by a specialized, state-funded transportation route which connects only border region islands (Table 5).

Additionally, the mean fare cost by ferry from Attica (Piraeus, Rafina) to each of the islands for the same time period was calculated in order to quantify the

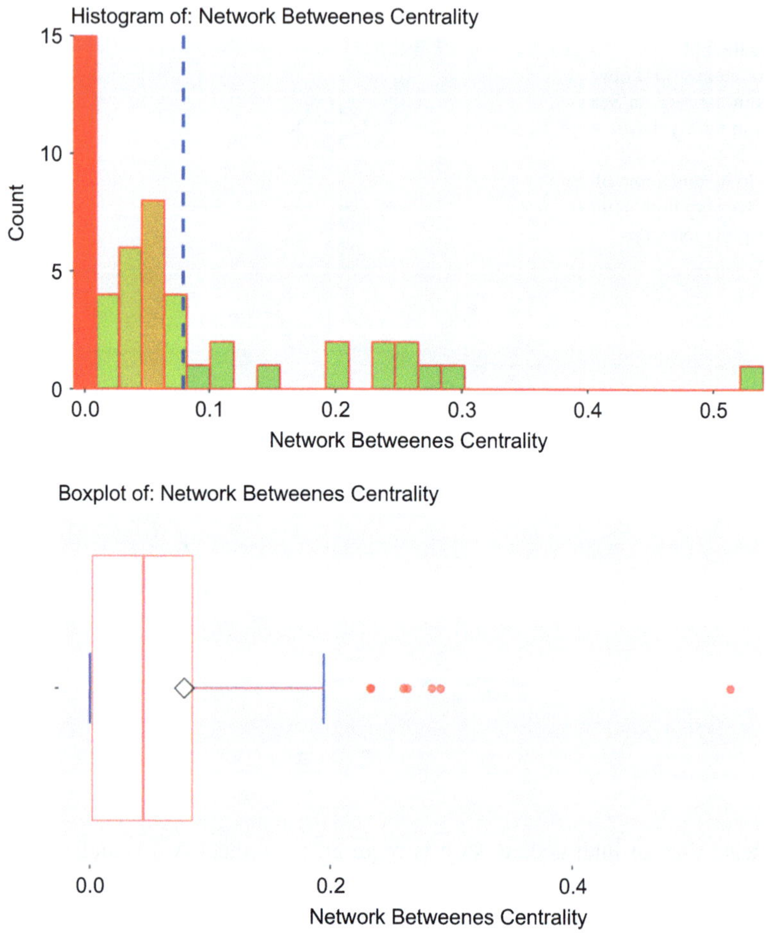

Fig. 5 Histogram and box-plot of distribution of "betweenness centrality measure" for the islands of the Aegean Sea. Right-skewed distribution with many small values

commuting cost from/to each island. This is a relative measure which in some cases is subsidized by the central government in order to reduce the cost of living is some islands. The "mcost" variable was calculated from ferry fare data and shows an almost uniform distribution with some increased frequency of values around the mean of the distribution (Fig. 6).

Furthermore, in order to quantify the spatial autocorrelation among the values of "weighted route frequency," the Moran's *I* index was used (Fig. 7). The neighborhood construction for this index was based on the four closest islands in order to quantify the relationship between distance and frequency of weekly scheduled routes. Map 8 depicts the spatial distribution of autocorrelation groups in the study area. This grouping of islands is based in the relationship of "weighted route frequency" between each island and its surrounding islands. As the following map

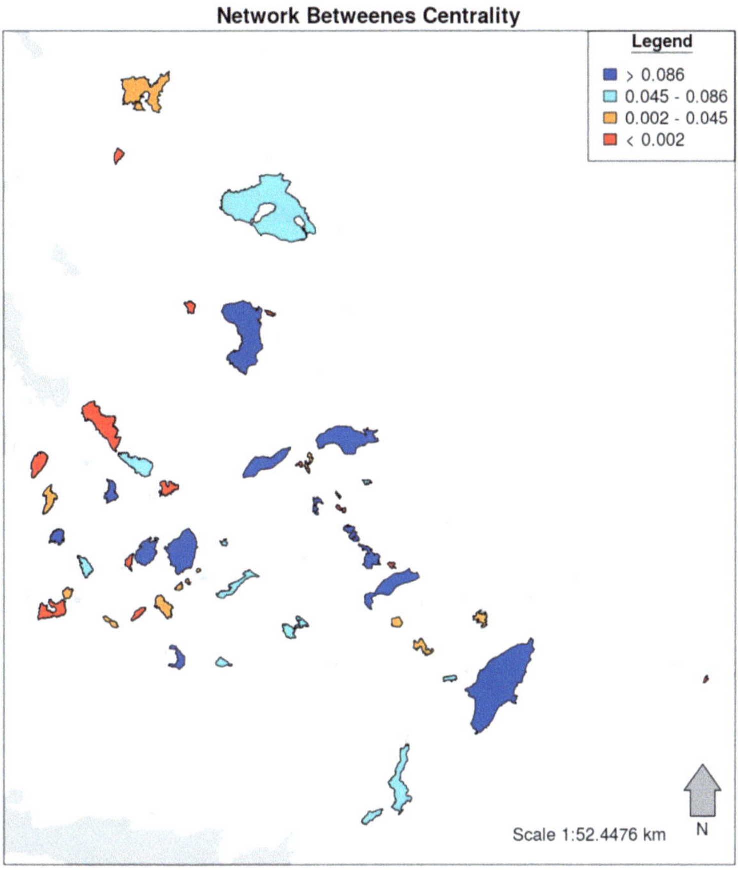

Map 7 Spatial distribution of betweenness centrality index for the islands of the Aegean Sea. Islands close to the border show relatively small network betweenness centrality

Table 5 Descriptive statistics of "network standardized betweenness centrality" variable

Mean	Median	Lower quartile	Upper quartile	Min	Max	Standard deviation
7.9%	4.4%	0.2%	8.6%	0%	53%	10.7%

shows, there are some areas with relatively small values of weekly frequency of scheduled routes surrounded by islands which also have small values of this variable (Low-Low). Islands in the Low-Low cluster are Samos and islands south of Samos as well as islands in the southern part of the Aegean Sea except for Santorini.

The Moran's I scatterplot shows that many islands of the central Aegean (Cyclades group) appear to have a relatively strong positive relationship with their surrounding islands, appearing in the upper right part of the scatterplot (High-High). In other words, the relation of the variable with respect to the values of that

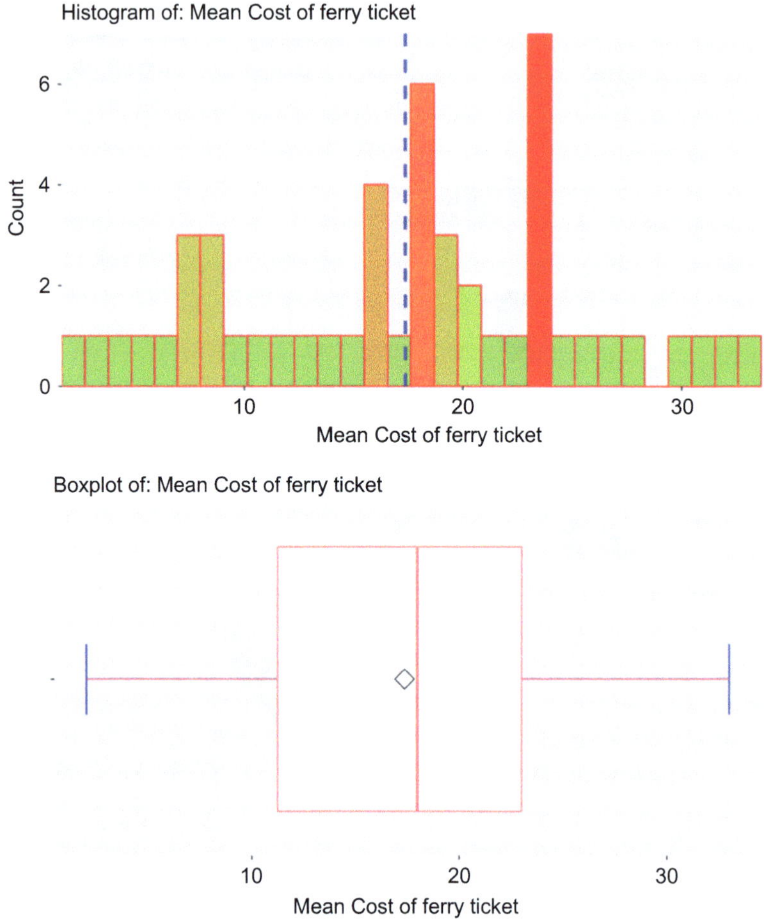

Fig. 6 Histogram and box-plot of distribution of "mean cost of ferry ticket" for the islands of the Aegean Sea. An almost uniform distribution with some increased frequency values around the mean of the distribution

variable in the neighboring locations is positive. As this area of the Aegean Sea is a very popular tourist destination in both winter and summer, this is expected to be true. The Cyclades is popular for bank holidays as well during winter as it is close enough to the city of Athens with relatively cheaper transportation options. Transportation companies tend to fulfill the increased demand for these islands by offering many and more frequent routes compared to other Aegean Sea islands.

In order to evaluate the contribution of the variables used in this study in grouping similar islands, the PAM method was used. The results are shown in Fig. 8. This is the result of partitioning around medoids based on two variables ("mean fare cost" and "cost and ferry root frequency") which explains 100% of the variability between

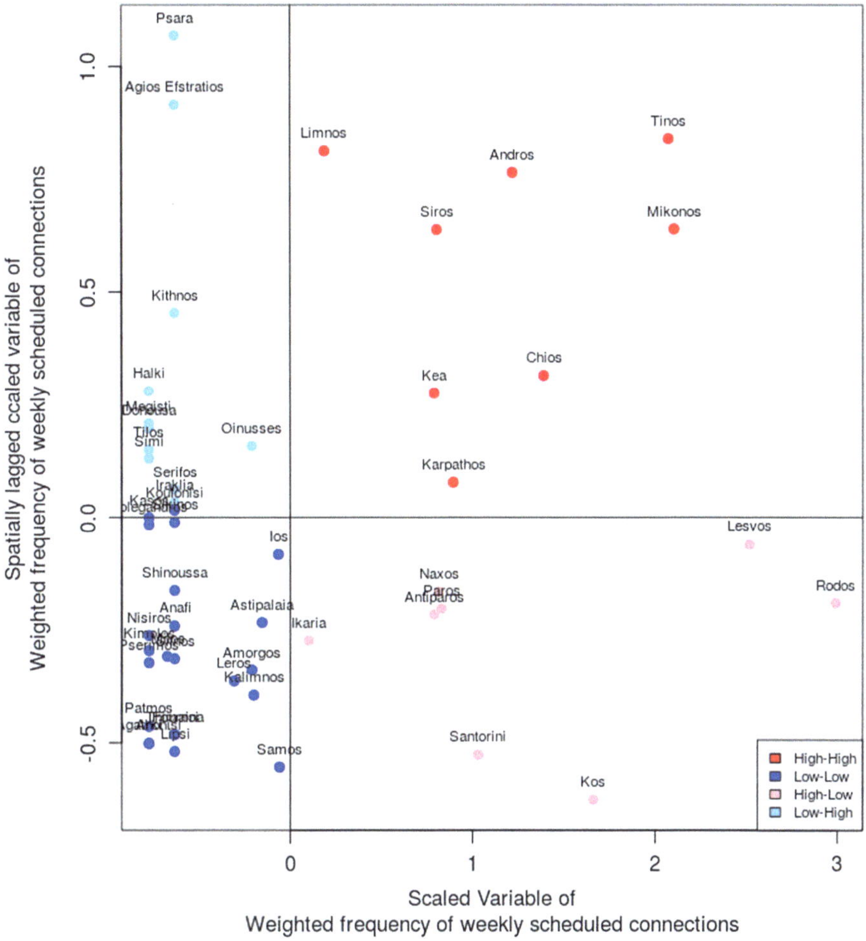

Fig. 7 Moran's *I scatterplot* of "weighted frequency of weekly scheduled connections"

observations (islands). These two variables can categorize all islands in five distinct groups of islands for policy-making regarding transportation routes and public funding in the North and the South Aegean Sea Regions.

The following table (Table 6) depicts the contents of each group:

The fifth group contains islands very close to Attica and with very frequent ferry connections mainly because they are located in the path of many ferry routes connecting other islands in the South Aegean Sea. The second group contains islands which are mainly indirectly connected to Attica and most of them have relatively smaller populations than most of the other islands in the area. Group number 1 includes islands with very frequent ferry routes (both in winter and summer months) with larger numbers of permanent residents.

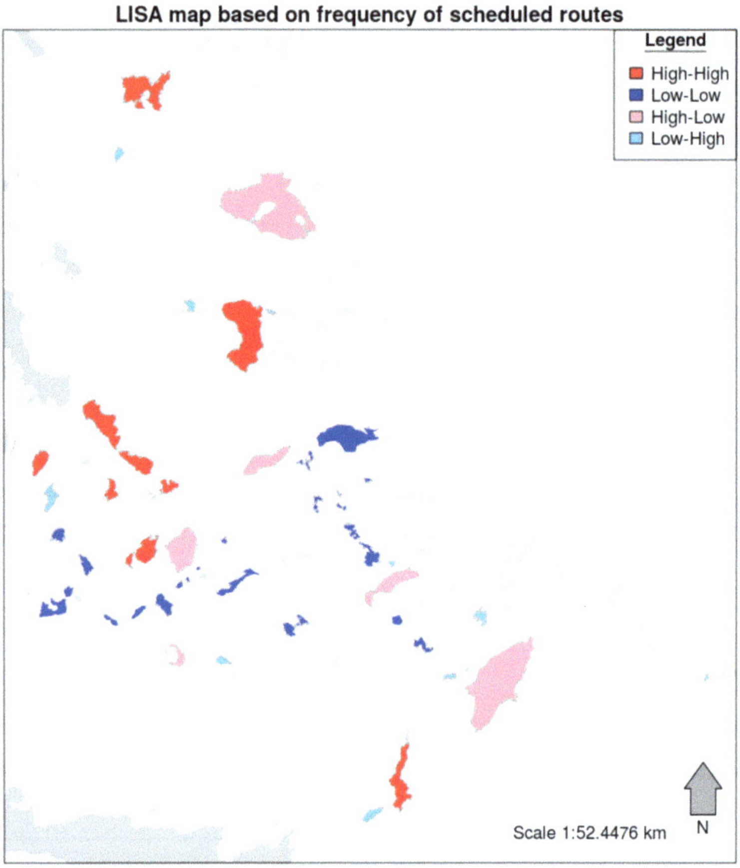

Map 8 Spatial distribution of Morans *I* values for the variable "weighted route frequency." Samos island is the larger island belonging to the Low-Low cluster

5 Discussion-Conclusions

The aim of this study was to identify and analyze the spatial patterns of accessibility characteristics of the North and South Aegean Sea islands with respect to Attica on Greece's mainland, the prefecture where Athens, the country's capital, is located. By producing new variables and calculating measures for each island in the study area, it was possible to depict and analyze measures related to transportation accessibility. The focus was to identify and examine islands with very high/low values of specific variables as well as to apply multivariate clustering techniques for grouping the islands. Classical approaches of spatial analysis were used, descriptive statistics as well as data clustering approaches for discovering relationships among variables

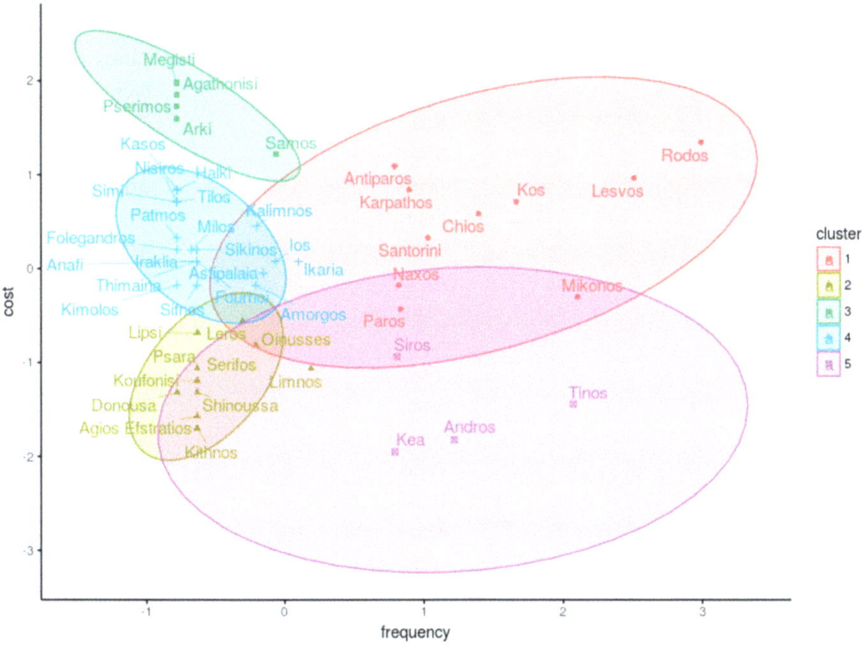

Fig. 8 Groups of islands after clustering with PAM method based on two variables (mean ticket cost and ferry root frequency)

obtained from public sources. The data sources were combined on the basis of theory and previous works in order to derive new variables related to island accessibility.

Overall, the results showed that there are some interesting cases of under and over-served islands which need further attention. The time frame of the data used is outside the touristic period, which implies that the results can only inform policies related to permanent residents of the islands and the provision of services during winter time. The values of the "passengers to population ratio" were particularly interesting, especially in the areas of the North Aegean and the Dodecanese islands. Although medium and large islands have increased values of the "weighted frequency of scheduled routes" and the "total weighted time of journey," they have smaller values of "passengers to population ratio." This is particularly evident in islands such as Oinousses, Arki, and Symi. It seems to be less applicable for other islands such as Psara, Agios Efstratios, Thirassia, Agathonisi, Nisyros, Tilos, Halki, and Megisti. On the contrary, this could be interpreted on the reverse for Lesvos and Rodos as these islands do not seem to have satisfactory levels of public and private services (administrative, health, commercial, etc.) which forces residents to move more often to Attica. Consequently, the "passengers to population ratio" is not always related with low or high accessibility but it may reflect lack or presence of public services and economic opportunities at the local level. There is also an indication of a relatively increased number of available transport options for some

Table 6 Grouping of islands based on two variables (mean ticket cost, cost, and ferry root frequency)

Cluster	Island	Cluster	Island
5	Andros	3	Samos
	Tinos		Pserimos
	Kea		Agathonisi
	Siros		Megisti
4	Ikaria		Arki
	Fournoi	2	Limnos
	Thimaina		Agios Efstratios
	Kalimnos		Psara
	Astipalaia		Oinousses
	Kasos		Leros
	Tilos		Lipsi
	Simi		Kythnos
	Nisiros		Serifos
	Patmos		Donousa
	Halki		Shinoussa
	Milos		Koufonisia
	Amorgos	1	Lesvos
	Ios		Chios
	Sifnos		Rodos
	Sikinos		Karpathos
	Anafi		Kos
	Folegandros		Naxos
	Iraklia		Paros
	Kimolos		Mikonos
			Santorini
			Antiparos

islands mainly due to increased demand for transportation (larger populations). This is the case for some medium and large islands with relatively small weighted total journey time even though they lie at longer distances from Attica.

Based on the results of the "weighted frequency of scheduled weekly connections," a very interesting finding is the existence of a "gap" between values 9 and 13, showing that there is no island that has a weighted frequency schedule from 8.8 up to 12.9 weekly connections. This indicates that the supply of weekly transportation connections (private companies) follows the demand which is related to population size. The values of "passengers to population ratio" is also interesting, especially in relation to the values of other variables for large and small islands. Islands such as Lesvos, Rodos, Samos, Limnos, and Chios appear in the lower part of the distribution indicating that there is a relatively weak correlation between passengers and total population of each island. This may happen because of the existence of many facilities in these islands (administrative facilities, health services, commercial businesses, and university facilities). Furthermore, islands such as Syros, where the headquarters of the Cyclades administration are located, shows increased values of connectivity with other islands (inter-islands connections). According to the results of PAM, high accessibility spots are located in a particular part of the Cyclades, specifically in the islands of Andros, Mykonos, Tinos, Syros, and Santorini. On the contrary, low access spots appear in the northern Dodecanese and include the islands of Fournoi, Thymena, Agathonisi, Lipsi, and Pserimos.

In conclusion, it can be said that the use of spatial analysis techniques could offer important support in formulating transportation policies in the North and South Aegean Sea Regions especially because of the fragmentation of this geographical area. The incorporation of space in such policies may help better identify hot spots that warrant specialized analysis. The results of data clustering are also helpful as the resulting island groups could assist in transportation policy-making in the study area by zoning areas or/and assigning new funded routes for under-served islands. The seasonality of economic activities in the North and South Aegean Sea islands is another problem that policy-making should address, as the demand for administration services and transportation services has a monthly and annual variation following the supply of private resources and the demand of population and tourist activities. In a very fragmented geographic space such as the Aegean Sea, where transportation between islands may be very difficult and costly (inter-island connectivity up to nine stops), new approaches in conceptualizing transportation services and public policy priorities are needed. The use of specialized spatial analysis methods along with detailed data may help policy-makers to identify areas which require attention in order to shape policies that balance equality and proportionality of services.

References

Ackerman, E. A. (1958). *Geography as a fundamental research discipline*. Chicago: University of Chicago.

Amato, N. M., Bayazit, O. B., Dale, L. K., Jones, C., & Vallejo, D. (1998). Choosing good distance metrics and local planners for probabilistic roadmap methods. In: *Proceedings of the 1998 I.E. International Conference on Robotics and Automation* (Cat. No.98CH36146) (Vol. 1, pp. 630–637).

Ankre, R., & Nilsson, P.-Å. (2016). Remote yet close: The question of accessibility in the Faroe Islands. In *Archipelago tourism: Policies and practices* (p. 137). Farnham: Ashgate.

Anselin, L. (2001). Spatial econometrics. In *A companion to theoretical econometrics* (pp. 310–330). Malden: Blackwell.

Bailey, T. C. (1994). A review of statistical spatial analysis in geographical information systems. In *Spatial analysis and GIS* (pp. 13–44). London: Taylor & Francis.

Cliff, A. D., & Ord, J. K. (1970). Spatial autocorrelation: A review of existing and new measures with applications. *Economic Geography, 46*, 269–292.

Cliff, A. D., & Ord, J. K. (1973). Spatial autocorrelation. In *Monographs in spatial and environmental systems analysis; 5*. London: Pion.

Cliff, A. D., & Ord, J. K. (1981). *Spatial processes: Models & applications*. London: Pion.

CRAN. (2017). *The Comprehensive R Archive Network [WWW Document]*. Retreived from http://cran.r-project.org/.

EU. (2010). *EUROISLANDS - The development of the islands – European Islands and Cohesion Policy | ESPON [WWW Document]*. Retrieved August 31, 2017, from https://www.espon.eu/programme/projects/espon-2013/targeted-analyses/euroislands-development-islands-%E2%80%93-european-islands.
Farrington, J. H. (2007). The new narrative of accessibility: Its potential contribution to discourses in (transport) geography. *Journal of Transport Geography, 15*, 319–330.
Goodchild, M., Haining, R., & Wise, S. (1992). Integrating GIS and spatial data analysis: Problems and possibilities. *International Journal of Geographical Information Systems, 6*, 407–423.
Greek Statistical Agency. (2011). *Census of Population [WWW Document]*. Census of Population. Retrieved from http://www.statistics.gr/portal/page/portal/ESYE/PAGE-themes?p_param=A1604&r_param=SAP03&y_param=2001_00&mytabs=0.
Haining, R. (1994). Designing spatial data analysis modules for geographical information systems. In *Spatial analysis in GIS* (pp. 45–63). London: Taylor & Francis.
Haining, R. (2003). *Spatial data analysis: Theory and practice*. Cambridge University Press, Cambridge.
Johnston, R. J. (1986). *Dictionary of human geography*. Oxford: Blackwell.
Karampela, S., Kizos, T., & Spilanis, I. (2014). Accessibility of islands: Towards a new geography based on transportation modes and choices. *Island Studies Journal, 9*, 293–306.
Karampela, S., Kizos, T., & Papatheodorou, A. (2015). Patterns of transportation for tourists and residents in the Aegean archipelago, Greece. In *Archipelago tourism: Policies and practices* (pp. 35–50). Aldershot: Ashgate.
Kaufman, L., & Rousseeuw, P. J. (2009). *Finding groups in data: An introduction to cluster analysis*. Chichester: Wiley.
Kizos, T. (2007). Island lifestyles in the Aegean islands, Greece: Heaven in summer, hell in winter? In *Seasonal landscapes* (pp. 127–149). New York: Springer.
Kizos, T., Spilanis, I., & Pralakidis, S. (2005). Attractiveness and less favored areas: The case of Lesvos. *Geographies, 10*, 80–107.
Kwan, M.-P., Murray, A. T., O'Kelly, M. E., & Tiefelsdorf, M. (2003). Recent advances in accessibility research: Representation, methodology and applications. *Journal of Geographical Systems, 5*, 129–138.
Makkonen, T., Salonen, M., & Kajander, S. (2013). Island accessibility challenges: Rural transport in the Finnish archipelago. *European Journal of Transport and Infrastructure Research, 13*, 274–290.
Moseley, M. J. (1979). *Accessibility: The rural challenge*. London: Methuen.
O'Sullivan, P. D., & Unwin, D. (2010). *Geographic information analysis* (2nd ed.). Chichester: Wiley.
Reynolds, A. P., Richards, G., de la Iglesia, B., & Rayward-Smith, V. J. (2006). Clustering rules: A comparison of partitioning and hierarchical clustering algorithms. *Journal of Mathematical Modelling and Algorithms, 5*, 475–504.
Schürmann, C., & Talaat, A. (2002). *The European peripherality index (No. ersa02p224), ERSA conference papers*. European Regional Science Association.
Scott, L. M. (2000). Evaluating intra-metropolitan accessibility in the information age: Operational issues, objectives, and implementation. In *Information, place, and cyberspace* (pp. 21–45). New York: Springer.
Shih, H.-Y. (2006). Network characteristics of drive tourism destinations: An application of network analysis in tourism. *Tourism Management, 27*, 1029–1039.
Spilanis, I., & Kizos, T. (2015). *Νησιωτικότητα και Βιωσιμότητα: Η περίπτωση των Νησιών του Αιγαίου*. Mytilini: University of the Aegean.
Spilanis, I., Kizos, T., Kondili, I., & Misailidis, N. (2005). Accessibility and attractiveness of Aegean Islands. *Aeihoros* (in Greek), *4*. Retrieved from http://www.aeihoros.gr/article/el/prospelasimotita-kai-elkustikotita-ton-nision-tou-aigaiou.
Spilanis, I., Kizos, T., & Petsioti, P. (2012). Accessibility of peripheral regions: Evidence from Aegean Islands (Greece). *Island Studies Journal, 7*, 199–214.

Spilanis, I., Kizos, T., & Giordano, B. (2013a). The effectiveness of European Regional Development Fund projects in Greece: Views from planners, management staff and beneficiaries. *European Urban and Regional Studies, 23*, 182–197.

Spilanis, I., Kizos, T., Vaitis, M., & Koukourouvli, N. (2013b). Measuring the economic, social and environmental performance of European Island Regions: Emerging issues for European and Regional Policy. *European Planning Studies, 21*, 1998–2019.

Weidenfeld, A., Butler, R., & Williams, A. W. (2011). The role of clustering, cooperation and complementarities in the visitor attraction sector. *Current Issues in Tourism, 14*, 595–629.

Zervakis, V., Georgopoulos, D., & Drakopoulos, P. G. (2000). The role of the North Aegean in triggering the recent Eastern Mediterranean climatic changes. *Journal of Geophysical Research: Oceans (1978–2012), 105*, 26103–26116.

Part II
Analysis and Modelling of Urban Transportation

Agent-Based Activity/Travel Microsimulation: What's Next?

Eric J. Miller

Abstract This chapter briefly summarizes and reviews the current generation of operational activity/tour-based model systems. These model systems are developed to varying degrees within an agent-based microsimulation (ABM) framework. ABM provides an extremely flexible, powerful, and efficient means for modelling complex spatial-temporal, socio-economic behaviour such as travel. A high-level definition of microsimulation in general and agent-based microsimulation in particular is presented. Overall, currently operational activity/travel model systems represent a sound "first generation" of such methods, but they are far from realizing the full potential of the ABM concept. A wide range of issues and challenges in advancing the ABM-based activity/travel modelling state of the art are discussed, leading to a few suggestions for key "next steps" in model development.

1 Introduction

Activity/tour-based models of urban travel demand are increasingly being used in operational planning practice (Castiglione et al. 2015). These are generally implemented within a microsimulation framework, in which out-of-home activity participation and the associated travel are modelled for individual trip-makers (agents). These operational model systems are the product of over 40 years of research and development, dating back at least to calls in the 1970s for an activity-based approach to modelling travel demand (Jones 1979; Hensher and Stopher 1979; Axhausen and Gärling 1992). They are also built upon the tremendous advances that have been made over this same time period in disaggregate, random utility choice modelling, computer hardware and software, and GIS-based spatial-temporal datasets, among other factors.

The rapidly growing availability of "big data" concerning travel behaviour from a variety of sources, continuing growth in computing capabilities, and ever-changing

E. J. Miller (✉)
Department of Civil Engineering, University of Toronto, Toronto, ON, Canada
e-mail: miller@ecf.utoronto.ca

H. Briassoulis et al. (eds.), *The Practice of Spatial Analysis*,
https://doi.org/10.1007/978-3-319-89806-3_6

(and increasingly challenging) policy issues (autonomous vehicles, new mobility services, increasing emphasis on active transportation, etc.) create both the opportunity and the need to continue to develop more advanced, robust travel demand modelling methods to help guide the continuing explosive growth of urban regions worldwide along more sustainable paths (Miller 2017).

Section 2 of this chapter briefly summarizes and reviews the current generation of operational activity/tour-based model systems. These model systems are developed to varying degrees within an agent-based microsimulation (ABM) framework. ABM provides an extremely flexible, powerful, and efficient means for modelling complex spatial-temporal, socio-economic behaviour such as travel (Kreibich 1979; Clarke et al. 1980; Mackett 1985; Goulias and Kitamura 1992; Ettema et al. 1993; Miller 1996, 2003). Section 3 provides a high-level definition of microsimulation in general and agent-based microsimulation in particular. Overall, currently operational activity/travel model systems represent a sound "first generation" of such methods, but they are far from realizing the full potential of the ABM concept. Section 4 discusses a wide range of issues and challenges in advancing the ABM-based activity/travel modelling state of the art. Section 5 concludes the chapter with a few suggestions for key "next steps" in model development.

2 Modelling Activity and Travel

2.1 Introduction

Many "activity-based" travel models currently exist worldwide. These can be loosely divided into two primary types: *tour-based* models and *activity-scheduling* models. These two classes of models are briefly discussed in the following two sub-sections.

2.2 Tour-Based Models

Tour-based models are the most common form of currently operational models. As their name implies, these models focus on predicting the most common forms of daily tours made by individuals. A *tour* (also often referred to as a *trip-chain*) is a connected set of trips in which the origin of each subsequent trip is the destination of the previous trip, with the origin of the first trip being the destination of the last trip. Figure 1a illustrates a simple tour consisting of three trips: from home to work, work to shopping, and shopping back home again. Tours generally are assumed to be *home-based* (as in Fig. 1a), with home being the start and end point of the tour (or the tour's "anchor point"), but non-home-based sub-tours are, of course, possible in which a non-home anchor point exists within the overall home-based tour. Figure 1b

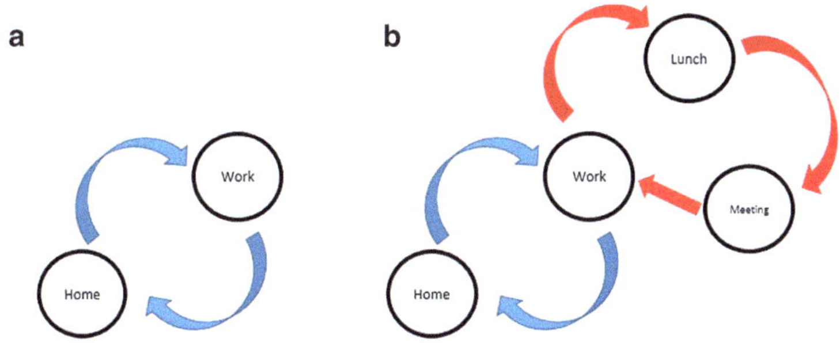

Fig. 1 Examples of tours. (**a**) A simple 2-trip home-work-home-tour. (**b**) 5-Trip home-based tour with a 3-trip work-based sub-tour

provides an illustration of this, in which a work-based sub-tour (work-lunch-meeting-work) exists within the overall home-based tour.

Tour-based models represent a major advance over conventional trip-based models in several important respects. Most fundamentally, they allow the logical interconnections between individual trips to be explicitly accounted for. Important examples of this include:

- The decision to drive to work in the morning is not independent of the decision to drive home in the evening. It may be, for example, that it would be feasible for a worker to take a commuter train to work during the morning peak period, but the worker is planning on staying at work late that evening and will miss the last train home. In such a case, the worker will drive to work so that she has the car available to drive home again that night. More generally, if a "car leaves the driveway", it must return again at some point.[1] Only a tour-based approach can ensure that this "tour-level" constraint is imposed on the individual trip mode choices.
- The existence of other trips in the tour may influence mode and/or location choices for other trips. Again consider a worker who might take the train to and from work if these are the only two trips being made that day (i.e. a simple home-work-home tour). But if he needs to do some shopping on the way home from work, and it is only feasible to execute this shopping trip by car (i.e. transit does not serve the shopping mall), then he will use the drive mode for the entire tour. Conversely, given that he has decided to drive to work he may choose to drive to a nearby mall to do his shopping on the way home, whereas, if he did take the train to/from work he might then do his shopping by taking the car once he had returned home to drive back and forth to this mall (or maybe he walks to and from a neighbourhood shopping street instead), in either case generating a second

[1]The same holds true for bicycles, which are also, in principle a "tour-based" travel mode.

home-shop-home tour by a different mode and perhaps to a different shopping location.

- In trip-based models, *non-home-based trips*, in which home is neither the origin nor destination of the trip, are notoriously difficult to model in any defensible way since little logical explanation for such trips exist with the model: why is this trip going from here to there by this mode at this time? In a tour-based formulation, as illustrated above (e.g. the work-shop trip), these trips sit within a logical/causal framework that considerably improves one's ability to model them in a sensible way.
- The "auto-drive to transit" ("park and ride") "mixed-mode" type of trip is also intrinsically a tour-based mode choice. The decision to drive to the train station in morning commits the commuter to return to that train station in the evening to retrieve the parked car and use it to drive home again from the station. The trip-maker in the model needs to "remember" at which station her car is parked and arrange her tour so that she returns at the end of the day to pick it up. The model can only be internally consistent in its prediction of auto access and egress trips to/from transit (and the transit trips originating and destined for each park and ride station) if the logical connections between the trips within the tour are explicitly maintained.
- As all of the examples above illustrate, a 24-h travel day can only be modelled in a logical, internally consistent way if a tour-based approach is used.

In order to model tours, a microsimulation approach must be adopted in which the tours (and their constituent trips) are explicitly modelled for each individual trip-maker. Aggregate, matrix-based approaches simply cannot deal with tours. As is discussed in Sect. 3, many reasons exist for adopting a microsimulation modelling approach, but, over and above these reasons, microsimulation is simply the only feasible means for modelling tours.

Within this microsimulation framework, many approaches are conceivable for modelling tours, including the activity-based models discussed in the next sub-section. In current operational practice, however, tour-based models very typically make use of random utility theory-based (deeply) nested logit models for their implementation. While details vary from one model to another, Fig. 2 presents a

Fig. 2 Example Tour-Based Model Structure. Adapted from Castiglione et al. (2015)

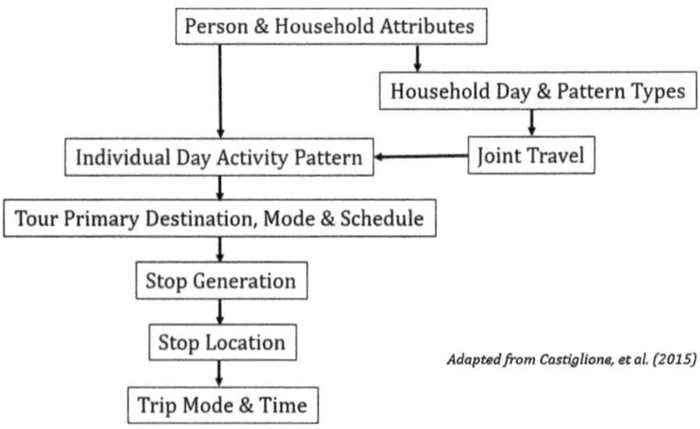

high-level representation of a typical tour-based model, which includes the following key components or steps (Castiglione et al. 2015):

- Given the synthesized attributes of persons within a given household, a daily activity pattern for each individual is predicted (e.g. number of tours).
- A primary destination, mode and overall timing for each tour is determined (e.g. a "mandatory work" tour, with auto drive as the travel mode).
- Additional stops by purpose and location may be added to each tour.
- Given the selected primary mode for the tour, modes for individual trips within the tour are assigned, as well as time of day for each trip.
- Optionally (in some models), household-level joint trips may be generated that influence the activity patterns for the individual household members involved.

This model structure derives in large part from the seminal work of Bowman and Ben-Akiva in Portland (Bowman and Ben-Akiva 1997). For a more complete discussion of such models, see Castiglione et al. (2015).

Operational tour-based model implementations are particularly increasingly common in the USA, usually involving the use of relatively standardized software developed by a handful of leading transportation consulting firms. Urban regions with currently operational or near-operational models include San Francisco, Columbus, Ohio, Atlanta, Denver, Sacramento, Los Angeles, and Jerusalem, among others (Bradley et al. 2010; Paleti et al. 2017; Vovsha et al. 2011).

2.3 Activity-Scheduling Models

This class of models focusses on predicting out-of-home activities and the associated travel required to execute these activities. That is, the primary focus is on predicting the start time, duration, location, and purpose (or type) of the out-of-home *activity episodes* in which a given person decides to engage in the day (or other time period) being modelled. Trips are then the emergent outcome of the need to travel from one activity location to another. Indeed, trips can themselves be thought of as another, special type of activity episode. Activity episodes and their associated trips need to be *scheduled*, generating a daily *activity pattern* for each person being modelled. Tours clearly emerge out of this scheduling process as a travel linkage among a sequence of out-of-home activity episodes.

Given this explicit focus on generating activity episodes, it is argued that these models are more truly "activity-based" than the tour-based models discussed in the previous sub-section, since these still focus on directly predicting trip-making, rather than activity participation. At some point, this distinction may be largely semantic in nature. What is more important is the extent to which an *activity-scheduling* approach may be a more behaviourally fundamental and/or more flexible/extensible one than the tour-based approach.

As with tour-based models, activity-scheduling models must use microsimulation, since it is impossible to develop a daily (or other time period) activity pattern for individuals in any other way, and the notion of somehow modelling activity patterns in some sort of aggregate, matrix-based way is simply inconceivable.

A much wider set of approaches are used in activity-scheduling models than for tour-based models, with no dominant method being apparent at this time. Random utility models (RUM) are often used, but examples of rule-based ("computational process") approaches, and hybrid RUM-rule based, also exist. Examples of fully operationally implemented activity-scheduling models are relatively rare compared to tour-based models. Many have been "quasi-operationally" used in various policy studies, but are not yet generally in day-to-day use by mainstream planning agencies.[2] Important examples of activity-scheduling models (in alphabetical order) include:

- ADAPTS (USA; Auld and Mohammadian 2012).
- ALBATROSS (The Netherlands; Arentze and Timmermans 2004).
- CEMDAP (USA; Bhat et al. 2004).
- C-TAP (Switzerland; Märki et al. 2014).
- CUSTOM (Canada; Habib 2018).
- FAMOS (USA; Pendyala et al. 2005).
- FEATHERS (Belgium; Arentze and Timmermans 2004).
- MATSIM (Multiple applications; Balmer et al. 2006).
- PCATS (Japan; Kitamura and Fuji 1998).
- TASHA (Canada; Miller and Roorda 2003).

3 Agent-Based Microsimulation Modelling

3.1 Introduction: Simulation and Complexity

Simulation is a widely used method for implementing models of a wide variety of systems and behaviours. It provides a computational/algorithmic mechanism for modelling such systems when simpler, typically analytical-based, methods are not able to deal with the complexity and/or detail of the process to be modelled. Simulation is a procedure for evolving a "system state" over time as a function of both exogenous and endogenous factors. It can also be thought of as providing a computer-based "laboratory", within which experimental investigations of a system's behaviour can be undertaken. Key characteristics of a simulation model include:

- It is a numerical algorithm (as opposed to analytical approach) for modelling the system or behaviour in question.

[2]A notable exception to this general statement is the TASHA model, which is the operational travel demand forecasting model for the City of Toronto.

- Dynamic changes in system behaviour are modelled over time, i.e. time is an explicit dimension within the model.
- The model is usually stochastic, i.e. random elements exist within the processes being modelled—"state outcomes" are not known with certainty.[3]
- The forecasted end state of the system being modelled is "evolved" rather than "solved for". Again, simulation involves the iterative stepping through time, with the system state incrementally evolving in each time step, rather than the end state being determined by, for example, analytically solving for an equilibrium state.

Simulation is used to model complex systems that cannot be modelled adequately by other means. Complexity arises due to combinations of:

- Dynamics
- Stochastic elements
- Complex (non-linear) behavioural processes
- Path (and initial condition) dependencies
- Multiplicity of heterogeneous actors/agents

Travel behaviour (as with most other urban socio-economic processes) clearly displays all of these characteristics.

Complexity theory has emerged over the past 50+ years as a new paradigm for understanding and modelling complex phenomena. The emergence of this new science is tied directly to the widespread access to high-speed digital computing that first occurred in the late 1950s and early 1960s, which allowed scientists and mathematicians for the first time to gather and analyse data concerning complex behaviours (weather patterns, stock market prices, fluid flows, etc.) in new and much more powerful and insightful ways (Gleick 1987). Also, the ability to simulate complex system behaviour within the computer as a means for exploring and understanding this behaviour was also integral to the emergence of this "complexity revolution". As illustrated in Fig. 3, Downey (2012) argues that complexity theory has led to a shift in the nature of models from analysis to computation and from equation-based to simulation-based in a variety of "dimensions" in terms of our representation of processes, populations, and spaces from:

- Homogeneous to heterogeneous/composite
- Linear to non-linear
- Deterministic to stochastic
- Continuous to discrete and so on.

Although not generally discussed in these terms, the evolution of, first trip-based, and, more recently, activity-based travel demand modelling has displayed at least some elements of this shift towards a "complexity paradigm". Travel demand modelling also emerged as a recognizable discipline in the mid-1950s with the advent of high-speed digital computing and the emergence of first-generation four-

---

[3] Although deterministic simulation models also exist.

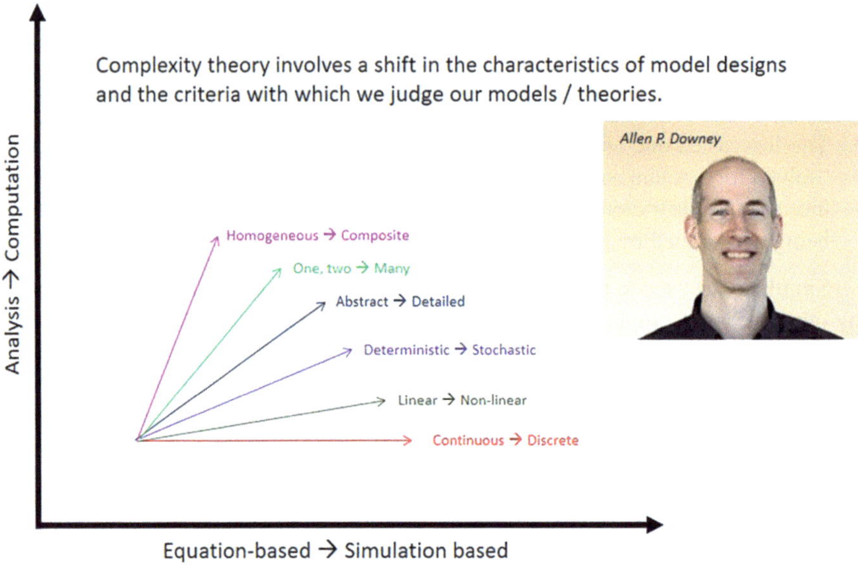

Fig. 3 Complexity Theory: A New Kind of Science. Complexity theory involves a shift in the characteristic of model designs and the criteria with which we judge our models/theories

step travel demand models in Detroit, Chicago, and elsewhere, and its evolutionary development has always been tied to computing capabilities and associated data availability (Meyer and Miller 2013). Since the early 1970s, it has grown increasingly discrete, stochastic, non-linear, and heterogeneous in its modelling approaches. This trend is continuing (and, indeed, accelerating) today as the availability of "big data", ever-increasing computing power and software, and new modelling methods, such as machine learning, continue to push the evolution of the field.

As already discussed in Sect. 2, activity-based travel demand modelling has also moved firmly into a microsimulation computational framework. Section 3.2 defines microsimulation and briefly discusses key motivations for its use in the construction of activity/travel demand forecasting model systems. Many of these model systems are also agent-based in their design. Section 3.3 defines what is meant by an agent-based microsimulation model system and why it is a useful approach to modelling activity and travel.

3.2 Microsimulation

O'Donoghuel (2014) defines microsimulation as "a simulation-based tool with a micro unit of analysis that can be used for ex-ante analysis. It is a micro-based methodology, utilizing micro units of analysis such as individuals, households, firms, and farms, using surveys or administrative datasets. It is a simulation-based methodology utilizing computer programmes to simulate public policy, economic or social changes on the micro population of interest". "Micro" implies simulating a system in a highly disaggregated way spatially, temporally, socio-economically (representation of actors), and in the representation of processes. Microsimulation modelling of socio-economic processes such as travel date at least to the seminal work of Orcutt in the late 1950s (Orcutt 1957, 1960).

Many important reasons exists for microsimulating travel demand (Miller 1996, 2018; Miller and Salvini 2002):

- Heterogeneity in trip-makers (attributes, preferences, contexts, history, behaviour, etc.).
- Identification of detailed impacts of policies across people and locations.
- Modelling complex behaviour. In addition to the modelling of tours and mixed-mode trips discussed in Sect. 2, this includes inter-personal interactions: within-household members (resolving competing demands for car usage, joint activities, etc.); among non-household alters (social network interactions, carpooling, etc.); and market interactions (buyers and sellers interacting within a market).
- Potential to capture memory, learning, and/or adaptation.
- Efficiency in data storage and processing.
- Emergent behaviour.

3.3 Agent-Based Modelling

Once one is microsimulating a socio-economic system it is a (relatively) small step to adopting a full agent-based microsimulation (ABM) approach. An agent is an "intelligent object" that:

- Perceives the world around it (monitors and receives information concerning its environment)
- Is able to control its actions in response to its environment, usually based on goal-oriented decision-making
- Acts into the world in an attempt to achieve its goals and objectives, thereby altering the world's state and its own environment

Persons, households, firms, etc. are obviously agents. Within the context of this chapter, they make decisions about their daily activity/travel patterns based on their personal goals and their perceptions of the transportation and urban activity systems (Manheim 1978). And by travelling through the transportation system, they interact with other trip-makers, change network congestion levels, etc.

While it has been noted that all tour/activity-based travel demand models are disaggregate microsimulation models that simulate the travel behaviour of individual trip-makers, it can be debated the extent to which the various models are "truly agent-based" or not. One need not get bogged down into too detailed a semantical debate about this, but the working assumption in this chapter is that an ABM is one in which the individual trip-makers are explicitly modelled as agent "objects" within which information concerning their attributes, their experiences in the system and their decision-making processes are encapsulated. Advantages of an explicit ABM formulation include:

- Agents are clearly a very natural, "high fidelity" representation for implementing any model of individual decision-making: as noted above, we are, literally, agents.
- Encapsulation of data and processes within the agent makes for "clean" coding and well-defined interfaces for information flow, agent interactions, etc. within the software.
- Facilitating model system modularity and extensibility.

Agent-based modelling is a very practical approach to decomposing the enormous complexity of modelling a system involving the multiple decisions and interactions in time and space of literally millions of heterogeneous agents with non-linear, context-dependent decision processes. Each agent is literally the "container" within which is stored all information concerning that agent, as well as the knowledge that the agent needs to make its daily activity/travel decisions, and wherein these decisions are made. In particular, as illustrated in Fig. 4, each agent is able to monitor its own context/environment, retain its memory of past events, its tastes and preferences, etc., keep track of its relationships and interactions with other agents, and make decisions about (and eventually execute) its daily activity/travel patterns. Within the confines of an individual agent, these decisions may be relatively simple and straightforward to model, given that the agent's decision-making context, memory, etc. are all explicitly "known" (at least within the virtual world of the simulation model). Complexity within the overall travel market still exists,

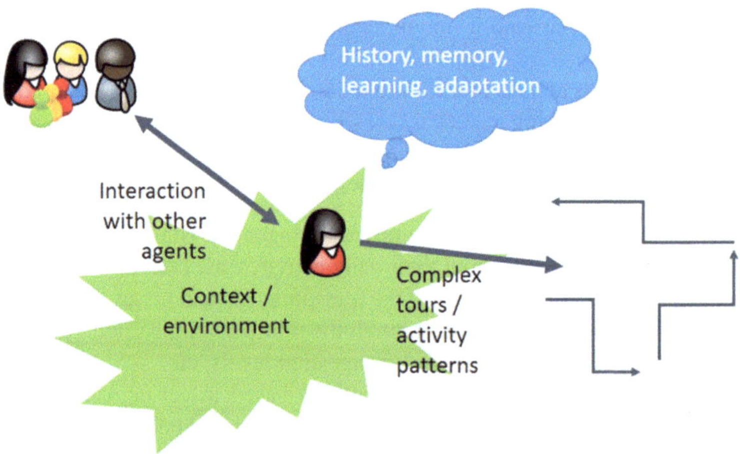

Fig. 4 Agent-Based Modelling

largely in terms of the interaction of all the trip-makers as they actually execute their activity/travel plans and compete for road space, seats on the bus, etc. But modelling the decisions of millions of individual trip-makers is, in principle, no more difficult than modelling a single individual.

4 Towards Next-Generation ABM Activity/Travel Models

4.1 Introduction

It is argued that ABM is a "logical", powerful, and practical approach to modelling the complexities of travel behaviour. But adopting an ABM approach does not automatically solve all modelling problems. ABM is a computational framework for modelling behaviour, rather than a model per se. Issues of model formulation, information flow, interactions among agents, data requirements for model construction and usage, computational practicality, etc. all remain to be addressed.

Indeed, there may be limitations in data, theory, or computational practicality that may limit the capabilities of ABM models in a variety of ways. Whether other modelling methods might better deal with at least some of these issues remains to be seen. It is arguable that ABM is the currently best available approach, but this remains to be demonstrated, especially as we try to expand beyond our current capabilities.

Research issues/challenges in developing improved activity-based model travel demand model systems include:

1. Dynamics and information flows (memory, inertia, state dependencies, adaptation, etc.)
2. Heterogeneity in both trip-makers and choice contexts
3. Issues in modelling spatial choice (activity episode locations)
4. Multi-dimensionality and sequencing of activity/travel decisions
5. Inter-agent interactions (within and between households; individual- vs. household-based models)
6. Activity episode utility (why do we travel?)
7. Modelling in-home activity
8. Moving beyond daily travel:
   (a) Modelling intercity (long-distance) travel
   (b) Multi-day (week-long) models
9. Computational efficiency
10. Statistical representativeness of microsimulation results
11. New/big data sources—how will these change our models?

The first eight of these issues all deal with modelling various elements of agent behaviour. Behavioural representation is the primary focus of this chapter, and so these issues are briefly discussed in turn in the following sub-sections. The remaining three issues are generic, technical issues with which all models (ABM or otherwise) must deal. They are largely "implementation issues" although they can never be totally excluded from the discussion of behavioural representation. The point is that these issues are "always with us" and act as constraints on what can/cannot be done, rather than as primary criteria for what we would "like" or "need" to do in terms of modelling activity/travel behaviour.

In the following sections, the intent in the discussions is not to solve problems in building activity/travel demand models, but simply to raise them in sufficient detail to show:

- Why there is a concern/need/issue.
- Identify major challenges/obstacles to model improvement, which inevitably involve some combination of limitations in data, theory, methods, and/or computational requirements
- Possible approaches for investigation (where this can be identified).

4.2 Behavioural Dynamics, Information, and Attitude Formation

In "real life", each person has memories of past experiences which provide an "information base" for current decisions as well as help shape our tastes, preferences, and attitudes. Our information and interactions with our social networks, media, etc. also shape our knowledge and our attitudes over time, and, hence, our current decision-making. We develop habits over the course of time (mode to work, favourite grocery store, etc.) that, once formed, are often not re-evaluated/changed over possibly considerable periods of time. The length of time that we have lived in a given neighbourhood affects our "awareness set" of potential destinations for shopping and other activities, as well as our experience with the transportation services connecting this neighbourhood to the rest of the urban area. People adapt to the circumstances within which they find themselves, which means that their preferences and behaviours evolve over time in response to their environment and their day-to-day experiences.

Memory/experience, inertia/habits, and adaptation are all examples of dynamic (temporally varying/evolving) factors that influence our day-to-day activity/travel behaviour. Essentially all activity/travel models, however, are static in that they model behaviour on a single, arbitrary "day in the life" with no consideration of memory or past experience. This massive "left censoring" is seemingly inescapable given that these models are inevitably built using similarly "left-censored" cross-sectional data collected from a 1-day survey.

Panel surveys are sometimes used to try to observe changes in behaviour over time. But these traditionally have consisted of conducting a conventional one- or two-day travel survey once a year. Arguably, this does not help much: observing a person's behaviour on one arbitrary day a year misses virtually all of the variability in day-to-day travel experience, trends in behaviour and evolution of habits that are actually of interest.

It is not clear, of course, how people actually do "internalize" their day-to-day travel experiences to evolve their "cognitive map" of the transportation system and their tastes/preferences over time, and, hence, how to model these processes. Much/most of this is likely to be intrinsically unobservable (latent), at least at a sampling scale that would be sufficient for generalized modelling purposes, i.e. beyond small sample, focussed research efforts, which are not easy/possible to generalize. Observing trip-makers over extended, continuous periods of time, however (from a minimum of 1 week up to several weeks) would, however, certainly be extremely beneficial in terms of observing both what behaviour varies over time and what is more stable, as well as providing valuable insights into at least short-run activity/travel dynamics.

"Outer loop" iteration/equilibration in all operational models arguably represents a crude form of "learning" in which agents' information concerning modal service levels is iteratively updated until they reach an equilibrium state in which their (aggregate) behaviour is consistent with (aggregate) service levels. While this may (very loosely) approximate the learning process of commuters, who undertake approximately the same set of trips each weekday, it clearly is not really representative of how people actually learn, gather information and evolve their behaviour over time. It also typically assumes that trip-makers are fully aware of the attributes of all feasible destinations, service levels for all modes for all trips to all destinations, etc., which clearly is not correct.

Another temporal-related problem with current modelling practice is the merging of survey travel records from different weekdays (i.e. Monday through Friday) into a single pooled dataset that is used to develop a model of travel on a "typical weekday". This represents a classic example of aggregation bias (in this case temporal aggregation), since it is clear that travel patterns vary by day of the week (Dianat et al. 2018a, b), and, hence, the "typical" day being modelled with the pooled data actually never occurs. Ideally travel on each day of the week should be modelled and then statistics useful for planning purposes (weekly averages, peak loads, etc.) could be computed. These would provide a far more realistic representation of "typical" system usage and travel behaviour for planning and decision-making purposes than the current biased pooled-data models.

The universal focus on weekday travel while ignoring weekend travel is also problematic. While planning for weekday peaks still dominates planning analyses, weekends also experience significant travel flows, which are quite different in their spatial-temporal patterns than weekday patterns and so pose their own challenges for the transportation system. Total greenhouse gas and air pollution emissions should include weekend effects in their calculation. And weekday–weekend trade-offs exist in activity episode scheduling, especially for shopping and other non-work/school activities.

While, as noted above, dynamically modelling trip-makers' tastes, preferences, and cognitive maps over time is an extraordinarily difficult task, it is arguable that activity-based models could be improved by:

- Collecting activity/travel participation data over extended periods of time, ranging from at least one full week up to several weeks. In the past, this has been an almost impossible thing to do, except for small-sample surveys involving considerable effort on the part of both surveyors and respondents. The emergence of smartphone apps and other large, continuous passive data streams concerning travel, however, makes multi-week, continuous tracking of travel by large samples of trip-makers an increasingly feasible proposition. Although not without their own issues, such new data sources offer exciting new possibilities for building more dynamic activity-based models (Habib 2018).
- Moving from conventional 1-day models to 1-week models offers the potential for significantly improved models. It is very arguable that 1 day is too limited a time period to fully exploit the potential of the activity-based paradigm for improved travel modelling. Too much day-to-day variability in activities and, perhaps more importantly, too many trade-offs within the week concerning when activities can be scheduled (especially for non-work/school activities, but the argument also holds for many workers and students as well), exist for a 1-day snapshot to be adequate. As noted above, week-long models would eliminate the aggregation bias of "typical day" models based on pooled data, capture weekday–weekend activity-scheduling trade-offs, and permit more complete accounting with respect to emissions, VKT, transit ridership and revenues, and other system performance measures of planning interest. Historically, week-long models were simply not practical due to lack of data and computational limitations. Again, new large data streams may address the data issue, while advanced high performance computing (HPC) capabilities (notably cloud computing) effectively eliminates computational constraints.
- In all surveys, asking how long the respondents have lived in their current residence, and, ideally, briefly gathering key retrospective information concerning their housing and labour market histories would provide some insight into where in "the learning curve" respondents are with respect to information about and experience with their neighbourhood and the local transportation system.

4.3 Heterogeneity

Over and above the dynamics/information issues discussed above, different people will tend to have different tastes and preferences due to their socio-economic characteristics. At least this is what we conventionally assume. It is an open question, however, how much of this heterogeneity in tastes and preferences is literally attributable to socio-economic factors (e.g. rich people behave differently than poor people simply because they are rich) and how much it is due to different dynamic/information contexts (rich people have different social influence networks than poor people and different experiences)? Certainly income, in particular, is a resource which enables/constrains what is feasible for people to do, and so it changes the context for decision-making in a very direct way. This then leads to different experiences, etc. A similar argument undoubtedly holds for age, and probably at least some other socio-economic attributes. Thus, experience (and hence, learning, memory, and preference formation), in general, is conditioned by socio-economic attributes, thus confounding the roles played by socio-economics *per se* versus the dynamics of social network information flows, day-to-day experience, etc. Perhaps as a result of this, in the absence of generally being able to dynamically model information acquisition and preference formation, modellers lean heavily on using socio-economic attributes to condition their choice models. This is certainly a practical approach, but it generally "locks into" the model current average behaviour by socio-economic category. Thus, the problem of allowing for changes over time in these behaviours is not solved in what remains a static model formulation.

Decision-maker heterogeneity is usually handled by some combination of categorization (different models for different socio-economic groups), inclusion of socio-economic attributes as explanatory variables in the model (e.g. dividing cost by income in mode choice models to allow for values of time to vary by income), and/or mixed-logit models (in which at least some parameters are allowed to vary randomly from person to person as a means of accounting for heterogeneity in these parameters across the population). Classification methods can include *ad hoc*, exogenously imposed categories (e.g. simply imposing a classification scheme on the model), a wide variety of statistical classification methods to find a "best-fit" classification scheme given the data, and latent class models (in which the allocation of agents into groups is performed by a statistically estimated stochastic model), among others.

To the extent that the classifier is a parameterized model with explanatory variables that can change over time, at least some dynamics are introduced into the model, since such classifiers permit agents to change category as their attributes (and/or their choice context) changes over time. The categorization scheme itself, however, remains static and can be susceptible to over-training to base case conditions (Badoe and Miller 1998).

Stochastic models such as mixed-logit and latent class models are not easy to apply in forecasting and have seen almost no application in operational applications to date. Latent class models are also generally computationally burdensome to estimate. Mixed-logit models indicate when significant variance in a parameter appears to exist in a population, but usually do not provide direct explanation of who is likely to have a greater or lower value of a given parameter. Hierarchical models of various kinds, in which utility function parameters are themselves parametric functions of explanatory variables, would generally appear to be a better approach to addressing heterogeneity in a more systematically forecastable manner.

Heterogeneity, of course, also exists in any practical characterization of activity alternatives, as well as in the attributes of these alternatives. "Shopping", for example, is often an activity type in both trip-based and activity-based models. But there are many different types of shopping (shopping for groceries is a very different activity than shopping for a new washing machine), with very different activity rates, costs, and feasible locations. Regardless of how detailed an activity categorization one might adopt, it is clear that these categories will always be aggregations of the myriad activities actually available in a complex economy/ society. Thus, significant heterogeneity in choice alternatives and their attributes is inherent in the problem. This issue is discussed further in the next section.

Better data, advanced econometric modelling methods and, increasingly, machine learning methods applied to activity/travel modelling can all help address heterogeneity issues. Access to very large samples of trips over extended time periods, in particular, should be extremely useful in obtaining a much better understanding of both the spatial and temporal variability in travel. Despite this, it is highly arguable, however, that heterogeneity as a modelling challenge is "here to stay": it is inherent in the complexity and diversity of the modern urban systems that we are attempting to model. Put another way, a fundamental "social Heisenberg Uncertainty Principle" exists, in that we are inevitably limited in our ability to accurately and completely observe a complex socio-economic system and to deterministically determine its behaviour. We are working in an inherently stochastic "world". This reinforces the case for microsimulation as the method of choice for modelling this world.

But this also means that we need to take the stochastic properties of this world and of our models' results more seriously. We need to work towards having the capability of routinely running many model replications for a single policy scenario to better account for this stochasticity. We also need to be able to "aggregate" over these replications to generate useful depictions of both the "average" predicted outcome and the "variance" in predictions around this average. Given the huge multidimensionality of even a single day's travel in a large urban area, such a "meta-simulation" approach is a non-trivial task. Current and ever-improving (and increasingly cost-effective) HPC hardware and software, as well as advanced visualization capabilities, however, make this a practical possibility.

4.4 Issues in Modelling Spatial Choice

A particularly important weakness in arguably all travel demand models is modelling location choice for non-work/school activity locations.[4] The predictive accuracy of such models is typically quite weak (Wang and Miller 2014). Many reasons for this exist, but, as noted in the previous section, considerable heterogeneity is inherent in the process in two ways:

[4]Work and school location choices/allocations for workers and students are also a major modelling challenge. For the purpose of this chapter, which is focussing on modelling day-to-day activity/ travel, we assume that work and school locations are determined exogenously to the activity/travel model, through longer-term labour market (place of residence—place of work) and school participation (place of residence—place of school) models.

1. There are a very large number of competing destinations for any given activity, with varying attributes: the choice set is very heterogeneous. Historically, data concerning the attributes of competing destinations has also often been limited.
2. Activity episode types (purposes) are very heterogeneous. In any practical model, we are all always dealing with broad classes of activities ("shopping", "recreation", etc.). Therefore, describing the attractiveness of one location versus another for a given episode (even if the attributes of the locations are known in great detail, which they usually aren't) is very difficult due to the aggregation of the episode types.

Time-space prisms are routinely touted as the "solution" to the first of these problems, and they certainly help, but they come with a host of practical problems:

- Prism calculation requires prior and posterior time-location "anchor points". This means that the set of feasible locations varies depending on what "gap" in the current (provisional) schedule is being considered for executing the given activity episode. Thus, location choice is interconnected with "gap" or episode start time choice.
- Episode duration inevitably also influences feasible prism calculations (and hence the location choice set), but durations may depend on the location chosen.
- Prisms are mode specific, so mode and location choices also become intertwined.
- Episode start/end times and durations are often somewhat flexible (fuzzy) and may be modified from originally desired or nominal values during the scheduling process in order to accommodate addition episodes or for other reasons. As a practical matter, we always observe realized episode start times and durations that are the outcome of the scheduling process, never possible "prior" nominal/desired start times and durations that may have been originally envisioned by the trip-maker at the time the "idea" of a given activity episode is first "generated".
- Even with crisp, well-defined prisms, the "feasible" choice set is still often very large (Wang and Miller 2014).

The issue of the interconnections between gap (start time), duration, location, and travel mode choices is discussed further in the next section.

With respect to the second issue of location attributes, modern GIS-based datasets (POI, etc.) are significantly improving the ability to characterize location attributes (at least for the base case; what they will be in the future is another matter), but even these have limitations in terms of detail and accuracy.

New/emerging, passive datasets (smartphone apps, third-party sources, smartcard data, etc.) generally seem to be compounding this problem since we often don't know/observe the purpose of the trip/activity. We, instead, impute it from the "land use" at the destination. This may introduce some circularity in the modelling logic (we choose this destination for a given purpose because of its attributes, but we impute the destination purpose based on these same attributes). This may cause us to redefine how we define activity type/purpose, and, hence, how we model activity generation and scheduling.

Activity episode location choice is also an important use case for returning to the Sect. 4.2 discussion of the role of information on decision-making. Time-space prisms are intended to identify the set of locations that are feasible to visit during a given time gap in a person's activity schedule. But what we also ideally need to know is what locations are known to/considered by the decision-maker—the so-called consideration or awareness set. Attempts have been made to model the probability that a person will have a particular choice set using a variety of methods, including latent class models, these are generally difficult to construct and typically become computationally intractable for operationally sized problems. Ideally, one would like to have a dynamic, simulation-based learning process in which individuals build up their awareness sets as they experience their environment over time. As discussed in Sect. 4.2, however, there may always be practical limits to what can be implemented in this regard.

4.5 *Multi-dimensionality and Sequencing of Activity/Travel Decisions*

Activity Scheduling Within the Overall Demand Modelling System

Over and above the challenges in modelling activity/travel behaviour posed by unobserved information flows, agent, and environment/context heterogeneity, and other issues discussed in the previous sections, the sheer multi-dimensionality of this behaviour poses major theoretical and practical problems in building behaviourally sound, operationally practical models. The traditional four-step model system represented a practical approach to this problem by decomposing it into a sequence of individual decisions (generation, distribution, etc.) which incrementally construct the entire set of trips and their attributes for an urban region. Although organized somewhat differently, activity-based models for practical reasons similarly decompose the overall decision process into a set of components or stages. Unlike the four-step paradigm, however, no strong consensus yet exists across the various models developed to date as to how best to do this.

It is standard practice to separate longer-term decisions from shorter-term ones. The so-called mobility tools—auto ownership and ownership of driver's licences and transit passes—are all longer-term decisions that do not vary from day to day and so are modelled separately and are treated as exogenous inputs into the activity/travel model *per se*. It is arguable that insufficient attention has generally been paid to mobility tool modelling, at least in the North American context, given the importance that these "resources" play in enabling travel. Indeed, in many models these are not modelled in detail at all, but simply taken as non-policy-dependent exogenous inputs, perhaps generated as part of the population synthesis process. These are, however, potentially important policy levers for influencing travel behaviour. Encouraging households to own one (or maybe even no) rather than two or more cars clearly is an important way of reducing auto usage and increasing transit and active transportation usage. Similarly, increasing transit pass ownership leads to increased transit usage. Further, as we enter an era of electric and/or autonomous vehicles, decisions concerning the number and type(s) of vehicle(s) to own—or whether to own any vehicles at all—are becoming even more important to understand and to include as an endogenous component of the overall travel demand modelling system.

Work and school participation and locations are also clearly longer-term decisions that should also be determined prior to modelling daily activity/travel. Some models, however, do include work and school location choice within the travel model. In terms of work locations, this may be a practical solution in the case where a more formal labour market or place of residence—place of work model is not available, but it must be recognized that is a mis-specification of the overall decision process since work locations are not determined on a day-to-day basis.[5] Students' school locations are generally poorly modelled in most models. While school location choice is a difficult process to model for both theoretical and data availability reasons, it is still surprising how little attention has been paid to this problem, given the importance of school-related travel (especially during peak travel periods), both in terms of the students' travel itself and the significant impacts that travel by younger children has on their parents' activity/travel schedules.

At the "other end" of the decision chain, route choices through the road and transit networks[6] (trip assignment) are generally modelled separately from the other components of trip-making, taking the origin-destination trips by mode and time and day generated by the rest of the demand model as inputs. This is a very reasonable and practical approach, for at least three reasons:

1. Route choice is a very complex and computationally intensive problem requiring specialized software to address. Enormous effort worldwide has been put into focussing on this problem and developing these models as stand-alone products.
2. Actual travel through the transportation network occurs within a time scale that generally is measured in minutes and seconds. Activity/travel-scheduling/planning is concerned with activity participation over the course of a day (or perhaps a week).[7] Thus, just as longer-term mobility decisions are separated from shorter-term travel planning decisions, so too does it make sense to separate daily planning decision-making from the modelling of even shorter-term trip execution.

[5]Exceptions, of course exist. Service workers (plumbers, etc.) and salespersons may have no fixed place of work, travelling each day to wherever their clients on a given day are located; construction workers move from site to site on a frequent basis, etc. But these special cases should be dealt with as such (and certainly are not in current operational models). For most workers, their primary work location is still determined by a longer-term work participation process.

[6]And, in principle, the bicycle and pedestrian networks as well, although bicycle and pedestrian route choices are rarely, if ever, explicitly modelled in operational models to date.

[7]Of course, this planning may go down to the level of the minute (or a few minutes), but the overall scale of the problem is at the level of the day.

3. Road and transit networks are physical systems that exist independently of the people using them (although their performance depends on this usage). In object-oriented terms, they are objects with their own attributes and behaviours (performance functions) that respond to travellers' usage of their services. Trip-makers clearly are different objects (agents) who make decisions within their individual minds about why, where, and when to travel given, among other factors, their perceptions of transportation network service levels. Network assignment models can be thought of as a "market" within which trip-makers' demand for transportation services is matched with the services that the transportation system is able to supply, with the "price" of these services (travel times, etc.) being determined through this demand–supply interaction (as in many other markets). Put another way, the trip assignment model simulates the execution of trip-makers' travel plans within the physical transportation system, whereas the "rest" of the demand model system represents the formation of these plans (activity/travel scheduling) in the trip-makers' minds. Although connected through the "feedback" of travellers' experiences when using the transportation system, there are quite distinct and separate processes.

Having said this, it must be recognized that there is a trend towards trying to integrate travel planning with route choice, so that trip plans (e.g. destinations and possibly even whether a trip will be made at all) are dynamically altered as the simulated day proceeds. That is, activity/travel decisions are continuously updated as the individual moves through the day and travels through the transportation system. Such dynamic rescheduling obviously can happen (I may cancel or reschedule an appointment because I am stuck in traffic) and may be important for modelling the effects of real-time information systems or responses to emergency situations (e.g. a road closure due to an accident). It is an open question, however, whether such models are appropriate/needed for all demand modelling applications, especially long-range forecasting, which remains the primary objective of most activity/travel models. Issues with very tight, within-day, dynamic integration between activity/travel planning and execution of these plans with simulation of the realized network flows include:

- Such tightly integrated models are computationally very burdensome.
- It is also questionable whether this level of within-day dynamics is necessary (or even appropriate) for long-range forecasting applications. Given the level of uncertainties concerning future year scenarios (population and employment levels and distributions, traffic signals, etc.), is this very fine level of modelling detail meaningful?
- While rescheduling obviously exists, it may not be critical to the determination of overall future travel patterns. Arguably, most people most of the time execute their daily travel plans more or less as planned. The extra complexity and computational cost of very detailed rescheduling may not be worth the effort.
- Since one almost always observes the final outcome of travel decisions (including route choice, if it is observed), it is very difficult in general to observe cases of rescheduling that might be used to estimate/calibrate such models.

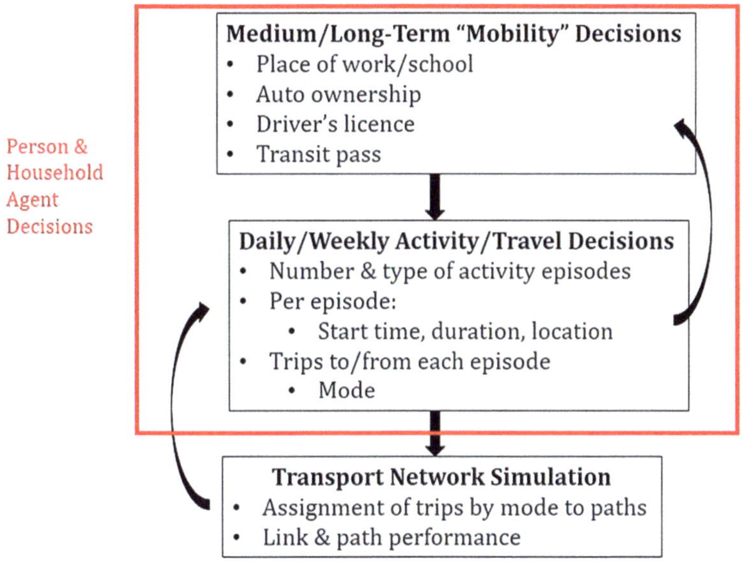

Fig. 5 High-Level Architecture of Activity/Travel Model Systems

- Unless traffic accidents, etc. are being explicitly modelled (which to date they never are in long-range forecasting models), there will be little in the model in the way of events to trigger rescheduling. In particular, it is generally impossible to differentiate such rescheduling from the overall equilibration of demand and supply through iterative feedback between the demand and network models that always occurs in such models.

Given these observations, it is argued that, for most applications, a fairly "classical" segmentation of processes (with appropriate feedbacks) such as sketched in Fig. 5, remains appropriate as a high-level architecture for developing activity/travel demand models (and, indeed, represents that vast majority of current model structures).

Activity-Scheduling Decision Structures

What remains to be determined is how to structure the actual generation and scheduling of daily/weekly activity episodes and their associated travel. This is also a multi-dimensional problem of determining:

- How many episodes[8] of what type in which to engage

[8]In this discussion, it is assumed that we are only concerned with out-of-home episodes. In-home episodes are briefly discussed in Sect. 4.8.

- For each episode determining its start time,[9] duration,[10] and location[11]
- For each trip determining its mode

It is fair to say that no consensus currently exists concerning how to structure these decisions, with different models varying considerably (indeed, fundamentally) in their approach to this problem. The closest that exists to a standard are the (primarily US-based) tour-based models briefly described in Sect. 2.2, which generally share a roughly similar structure. Activity-scheduling models, on the other hand, display considerable diversity in terms of the extent of joint versus sequential choices across the various dimensions of the problem, the ordering of decisions when being sequentially made, and the actual modelling methods (random utility, rule-based, etc.) to be used. It is beyond the scope of this chapter to enter into a detailed review and discussion of these various designs and their strengths and weaknesses, but a few observations/"propositions", based largely on the author's experiences with such models are presented for consideration below.

First, it is important to note that the basic scientific challenge in developing such models is the intrinsic latency (unobservability) of this decision process. We can only hypothesize what we hope is a "reasonable" model, at best based on well-argued axioms that are then empirically validated as best as possible. Psychological theories of decision-making, such as Maslow's Hierarchy of Needs (Maslow 1970), behavioural economics (Kahneman 2011), and standard microeconomic theory (random utility theory, etc.), all provide useful insights, but only take us so far. At the end of the day, a model must be hypothesized and tested against observed behaviour in order to assess its reasonableness and usefulness.

Second, continuing the theme of decomposition as a practical means for modelling complex decision bundles, it is suggested that the activity/travel-scheduling/planning process can be divided into three main components:

- Activity episode generation
- Episode scheduling (includes the generation of travel episodes—trips—to travel to and from the episode)
- Trip mode choice

Episode Generation: The "idea" of the desire to participate in an activity must "come from somewhere". Generation is the model component that instantiates specific episodes for possible scheduling and eventual execution. As part of the generation process, at least some attributes of each episode need to be determined—

[9]And/or "gap" in the current provisions schedule, as briefly discussed in Sect. 4.4.

[10]An episode's end time equals its start time plus duration, so only two of these three attributes are independently determined. While an ending time may sometimes be the primary consideration for scheduling a given episode (I need to be home by 5 pm), it is generally assumed that start time and duration are generally the more "primary" attributes.

[11]As per above, it is assumed herein that work and school locations are known. Locations for non-work/school locations, however, are assumed to be chosen as part of the activity episode generation/scheduling process.

at a minimum the activity episode type (shopping, social, etc.). It is suggested that for work and school episodes start time and duration should also be determined as part of the generation process, while for non-work/school (NWS) episodes at least duration should also be (at least provisionally) determined during generation. It is arguable that it is difficult to think of participating in an activity episode without at the same time having some idea of how long it is likely to take. Work and school start times arguably should also be included within the generation process, given that these are largely determined by the "logic" of the job or the education programme, rather than by scheduling considerations (this is discussed further below). NWS episode start times may vary in terms of their flexibility, but arguably from practical modelling considerations, may generally be handled within the scheduling process, since these start times are often influenced by scheduling considerations.

Episode Scheduling: Once a person has thought of possibly engaging in an activity episode, the feasibility of scheduling this episode—and where in the person's schedule this episode will be placed, needs to be determined. It is arguable that separating episode generation from scheduling is computationally attractive and behaviourally plausible in that it permits the episode generator to be only concerned with the logic of the given episode (I'd like to go golfing this weekend) without being concerned with trade-offs with other possible episodes (the lawn needs mowing this weekend). These trade-offs (and choices among them) reside within the scheduler, which "sees" all the proposed episodes and assesses which can be scheduled when and which might need to be rejected, or at least have their attributes (e.g. duration) modified to feasibly fit into the overall schedule that is being planned. The scheduler can be thought of as the person's time budget manager that allocates the resource of time to competing activities. Determination of NWS episode start times (and/or allocation to "gaps" in the schedule), as well as generation of the trips needed to travel to/from each scheduled episode also logically and conveniently should reside within the scheduler, since it is generally not possible to assess the feasibility of participating in a given episode without determining when it is to occur and how one is going to travel to/from it.

Trip Mode Choice: A trip is a special type of activity episode that is generated every time we decide to participate in an out-of-home activity. Like any episode, trips have start time and duration (travel time). They also have mode (auto drive, transit, walk, etc.), which determines the trip's travel time.[12] Trip mode choice, therefore, is an intrinsic component of the activity/travel-scheduling process. Trip mode choice models are themselves typically sophisticated choice models and so treating these as a separate (but interacting) component within the scheduling process is a practical approach. Ideally, this model is framed as a tour-based model to capture the tour-level interactions discussed in Sect. 2.2. This, however, introduces complications

[12]In such models, trip start time is usually determined by either subtracting the expected travel time from the desired start time of the activity episode being travelled to, or adding the travel time to the activity episode end time if one is leaving this episode to travel to another location.

into the overall scheduling process that need to be addressed in any practical implementation.

Implicit in the discussion above is the assumption that work and school activity generation and scheduling are treated somewhat differently than other activities. While debate about this approach exists, and some models treat work and school as "just another activity" (Doherty et al. 2001; Habib 2018), strong arguments exist that they are, in fact, "primary" activities, that generally take precedence over other activities in the activity generation/scheduling process (Dianat et al. 2018a, b). This largely derives from the "contracts"/"commitments" that they involve with other agents (the worker's employer; the legal requirement for students under the age of 16 to attend school; the need for university students to attend classes so that they can pass their courses; etc.) and that the location, timing, and duration of these episodes is typically determined by the arrangements of these contracts (I must work an 8-h shift from 8 am to 4 pm at the factory; my Chem Lab starts at 9 am in the Chemistry Building on campus; etc.) rather than by personal preferences or activity-scheduling considerations. Indeed, work and school episodes are generally assumed to be scheduled first in the scheduling process and thereby provide "anchor points" or a "skeleton schedule" around which other, lower-priority (and/or higher-flexibility) activity episodes can then be scheduled.

The final observation to be made here is that NWS location choice is a particularly problematic issue to deal with in model design, for at least two reasons. First, as noted in the previous section, it is easily argued that location choice is often intertwined with duration, start time (gap), and/or mode choice. Jointly modelling all these attributes simultaneously is very challenging, however, especially if duration is treated as a continuous variable. Thus, some structuring of these decisions is generally a practical necessity.

Second, for some activities, location may well be predetermined by longer-term processes, similar to work or school (I don't randomly pick a doctor for my annual check-up; my grandmother whom I want to visit this Sunday lives at a known location; my son's hockey game is at a pre-scheduled arena; etc.), while others are much more dynamically determined on an episode-by-episode basis (Where should we eat tonight? Stopping at a hardware store that is "on the way" home from work, etc.). But longer-term location choices are generally not observable within typical practical data collection methods, and, even if they were, predicting them in future years would be challenging. As a result, it is usually necessary to include them with episode-specific location choices in a single model.

How best to improve our models of activity location mode choice is perhaps the single biggest challenge in terms of improving the practical utility of these models for operational planning purposes. Bringing better data to bear on the problem that has often not been available in the past will certainly help. In particular, the widespread availability of high-quality GIS-based datasets greatly improves the set of explanatory variables potentially available for describing the attractiveness of different locations for different purposes. But finding a computationally attractive, behaviourally robust modelling structure for dealing with location, gap, and mode choice is also essential if significantly improved, practical location choice models are to be developed.

4.6 Inter-Agent Interactions

Over and above social network information flows, inter-agent interactions and constraints are clearly extremely important in determining activity/travel behaviour. To begin, it is fundamentally essential to observe that all individual person travel occurs within a household context, with associated constraints, collaboration, joint activities, etc. While including household-level variables (auto availability measures, etc.) in person-level models is commonplace, explicitly modelling activity/travel within an explicit household-level model is still not common. An exception to this rule is the TASHA ABM, which is fully household based and which is the operational travel demand forecasting system used by the City of Toronto since 2016 (Miller et al. 2015). TASHA endogenously models within-household car allocation, ride-sharing, and joint activity participation in a parsimonious and computationally efficient manner (Miller et al. 2005).

A particularly important, but usually neglected household-level interaction is that of "serve-dependent" activities in which one or more (adult) person within the household is responsible for taking care of one or more dependents (young children, the very elderly, physically/mentally disabled persons) who require assistance in travelling (taking one's child to school or daycare) or in-home assistance/supervision (e.g. young children are legally not allowed to be at home without adult supervision). These interactions can have very significant effects on the supervising adults' activity/travel schedules, but are rarely considered in individual-based models.

A concern with increasing use of smartphone app activity/travel tracking, cellular data records, transit smartcard records, and other large-scale passive data streams is that these all track individual trip-making, usually anonymously (so that attributes of the trip-maker are not known) and, without any linkage to the trip-maker's household. This may pose a significant challenge to household-based modelling, despite the case presented above for the need for such models. An important research challenge is to investigate ways in which household-level effects can be retained within models based on such individual-based datasets.

Inter-person interactions with non-household members are, of course, also important in determining activity/travel behaviour. Visiting friends and relatives, social outings, carpooling, etc. are all of importance. "Carpooling" (ride-sharing between non-household members) is a particularly challenging problem, which arguably is not handled well by <u>any</u> current travel demand model, trip-based or activity-based. Historically, lack of appropriate data and major computational issues have severely limited what could be done. As new mobility services continue to emerge, the modelling of ride/car/bike-sharing in a wide variety of service designs is going to be a major challenge. If access to data from such services becomes available this may facilitate the development of improved models. Given that these data, however, are invariably privately held, their availability for academic or public use is not clear at this time.

4.7 Activity Episode Utility and the Treatment of Time

Adopting a utility-based approach to activity scheduling is very attractive for a variety of theoretical and practical reasons. Indeed, it is not clear how trade-offs among activity episodes "competing" for scarce time slots in a person's schedule can be systematically assessed outside of some form of utility calculation. Utility-based models are still relatively new, but are evolving fairly rapidly. A significant literature on "time-use" also exists which is predominantly utility based. Utility-based models (or at least parameterized functions with policy- and schedule-sensitive explanatory variables) are also essential if elastic activity/travel generation models are to be developed.

Time is a resource to be spent, like money. To the extent that we try to construct episode utilities, these are usually expressed as a function of episode duration (time spent). This is probably appropriate for some activities, but not others. For example, the utility that I get from a visit to the doctor has almost nothing to do with the duration of the visit, nor is this duration in any way under my control, i.e. it is not the outcome of my choice involving trading the marginal utility of time spent at the doctor's versus competing possibilities. Much remains to be understood concerning how utility is derived from episode participation and the relationship of this utility to the time spent. Why is a minute spent shopping valued differently than a minute spent visiting friends? How does it change with duration, context, person, time of day, etc.?

4.8 In-Home Activities

Although literature exists concerning in-home activities and the interplay between in-home and out-of-home activities, the preponderance of activity/travel models understandably have focussed on out-of-home activity and associated trip-making. The basic assumption of this approach, obviously, is that little substitutability exists between in-home and out-of-home activities, or, at least, that this substitutability is essentially stable over time and so can be treated as a constant, implicit contextual factor rather than an explicit, endogenous component of the decision-making being modelled.

The considerable growth in online shopping, home entertainment options, shopping and meal delivery services, etc. all call this assumption into question. In addition, two in-home activities that are of particular importance to constraining out-of-home activities are:

1. As noted in Sect. 4.6, in-home supervision of dependents (young children, etc.) represents a major constraint on the out-of-home activities of supervising adults. This constraint is almost never recognized in current models.

2. Night-sleep is a mandatory activity[13] that consumes a significant portion of most people's days. It also heavily conditions (interacts with) both morning activity start times and evening/night activity end times. One can generally get away with not explicitly modelling night-sleep in a 1-day model, but it becomes essential to model it explicitly in multi-day/week-long models. Night-sleep arguably should join work and school activities as elements of a person's skeleton schedule (Dianat et al. 2018a, b).

Traditional time-use surveys as well as potentially new data sources, such as credit card transaction data, would appear to provide a sound starting point for modelling in-home activities in conjunction with out-of-home activities, as this becomes necessary to do. Further, in anticipation of eventually needing to model both in- and out-of-home activities, any activity-scheduling model conceptual design should be general/flexible enough to be extended to in-home activities. Opportunities and methods for obtaining both in-home and out-of-home activities in new data collection efforts should also be actively pursued.

4.9 *Beyond Daily Urban Travel*

Multi-Day (Week-Long) Activity/Travel Models

As noted in Sect. 4.2, multi-day (e.g. week-long) models of travel behaviour may provide significant improvements on our ability to understand activity/travel behaviour. Dianat et al. (2018a, b), for example, demonstrate that activity/travel patterns differ significantly from 1 day within the week to another. They also demonstrate the practicality of constructing a week-long model, providing that suitable data are available for model estimation.

Steadily improving methods for collecting large-sample activity/travel data for a week or more (e.g. smartphone apps, cellular data records), combined with ever-increasing computing power (notably cloud computing) make the possibility of week-long activity/travel models a much more practical possibility than has previously been the case. The major practical limitation at this point in time is the question of how to model road and transit assignments within a week-long model, given the still-too-long run times for these models for large urban areas. Even this concern, however, is not an insurmountable problem, again given cloud computing options, as well as the possibility of a "next generation" of highly parallelized, quick-running assignment models being developed in the coming years.

The other question concerning week-long models is how they would be used in operational practice by planning agencies. At a minimum, the results for the 5 weekdays could be aggregated to provide a true "average" weekday forecast that is free of

[13]The occasional "all-nighter" getting a term paper done, "partying 'till the cows come home", etc. notwithstanding.

the “temporal aggregation bias” of current “typical weekday” models that are based on pooled data from different weekdays naively combined in the development of a single weekday model.

A week-long model would also provide (for the first time in virtually all agencies worldwide) information on weekend travel, which should be of policy interest for a variety of reasons. A week-long model also would provide much better estimates of greenhouse gas and air pollution emissions than is possible from even a 24-h single-day model. It would also provide a much better basis for expanding model results to the annual totals which are often required for economic evaluation and other policy analysis purposes.

Intercity/Long-Distance Travel

Models of intercity and/or long-distance travel have always lagged behind the urban travel demand modelling state of the art and practice for a variety of reasons (Miller 2004). Increasing attention is now being paid to modelling longer-distance, non-intra-urban travel (Aultman-Hall et al. 2015; LaMondia et al. 2016). These efforts are hindered by a general lack of data to support advanced modelling, as well as lack of consensus on even very basic concepts such as the definition of a “long-distance” trip.

The extent to which ABM is a promising approach for developing the next generation of long-distance travel models is an open question, but one well worth investigating. The fact that long-distance travel is relatively “sparse” compared to urban trip-making actually is a strong argument for a microsimulation approach in which such trips are randomly generated as discrete events. Long-distance travel is also extremely heterogeneous in nature (think of all the different reasons that people generate business trips or possible destinations for “visit friends and relatives” and vacation trips), again arguing for a microsimulation approach.

5 Summary Comments

This chapter has discussed a broad range of issues and options for developing next-generation activity/travel models. Although it has not spoken very directly to the emerging major challenges associated with autonomous vehicles and Mobility-as-a-Service (MaaS) concepts, these very much provide an explicit subtext for this discussion. The assumption motivating this discussion is that we need to significantly improve our general travel demand modelling capability so as to be better able to model the impacts of these new, disruptive technologies and services. Current models make many assumptions that may have been acceptable given past circumstances, but these assumptions are being aggressively challenged by the “brave new world” that is rapidly approaching.

At the same time, new data collection methods and technologies are opening doors for new "views" of activity/travel behaviour, including much larger samples of behaviour, over extended periods of time, than have been available in the past. This is creating the possibility of building new models in new ways that perhaps will transcend some of the data limitations that we have experienced in the past.

Without trying to summarize the long list of items discussed in the chapter, a few key observations that emerge from this discussion include the following:

- A strong case exists for moving from models of one "typical" day's activities and travel to models of 1 week's activities. Such models arguably would be both more behaviourally representative and useful for policy analysis. They are also much more feasible to construct given current and emerging data, modern computing capabilities and improving theory concerning how to model activity and travel over a 1-week planning period.
- Heterogeneity in both agents and choice contexts is a continuing challenge that will always be a major issue in model design and application.
- Similarly, difficulties in observing and modelling information acquisition and usage by trip-makers is another major modelling challenge. It is easy to talk about building more dynamic models of agent learning and adaptation, but implementation of significantly improved models will continue to be difficult.
- While the new, "big" data streams are generally very promising for the development of next-generation models, one potential problem is that these are inevitably individual based and do not directly collect household-level behaviours and interactions. Being able to explicitly model household-level resources, constraints, and interactions, however, arguably is very important in order to understand individual-level travel. How we can "retain the household" in future models based on new data sources is a significant challenge going forward.

Finally, much of this chapter has, one way or another, wrestled with how one might best design the decision structure for modelling activity/travel decisions. This issue of "model architecture" is critical to the overall behavioural fidelity, computational efficiency, data requirements, and practical ability to implement the models in operational settings. No standard architecture currently exists, especially for activity-scheduling models. This chapter provides some "hints" as to where model design might usefully head, but much work remains to examine and test alternative designs, hopefully in a way that does lead to the emergence of new models that are both behaviourally sound and operationally useful.

References

Arentze, T. A., & Timmermans, H. J. P. (2004). A learning-based transportation oriented simulation system. *Transportation Research Part B, 38*, 613–633. https://doi.org/10.1016/j.trb.2002.10.001.

Auld, J., & Mohammadian, A. (2012). Activity planning processes in the Agent-based Dynamic Activity Planning and Travel Scheduling (ADAPTS) model. *Transportation Research Part A, 46*, 1386–1403. https://doi.org/10.1016/j.tra.2012.05.017.

Aultman-Hall, L., Harvey, C., LaMondia, J., & Ritter, C. (2015). Design and response quality in a one-year longitudinal survey of overnight and long-distance travel. *Transportation Research Procedia, 11*, 136–153. https://doi.org/10.1016/j.trpro.2015.12.012.

Axhausen, K. W., & Gärling, T. (1992). Activity-based approaches to travel analysis: Conceptual frameworks, models and research problems. *Transport Reviews, 12*(4), 323–341.

Badoe, D. A., & Miller, E. J. (1998). An automatic segmentation procedure for studying variations in mode choice behavior. *Journal of Advanced Transportation, 32*(2), 190–215.

Balmer, M., Axhausen, K., & Nagel, K. (2006). Agent-based demand-modeling framework for large-scale microsimulations. *Transportation Research Record, 1985*, 125–134. https://doi.org/10.3141/1985-14.

Bhat, C., Guo, J., Srinivasan, S., & Sivakumar, A. (2004). Comprehensive econometric microsimulator for daily activity-travel patterns. *Transportation Research Record, 1894*, 57–66. https://doi.org/10.3141/1894-07.

Bowman, J. L., & Ben-Akiva, M. (1997). Activity based travel forecasting. In *Activity-based travel forecasting conference, June 2–5, 1996: summary, recommendations and compendium of papers, New Orleans, LA* (USDOT Report #DOT-T-97-17).

Bradley, M., Bowman, J. L., & Griesenbeck, B. (2010). SACSIM: An applied activity-based model system with fine-level spatial and temporal resolution. *Journal of Choice Modelling, 3*, 5–31. https://doi.org/10.1016/S1755-5345(13)70027-7.

Castiglione, J., Bradley, M., & Gleibe, J. (2015). *Activity-based travel demand models: A primer, report S2-C46-RR-1*. Washington, DC: Transportation Research Board.

Clarke, M. C., Keys, P., & Williams, H. C. W. L. (1980). Micro-analysis and simulation of socio-economic systems: Progress and prospects. In R. J. Bennett & N. Wrigely (Eds.), *Quantitative geography in Britain: Retrospect and prospect*. London: Routledge and Kegan Paul.

Dianat, L., Habib, K. N., & Miller E. J. (2018a, January). *Investigating the influence of assigning a higher priority to scheduling work and school activities in the activity-based models on the simulated travel/activity patterns*. Presented at the 97th annual meeting of the Transportation Research Board, Washington, DC.

Dianat, L., Habib, K. N., & Miller, E. J. (2018b, January). *Modeling and forecasting daily non-work/school activity patterns in an activity-based model using skeleton schedule constraints*. Presented at the 97th Annual Meeting of the Transportation Research Board, Washington, DC.

Doherty, S. T., Miller, E. J., Axhausen, K. W., & Gärling, T. (2001). A conceptual model of the weekly household activity-travel scheduling process. In E. Stern, I. Salomon, & P. Bovy (Eds.), *Travel behaviour: Patterns, implications and modelling* (pp. 148–165). Cheltenham: Elgar Publishing Ltd.

Downey, A. B. (2012). *Think complexity*. Needham: Green Tea Press. Retrieved from http://greenteapress.com/complexity/thinkcomplexity.pdf.

Ettema, D., Borgers, A., & Timmermans, H. (1993). Simulation model of activity scheduling behavior. *Transportation Research Record, 1413*, 1–11.

Gleick, J. (1987). *Chaos*. New York: Viking Penguin.

Goulias, K. G., & Kitamura, R. (1992). Travel demand forecasting with dynamic microsimulation. *Transportation Research Record, 1357*, 8–17.

Habib, K. N. (2018). A comprehensive utility-based system of activity travel scheduling options modelling (CUSTOM) for worker's daily activity scheduling processes. *Transportmetreica A: Transport Science, 14*(4), 292–315.

Hensher, D. A., & Stopher, P. R. (1979). *Behavioral travel demand modelling (editorial)*. London: Croom Helm.

Jones, P. M. (1979). New approaches to understanding travel behaviour: The human activity approach. In D. Hensher & P. Stopher (Eds.), *Behavioral travel modeling* (pp. 55–80). London: Redwood Burn Ltd.

Kahneman, D. (2011). *Thinking, fast and slow*. New York: Farrar, Straus and Giroux.

Kitamura, R., & Fuji, S. (1998). Two computational process models of activity-travel behavior. In T. Garling, T. Laitila, & K. Westin (Eds.), *Theoretical foundations of travel choice modeling* (pp. 251–279). Oxford: Pergamon.
Kreibich, V. (1979). Modeling car availability, modal split and trip distribution by Monte-Carlo simulation: A short way to integrated models. *Transportation, 8*, 153–166.
LaMondia, J. J., Moore, M., & Aultman-Hall, L. (2016). Matching household lifecycle characteristics to clustered annual schedules of long-distance and overnight travel. *Transportation Research Record, 2594*, 11–17. https://doi.org/10.3141/2594-02.
Mackett, R. L. (1985). Micro-analytical simulation of locational and travel behaviour. In *Proceedings PTRC Summer Annual Meeting, Seminar L: Transportation Planning Methods* (pp. 175–188). London: PTRC.
Manheim, M. L. (1978). *Fundamentals of transportation systems analysis, Volume 1: Basic concepts*. Cambridge: MIT Press.
Märki, F., Charypar, D., & Axhausen, K. W. (2014). Agent-based model for continuous activity planning with an open planning horizon. *Transportation, 41*, 905–922. https://doi.org/10.1007/s11116-014-9512-y.
Maslow, A. H. (1970). *Motivation and personality* (2nd ed.). New York: Harper & Row.
Meyer, M. D., & Miller E. J. (2013). *Urban transportation planning, a decision-oriented approach* (3rd ed.). web-published: http://mtsplan.com/services.html.
Miller, E. J.. (1996). Microsimulation and activity-based forecasting. In Texas Transportation Institute (Ed.) *Activity-Based Travel Forecasting Conference, June 2-5, 1996 Summary, Recommendations, and Compendium of Papers*, (pp. 151–172). Washington, DC: Travel Model Improvement Program, U.S. Department of Transportation and U.S. Environmental Protection Agency.
Miller, E. J. (2003). Microsimulation, Chapter 12. In K. G. Goulias (Ed.), *Transportation systems planning methods and applications* (pp. 2-1–12-22). Boca Raton: CRC Press.
Miller, E. J. (2004). The trouble with intercity travel demand models. *Transportation Research Records, Journal of the Transportation Research Board, 1895*, 94–101.
Miller, E. J. (2017). Modeling the demand for new transportation services & technologies. *Transportation Research Record, Journal of the Transportation Research Board, 2658*, 1–7.
Miller, E. J. (2018). The case for microsimulation frameworks for integrated urban models. *Journal of Transport and Land Use* (submitted for publication).
Miller, E. J., & Roorda, M. J. (2003). Prototype model of household activity-travel scheduling. *Transportation Research Record, 1831*, 114–121. https://doi.org/10.3141/1831-13.
Miller, E. J., & Salvini, P. A. (2002). "Activity-based travel behavior modeling in a microsimulation framework", invited resource paper, Chapter 26. In H. S. Mahmassani (Ed.), *In perpetual motion, travel behavior research opportunities and application challenges* (pp. 533–558). Amsterdam: Pergamon.
Miller, E. J., Roorda, M. J., & Carrasco, J. A. (2005). A tour-based model of travel mode choice. *Transportation, 32*(4), 399–422.
Miller, E. J., Vaughan, J., King, D., & Austin, M. (2015). Implementation of a "next generation" activity-based travel demand model: The Toronto Case. In: *Presentation at the Travel Demand Modelling and Traffic Simulation Session of the 2015 Conference of the Transportation Association of Canada*, Charlottetown, PEI, Canada.
O'Donoghuel, C. (2014). "Introduction", Chapter 1. In C. O'Donoghuel (Ed.), *Handbook of microsimulation modelling* (pp. 13–31). Bingley: Emerald Group.
Orcutt, G. H. (1957). A new type of socio-economic system. *Review of Economic Studies, 39*(2), 116–123.
Orcutt, G. H. (1960). Simulation of economic systems. *The American Economic Review, 50*(5), 894–907.
Paleti, R., Vovsha, P., Vyas, G., Anderson, R., & Giaimo, G. (2017). Activity sequencing, location, and formation of individual non-mandatory tours: Application to the activity-based models for Columbus, Cincinnati, and Cleveland, OH. *Transportation, 44*, 615–640.

Pendyala, R., Kitamura, R., Kikuchi, A., Yamamoto, T., & Fuji, S. (2005). Florida activity mobility simulator: Overview and preliminary validation results. *Transportation Research Record, 1921*, 123–130. https://doi.org/10.3141/1921-14.

Vovsha, P., Freedman, J., Livshits, V., & Sun, W. (2011). Design features of activity-based models in practice. *Transportation Research Record, 2254*, 19–27. https://doi.org/10.3141/2254-03.

Wang, J. A., & Miller, E. J. (2014). A prism- and gap-based approach to shopping destination choice. *Environment & Planning B, 41*, 977–1005.

Bottleneck Models and Departure Time Problems

André de Palma and Claude Lefèvre

Abstract This chapter is concerned with the analysis of departure times in dynamic bottleneck models. First, the case of a set of individual drivers is discussed through both deterministic and stochastic approaches. The analysis is then extended to a new model that combines small and large agents. In the stochastic setting, the focus is mainly on model building. Simulations are planned for the near future.

1 Introduction

Dynamic bottleneck models have become very popular in transportation research. This has not always been the case. In the early 1980s, most transportation models were static, in the sense that congestion was assumed to be independent of the time of the day (see, e.g., Sheffi 1985). The static models were developed in different directions accounting for heterogeneity, multiple types of vehicles, elastic demand and stochastic equilibrium, among others. The treatment of congestion remained rather inappropriate because somewhat ad hoc peak and off-peak periods were introduced as a surrogate to take into account the time-dependent nature of congestion. In reality, the choice between peak and off-peak must be endogenous as it depends on the level of congestion. This is also true with several policies, road pricing, and change of the peak period lengths for instance, which have to be determined endogenously. Cross-elasticities between the periods may be assumed, but these elasticities should not be constant.

Intermediary models have been proposed by Nesterov and de Palma (2003) for a stable dynamic approach between the static and dynamic cases. These models are popular in operations research since they can be formulated through some convex

A. de Palma (✉)
Ecole Normale Supérieure (ENS) Paris-Saclay, CREST, Cachan, France
e-mail: andre.depalma@ens-cachan.fr

C. Lefèvre
Département de Mathématique, Université Libre de Bruxelles (ULB), Bruxelles, Belgium
e-mail: clefevre@ulb.ac.be

H. Briassoulis et al. (eds.), *The Practice of Spatial Analysis*,
https://doi.org/10.1007/978-3-319-89806-3_7

problem. In the present work, the focus is on bottleneck models whose construction is fully dynamic.

In dynamic modeling, congestion depends endogenously on the time of the day. Such models extend the dynamic assignment models considered for problems of route choice. The initial dynamic congestion model is due to economist Vickrey (1969). In this deterministic model, there is an origin, a destination and a single bottleneck of finite capacity. The cost function for the driver is the sum of the travel time and the schedule delay cost (i.e., the cost due to an early or late arrival); it is piece-wise linear.

Variants of this dynamic model were developed, independently, by the team Moshe Ben-Akiva, Nicolas Litinas, Pavlos Kanaroglou, and the two authors of this chapter. The reader is referred to de Palma et al. (1983), Ben-Akiva et al. Ben-Akiva et al. (1986a, b, 1987), and de Palma et al. (1987). These authors generalized the deterministic model à la Vickrey in two ways. First, the demand side is represented by a discrete probabilistic model, which is obtained from a continuous logit choice model. Second, the supply side is an extended bottleneck model inspired by Robert Herman, the pioneer of Transportation Science, who developed it with André de Palma during a visit at Austin, Texas.

In static bottleneck models, a major decision variable that was missing is the choice of departure time. This variable is encapsulated in the fully dynamic models. The first computational model was developed with Pavlos Kanaroglou who was then researcher with André de Palma at McMaster University. A computer program was developed to compute a stochastic equilibrium in a small network. It was based on a day-to-day adjustment process (an iteration corresponds to one day). At the beginning of a day, the driver combines the experienced travel time with the expected travel time. At the end of the day, the driver has experienced another travel time. This allows the driver to revise their expectation of travel time for the next day according to a moving average process. That iteration procedure converges in general after about 30 steps.

This line of research was continued and amplified in many papers. The early work in the economic literature is a seminal paper by Arnott et al. (1993); for a review, see, e.g., de Palma and Fosgerau (2011). The literature on dynamic bottleneck models is now rich and diversified. In particular, a large-scale dynamic model, METROPOLIS, was used in many cities all around the world; see de Palma and Marchal (1996) and de Palma et al. (1997). Recently, the driver simulation models DynaMit and MATSim have been developed for a similar purpose by other researchers.

A first goal of this chapter is to provide both deterministic and stochastic versions of this dynamic setting. There is an enormous literature relying on a stochastic approach but it often neglects to describe endogenously the arrival rates. Besides that reason, the aim is to consider a stochastic (instead of deterministic) bottleneck model because there exists no simple deterlinistic day-to-day adjustment process so far, which always leads to stable equilibria. Probabilistic birth-and-death master equations provide us with another angle of view to stability since they usually yield a stationarity distribution. As it is known, a deterministic macroscopic stable solution is not necessarily the most likely outcome to the corresponding stochastic process. The reader is referred, e.g., to the discussion in de Palma and Lefèvre (1983). It is also mentioned that stochasticity may play other roles in the bottleneck models. This is the case when considering stochastic capacities, as in Xiao et al. (2015), or stochastic travel times, as in Xiao et al. (2017).

The second goal of this chapter is to introduce in the model a new combination of *small and large* transport agents. A driver is considered as small when its departure time has no influence on the aggregate level of congestion. We consider a mix of small drivers and of a large fleet of vehicles, which have similar sizes (the case of variable sizes could be discussed similarly). The departure time choices in the fleet are assumed to be taken by a single manager whose objective is to minimize his various costs by taking into account the consequences of his choices. This decision process is assumed here to occur continuously in time, so that a small number of vehicles of the fleet can be reallocated at each instant. The fleet is large to guarantee that the cost per vehicle is of the same order of magnitude as the externality. The case where the manager decides periodically, at discrete times, is left for a future research.

The structure of the chapter is as follows. In Sect. 2, the basic deterministic model is constructed: the supply side is the bottleneck model such as introduced by Vickrey (1969), while the demand side is a continuous logit discrete choice model. In Sect. 3, the stochastic version of this model (stochastic demand) is examined by resorting to birth-and-death master equations (see, e.g., Ross 1996). In Sect. 4, a deterministic model is introduced, which combines small and large agents. The equilibrium state is computed by showing that the individual and fleet vehicles behave rather differently. In Sect. 5, this model is formulated in a stochastic framework, with a special attention to the choice problem between two departure times.

2 Starting Deterministic Model

A deterministic dynamic model for congestion is presented first. In this model, the traditional assumption is made that all drivers have the same importance for the traffic. It is also assumed that each driver is arbitrarily small so that the Nash equilibrium concept applies with a continuum of users.

2.1 Definitions and Notation

The key parameters of the model are:

t, the departure time ($t \in [0, T]$, continuous or not),
ω, the iteration (day, $\omega = 0, \ldots, \Omega$),
$r_\omega(t)$, the number of drivers departing from the origin at time t at iteration ω,
$R_\omega(t)$, the cumulated inflow at time t at iteration ω, i.e., $R_\omega(t) = \int_0^t r_\omega(u)du$,

N, the total number of drivers ($\int_0^T r_\omega(t)dt = N$),
s, the capacity of the bottleneck (number of vehicles per hour),
$Q_\omega(t)$, the number of users in the queue at time t at iteration ω,
$tt_\omega(t) = Q_\omega(t)/s$, the waiting time of a driver at the bottleneck at time t at iteration ω.

Without loss of generality, the free-flow travel time is assumed to be 0. Let $\tau_\omega(t)$ denote the last time before t when congestion started (provided it occurs) at iteration ω. So,

$$\tau_\omega(t) = \sup\{u \leq t : Q_\omega(u) > 0\}.$$

When there is congestion, the queue length is then given by

$$Q_\omega(t) = \int_{\tau_\omega(t)}^t r_\omega(u)du - s[t - \tau_\omega(t)]. \tag{1}$$

The cost for a driver departing at time t at iteration ω is defined by

$$C_\omega(t) = \alpha tt_\omega(t) + \beta\left[t^* - t_\omega^a(t)\right]^+ + \gamma\left[t_\omega^a(t) - t^*\right]^+, \tag{2}$$

in which

t^* is the desired time of destination,
$t_\omega^a(t)$ is the arrival time of a driver departing at time t at iteration ω, i.e., $t_\omega^a(t) = t + tt_\omega(t)$.

2.2 Departure Pattern

Denote by $P_\omega(t \mid t')$ the transition probability from departure choice t' at iteration $\omega - 1$ to departure time t at iteration ω. Note that t may be larger or smaller than t'. That probability t' is assumed to be of the form

$$P_\omega(t \mid t') = \frac{h_\omega(t \mid t')}{H_\omega(t')}, \tag{3}$$

where $H_\omega(t') = \int_0^T h_\omega(t \mid t')dt$ is the aggregator at iteration ω and yields, of course,

$$\int_0^T P_\omega(t \mid t')dt = 1.$$

Then, the number of drivers departing at time t at iteration ω is given by

$$r_\omega(t) = \int_0^T P_\omega(t \mid t')r_{\omega-1}(t')dt'. \quad (4)$$

The drivers are assumed to be small enough so that they may consider that their departure time decisions do not affect the value of the aggregator. That is, during each period, the value of the aggregator is constant from the driver point of view. This is an assumption similar to the monopolistic competition; see Dixit and Stiglitz (1977).

2.3 Logit Probabilities

In (3), the function $h_\omega(t \mid t')$ depends on the cost incurred at iteration $\omega - 1$ at time t' and on the cost incurred at iteration ω at time t.

A logit specification for the transition probabilities is considered. The logit model is a classical tool in Transportation (McFadden 2001) and in Industrial Organization (Anderson et al. 1992). Recently, it was also widely used in the study of Big Data, especially for sentiment analysis (Yessenalina and Cardie 2011). Specifically,

$$h_\omega(t \mid t') = \mathrm{e}^{-[C_\omega(t)-C_{\omega-1}(t')]}.$$

In this case, (3) is given by the continuous logit formula

$$P_\omega(t \mid t') = \frac{\mathrm{e}^{-C_\omega(t)}}{\int_0^T \mathrm{e}^{-C_\omega(u)du}} \equiv P_\omega(t). \quad (5)$$

It is worth recalling that when the variance of the error terms in the logit model is zero, this model reduces to the deterministic case. From (5), it is observed that the transition probability becomes independent of t', which makes the model analysis easier. Using (4), it is obtained

$$r_\omega(t) = \int_0^T P_\omega(t)r_{\omega-1}(t')dt' \equiv NP_\omega(t). \quad (6)$$

Let us check that this system has $r_\omega(t)$ as a fixed point. Indeed, from (5) and (6),

$$r_\omega(t) = N\frac{\mathrm{e}^{-C_\omega(t)}}{\int_0^T \mathrm{e}^{-C_\omega(u)du}}.$$

It is postulated that the equilibrium solution has the following intuitive pattern over time: first, departures occur early without congestion (arrival before t^*); then, departures occur early with congestion; next, departures occur late with congestion; finally, departures occur late without congestion. Let t_ω^i be the first early departure with congestion and let t_ω^f be the first late departure without congestion. For early arrivals,

$$C_\omega^E(t) = (\alpha - \beta)tt_\omega(t) + \beta(t^* - t), \quad t < \widehat{t}_\omega,$$

where $t_\omega^a(\widehat{t}_\omega) = t^*$. Likewise, for late arrivals,

$$C_\omega^L(t) = (\alpha + \gamma)tt_\omega(t) + \gamma(t - t^*), \quad t > \widehat{t}_\omega.$$

When there is no congestion, the travel time is constant (normalized to 0 with wlog). When there is congestion, the travel time is given by

$$tt_\omega(t) = \left[\int_{t_\omega^i}^t r_\omega(u)du - s\left(t - t_\omega^i\right)\right]\frac{1}{s}, \quad t_i \le t \le t_\omega^f.$$

So, the choice probabilities $P_\omega(t)$ depend on the cost function, which depends on the travel time, which itself depends on the departure time schedule $\{r_\omega(u)\}$. This shows that the departure rate is a solution of the fixed point problem.

It has been assumed above that each driver at iteration ω selects their departure time at the beginning of the iteration with full knowledge of travel time for each departure time and solves the fixed point accordingly. Alternatively, it is assumed that a driver has not a perfect foresight and builds an estimate of the travel times at the beginning of iteration ω for the whole period. The estimation made for period ω depends on the estimation made for period $\omega - 1$ and on the observed travel times during period $\omega - 1$, i.e.,

$$tt_\omega^{\text{ap}}(t) = g\left[tt_{\omega-1}^{\text{ap}}(t), tt_{\omega-1}^{\text{ob}}(t)\right].$$

In the general case, the transition probabilities might depend explicitly on the previous choice t' at iteration $\omega - 1$.

Estimation of departure time choices is standard using an ordered probit model (see, e.g., Small 1982). A logit model for the departure time choices is possible too, on the basis of stated or revealed preference data. In the next section, a stochastic approach is proposed which is more suitable to describe the departure time process and which has not been considered so far in the literature.

3 A Stochastic Framework

In this case, it is assumed that the departure times are discrete ($t = 0, \ldots, T$) and the iteration process is Markovian in continuous time ($\omega \in \mathbb{R}_+$). This provides a framework which is standard and convenient in probability theory (see, e.g., Ross 1996). Note that considering a continuous time ω yields an approximation to the daily decision process which is reasonable in the long term.

Let $X_\omega(t)$ be the number of drivers departing at time t at iteration ω. The aim is to derive the joint distribution of the vector $\mathbf{X}_\omega = \{X_\omega(t), t = 0, \ldots, T\}$. Let $p_\omega(\mathbf{x}) = P(\mathbf{X}_\omega = \mathbf{x})$, with $p_0(\mathbf{x})$ assumed to be fixed.

Using the Markovian property the master equation is obtained:

$$\begin{aligned}\frac{dp_\omega(\mathbf{x})}{d\omega} = &\sum_{j|x_j\geq 1}\ \sum_{i\neq j|x_i\leq N-1}(x_i+1)p_\omega(\mathbf{x}+\mathbf{e}_i-\mathbf{e}_j)g_\omega^{j|i}(\mathbf{x}+\mathbf{e}_i-\mathbf{e}_j)\\ &-p_\omega(\mathbf{x})\sum_{j|x_j\geq 1}\ \sum_{i\neq j|x_i\leq N-1}x_i g_\omega^{j|i}(\mathbf{x}),\end{aligned} \tag{7}$$

where $\mathbf{e}_i$ is the vector with 1 at the ith position and 0 elsewhere, and where $g_\omega^{j|i}(\mathbf{x})$ denotes the transition rate from departure time i to departure time j at iteration ω when the present state is $\mathbf{x}$. This rate is a function of the form

$$g_\omega^{j|i}(\mathbf{x}) = f[C_\omega(i), C_\omega(j)], \tag{8}$$

where the cost $C_\omega(t)$ is defined as before by (2) and depends on t and $\{X_\omega(u), u = 0, \ldots, t\}$. The latter property comes from the fact that the cost at time t depends on travel times at time t, which depend on departures before time t. Thus, $g_\omega^{j|i}(\mathbf{x})$ depends in fact on $\mathbf{x}_{i\vee j} = (x_1, \ldots, x_{i\vee j})$ only. In (8), the function $f(x, y)$ is expected to be increasing in x and decreasing in y.

The process $\{\mathbf{X}_\omega, \omega \in \mathbb{R}_+\}$ corresponds to a multi-dimensional birth-and-death process on $[0, N]^{T+1}$. As the state space is finite, the equations (7) admit, for all ω, a unique solution $p_\omega(\mathbf{x})$ under certain mild conditions on the transition rates. Moreover, this solution can then converge towards a stationary distribution $p(\mathbf{x})$. Further details can be found in, e.g., Zeifman (1995).

Note that a single driver might be considered as small with respect to the total mass of drivers. In such a case, they can neglect the impact of their decision on the travel pattern. Mathematically, this means that in (7), the transition rate $g_\omega^{j|i}(\mathbf{x}+\mathbf{e}_i-\mathbf{e}_j)$ can be approximated by $g_\omega^{j|i}(\mathbf{x})$.

4 Small and Large Agents: A Deterministic Setting

In this section, a more general model is examined where the drivers can have a variable importance. This is the deterministic approach, directly at the stationary state (so that the index ω becomes superfluous).

There are a number N of individual drivers and a large fleet of M vehicles that is managed by a single decision-maker. The manager of the fleet takes into account the congestion induced by their decision. The choice process is made in two stages. In the first stage, all the individual drivers may choose their departure time according to the method described in Sect. 3. At the next stage, the manager decides to reallocate the vehicles of their fleet by estimating the generated congestion. The fleet is large so that the cost per vehicle is approximately of the same level as the externality (see below).

4.1 Binary Choice

First, the case where there are only two possible departure times 1 and 2 is discussed. The cost per vehicle in the fleet is given at time 1 by

$$C_1^A = h_1(X_1 + Y_1),$$

and at time 2 by

$$C_2^A = h_2(\theta X_1 + \theta Y_1 + X_2 + Y_2),$$

where $0 < \theta < 1$. The functions h_1 and h_2 are increasing since the costs increase with the number of vehicles. The parameter θ represents the fact that the vehicles departing at time 1 are linearly less costly than those departing later at time 2. The cost of the whole fleet of vehicles is thus equal to

$$C^A = Y_1 C_1^A + Y_2 C_2^A.$$

(where Y is the number of vehicles in the fleet at time i, $i = 1, 2$)

Next, suppose that δ_Y vehicles from the fleet change their previous choice and decide to depart at time 2 instead of time 1. The direct cost generated by these δ_Y vehicles is:

$$\begin{aligned}\Delta^D\left(C^A\right) &= \delta_Y[-h_1(X_1 + Y_1 - \delta_Y) + h_2(\theta X_1 + \theta Y_1 - \theta\delta_Y + X_2 + Y_2 + \delta_Y)] \\ &= \delta_Y[-h_1(X_1 + Y_1) + h_2(\theta X_1 + \theta Y_1 + X_2 + Y_2)] + O\left(\delta_Y^2\right).\end{aligned}$$

Therefore,

$$\Delta^D\left(C^A\right) = \delta_Y\left[-C_1^A + C_2^A + O(\delta_Y)\right],$$

and as δ_Y is small with respect to the costs C_i^A, $i = 1, 2$,

$$\Delta^D\left(C^A\right) \approx \delta_Y\left(-C_1^A + C_2^A\right). \tag{9}$$

Using as before first-order approximations, the impact of the change made by the δ_Y vehicles on the other vehicles of the fleet is estimated. The generated indirect cost is:

$$\begin{aligned}\Delta^I(C^A) &= (Y_1 - \delta_Y)[h_1(X_1 + Y_1 - \delta_Y) - h_1(X_1 + Y_1)] + (Y_2 + \delta_y) \\ &\quad \times [h_2(\theta X_1 + \theta Y_1 - \theta\delta_Y + X_2 + Y_2 + \delta_Y) - h_2(\theta X_1 + \theta Y_1 + X_2 + Y_2)] \\ &= [-Y_1 h_1'(X_1 + Y_1) + Y_2(1 - \theta)h_2'(\theta X_1 + \theta Y_1 + X_2 + Y_2)]\delta_Y + O(\delta_Y^2),\end{aligned}$$

where f' denotes the derivative of a function f. Therefore,

$$\Delta^I\left(C^A\right) \approx \delta_Y\left[Y_1 C'^A_1 + Y_2(1 - \theta)C'^A_2\right]. \tag{10}$$

The total impact due to the δ_Y changes is $\Delta(C^A) = \Delta^D(C^A) + \Delta^I(C^A)$. From (9) and (10), it is obtained;

$$\Delta\left(C^A\right) \approx \delta_Y\left\{C_2^A - C_1^A + \left[Y_2(1-\theta)C'^A_2 - Y_1C'^A_1\right]\right\}. \tag{11}$$

At this point, it is possible to make precise the assumption that the fleet is large enough. This means that

$$\text{Assumption}: C_i^A \text{ is of the same order of magnitude as } Y_iC'^A_i, \quad i = 1, 2.$$

Recall that the externality, denoted by e_i^A, is the difference between the marginal social cost, $\left(Y_iC_i^A\right)' = Y_iC'^A_i + C_i^A$, and the individual cost C_i^A; so, $e_i^A = Y_iC'^A_i$. Therefore, the assumption above states that the cost C_i^A per vehicle in the fleet is of the same order of magnitude as the externality e_i^A.

4.2 *Equilibrium*

The choice process begins with the individual drivers. They select between the departure times 1 and 2 and, at the equilibrium, the associated costs are equal. A Nash equilibrium is assumed, so that each driver considers that the decision of the others is fixed. The costs per individual are given at time 1 by

$$C_1^I = g_1(X_1 + Y_1),$$

and at time 2 by

$$C_2^I = g_2(\theta X_1 + \theta Y_1 + X_2 + Y_2),$$

where the parameter θ is defined as in the case of the fleet vehicles. The functions g_1 and g_2 are again increasing.

At equilibrium,

$$C_1^I = g_1(X_1 + Y_1) = C_2^I = g_2(\theta X_1 + \theta Y_1 + X_2 + Y_2). \tag{12}$$

Since $X_1 + X_2 = N$, (12) becomes

$$g_1(X_1 + Y_1) = g_2(\theta X_1 + \theta Y_1 + N - X_1 + M - Y_1). \tag{13}$$

Assuming that there is an interior solution, it is unique since the left-hand side is increasing in X_1, while the right-hand side is decreasing in X_1. That solution is denoted by

$$X_1^{\text{br}} = \gamma(Y_1; N, M), \quad \text{with } dX_1^{\text{br}}/dY_1 < 0.$$

Then, it is the turn of the drivers in the fleet. The fleet manager minimizes the total cost of the fleet taking constant the decisions of the individual drivers. At the optimum, it is necessary and sufficient that any change has no impact on the cost. This means that $\Delta(C^A) = 0$, i.e., from (11),

$$\delta_Y\left(-C_1^A + C_2^A\right) + \delta_Y\left[-Y_1 C'^A_1 + Y_2(1-\theta)C'^A_2\right] = 0.$$

After simplification by δ_Y, we get

$$C_1^A + Y_1 C'^A_1 = C_2^A + Y_2(1-\theta)C'^A_2. \tag{14}$$

Equation (14) is equivalent to

$$\begin{aligned} h_1(X_1+Y_1) + Y_1 h_1'(X_1+Y_1) &= h_2(\theta X_1 + \theta Y_1 + N - X_1 + M - Y_1) \\ &\quad + Y_2(1-\theta)h_2'(\theta X_1 + \theta Y_1 + N - X_1 + M - Y_1). \end{aligned} \tag{15}$$

As before, assuming that there is an interior solution, it is unique in this case too and denoted by $Y_1^{\text{br}} = \mu(X_1; N, M)$, with $dY_1^{\text{br}}/dX_1 < 0$.

From (13) and (15), a Nash equilibrium state $\left(X_1^*, Y_1^*\right)$ will satisfy the following system

$$\begin{cases} X_1^* = \gamma\left(Y_1^*; N, M\right), \\ Y_1^* = \mu\left(X_1^*; N, M\right). \end{cases}$$

Of course, $X_2^* = N - X_1^*$ and $Y_2^* = M - Y_1^*$.

A more analytical formulation follows. It is assumed that the costs are linear with, for instance, $g_1 = g_2 = h_1 = h_2$ the identity function. Then, the individual condition (13) becomes

$$(X_1 + Y_1) = (\theta X_1 + \theta Y_1 + X_2 + Y_2),$$

which can be rewritten as

$$X_1(2-\theta) + Y_1(2-\theta) = M + N. \tag{16}$$

Moreover, the fleet condition (15) becomes

$$(X_1+Y_1) + Y_1 = \theta X_1 + \theta Y_1 + N - X_1 + M - Y_1 + (M - Y_1)(1-\theta),$$

which reduces to

$$X_1(2-\theta) + Y_1(4-2\theta) = N + M(2-\theta). \tag{17}$$

Combining (16) and (17), a system of two linear equations with two unknowns is obtained. Its solution is found to be

$$\begin{cases} X_1^* = (N+\theta)/(2-\theta), \text{and thus } X_2^* = [N(1-\theta)-\theta]/(2-\theta), \\ Y_1^* = (M-\theta)/(2-\theta), \text{and thus } Y_2^* = [M(1-\theta)+\theta]/(2-\theta). \end{cases}$$

Note that $0 < Y_1^*, Y_2^* < M$, while $0 < X_1^*, X_2^* < N$ if $\theta < N/(N+1)$, which will be often satisfied in practice.

4.3 Extension

In this part, a more general model with K possible departure times is examined. The vehicles departed before any time j, $1 \leq j \leq K$, influence the costs for the fleet vehicles and the individual drivers through the polynomial function

$$\sum_{r=0}^{j-1} \theta^{j-r} x_r.$$

A similar reasoning can be used to obtain the equilibrium state by recursion. For instance, take $K = 3$. The procedure starts by considering the first two departure times. Using the previous method, a solution $\{X_1^*, Y_1^*, X_2^*, Y_2^*\}$ is obtained. For the individual drivers, the common cost, denoted by C^I, satisfies (12), i.e.,

$$g_1(X_1^* + Y_1^*) = C^I = g_2(\theta X_1^* + \theta Y_1^* + X_2^* + Y_2^*).$$

At equilibrium, the costs for the individuals are the same for the K times, which yields

$$g_3(\theta^2 X_1^* + \theta^2 Y_1^* + \theta X_2^* + \theta Y_2^* + X_3 + Y_3) = C^I. \tag{18}$$

For the vehicles in the fleet, the common cost, denoted by MSC^A (marginal social cost), is given by

$$\text{MSC}^A = h_1(X_1^* + Y_1^*) + Y_1^* h_1'(X_1^* + Y_1^*),$$

and from (15),

$$\text{MSC}^A = h_2(\theta X_1^* + \theta Y_1^* + X_2^* + Y_2^*) + Y_2^*(1-\theta) h_2'(\theta X_1^* + \theta Y_1^* + X_2^* + Y_2^*).$$

Here again, at equilibrium, the marginal social costs are identical for the K times, so that

$$\begin{aligned} & h_3(\theta^2 X_1^* + \theta^2 Y_1^* + \theta X_2^* + \theta Y_2^* + X_3 + Y_3) \\ & \quad + Y_3 h_3'(\theta^2 X_1^* + \theta^2 Y_1^* + \theta X_2^* + \theta Y_2^* + X_3 + Y_3) = \text{MSC}^A \end{aligned} \tag{19}$$

The system (18) and (19) will have a unique solution X_3^* and Y_3^*, under some regulatory conditions. The procedure can then continue recursively.

5 Small and Large Agents: A Stochastic Setting

In this section, a Markovian model in continuous time is used (see, e.g., Ross 1996) to describe the decision process of the individual drivers and of the fleet manager.

5.1 Binary Choice

The procedure starts by considering two possible departure times 1 and 2. Let (X_ω, Y_ω) be the numbers of individual drivers and vehicles in the fleet that choose to depart at time 2 at iteration ω. Note that the number of departures at time 1 is $(N - X_\omega, M - Y_\omega)$. The focus is on the joint distribution of the vector (X_ω, Y_ω). Let $p_\omega(x, y) = P(X = x, Y = y)$, with $p_0(x, y)$ assumed to be fixed.

From the Markovian assumption, the master equations are obtained:

$$\begin{aligned}\frac{dp_\omega(x,y)}{d\omega} &= p_\omega(x,y+1)f_\omega^{2|1}(x,y+1) + p_\omega(x,y-1)f_\omega^{1|2}(x,y-1) \\ &\quad + p_\omega(x+1,y)(x+1)g_\omega^{2|1}(x+1,y) \\ &\quad + p_\omega(x-1,y)(N-x+1)g_\omega^{1|2}(x-1,y) \\ &\quad - p_\omega(x,y)\left[f_\omega^{2|1}(x,y) + f_\omega^{1|2}(x,y)\right. \\ &\quad \left. + xg_\omega^{2|1}(x,y) + (N-x)g_\omega^{1|2}(x,y)\right],\end{aligned} \tag{20}$$

where, for instance, $f_\omega^{2|1}(x,y)$ (resp. $g_\omega^{2|1}(x,y)$) is the fleet transition rate (resp. individual transition rate) from departure time 1 to departure time 2 at iteration ω when the state is (x, y) at iteration $\omega - 1$. The transition rates at state (x, y) equal 0 when x is outside $[0, N]$ or y is outside $[0, M]$.

The transition rates can be constructed by means of a logit model. The argument ω is omitted to simplify the notation. For the fleet, arguing as for (10), this leads to

$$f^{2|1}(x,y) = \lambda \frac{\mathrm{e}^{-\left[C_2^A(x,y)+y(1-\theta)C'^A_2(x,y)\right]}}{\mathrm{e}^{-\left[C_1^A(x,y)+(M-y)C'^A_1(x,y)\right]} + \mathrm{e}^{-\left[C_2^A(x,y)+y(1-\theta)C'^A_2(x,y)\right]}}, \tag{21}$$

where λ is the decision-making rate of the manager and $C_i^A(x, y)$, $i = 1, 2$, is the cost of a vehicle in the fleet departing at time i. For the individual drivers, we have

$$g^{2|1}(x,y) = \mu \frac{e^{-C_2^I(x,y)}}{e^{-C_1^I(x,y)} + e^{-C_2^I(x,y)}}, \tag{22}$$

where μ is the individual decision-making rate and $C_i^I(x,y)$, $i = 1, 2$, is the cost of a driver departing at time i.

The process $\{(X_\omega, Y_\omega), \omega \in \mathbb{R}_+\}$ corresponds to a bivariate birth-and-death process on $[0, N] \times [0, M]$. The associated equations (20) obtain a unique solution $p_\omega(x, y)$ for all ω and a stationary solution $p(x, y)$, under some regularity conditions on the rates.

5.2 General Case

Let $X_\omega(t)$ be the number of individual drivers departing at time t at iteration ω, and let $Y_\omega(t)$ be the number of vehicles of the fleet departing at time t at iteration ω. The aim is to derive the joint distribution of the vector $(\mathbf{X}_\omega, \mathbf{Y}_\omega) = \{[X_\omega(t), t = 0, \ldots, T], [Y_\omega(t), t = 0, \ldots, T]\}$. Let $p_\omega(\mathbf{x}, \mathbf{y}) = P(\mathbf{X}_\omega = \mathbf{x}, \mathbf{Y}_\omega = \mathbf{y})$, with $p_0(\mathbf{x}, \mathbf{y})$ supposed to be fixed.

In this case, the master equations become

$$\begin{aligned} \frac{dp_\omega(\mathbf{x},\mathbf{y})}{d\omega} &= \sum_{\mathbf{z}\neq\mathbf{y}} p_\omega(\mathbf{x},\mathbf{z}) f_\omega^{\mathbf{y}|\mathbf{z}}(\mathbf{x},\mathbf{z}) \\ &+ \sum_{j|x_j\geq 1} \sum_{i\neq j|x_i\leq N-1} (x_i+1) p_\omega(\mathbf{x}+\mathbf{e}_i-\mathbf{e}_j,\mathbf{y}) g_\omega^{j|i}(\mathbf{x}+\mathbf{e}_i-\mathbf{e}_j,\mathbf{y}) \\ &- p_\omega(\mathbf{x},\mathbf{y}) \left[\sum_{\mathbf{z}\neq\mathbf{y}} f_\omega^{\mathbf{y}|\mathbf{z}}(\mathbf{x},\mathbf{y}) + \sum_{j|x_j\geq 1} \sum_{i\neq j|x_i\leq N-1} x_i g_\omega^{j|i}(\mathbf{x},\mathbf{y}) \right], \end{aligned} \tag{23}$$

using the same notation as for (20). To the best of these authors knowledge, such a birth-and-death process is a new framework for the congestion problems. It would be interesting to simulate this stochastic process, assuming a smooth behavior for the transition rates.

6 Conclusions

Two main ideas have been developed in this chapter. The first one is the use of a stochastic approach to tackle dynamic transportation bottleneck models. Equilibrium solutions can be determined in various situations but very little is known concerning the adjustment process and the stability. We believe that the problem is more amenable to numerical simulations when it is formulated in a truly stochastic environment (i.e., not only with discrete choice models but with birth-and-death master equations).

Secondly, the possible presence of a fleet of vehicles has been taken into account in the model. The technology of tomorrow suggests that fleets of vehicles will become more frequent in the real world (e.g., vehicles on demand that are centrally managed). These issues were rarely addressed in the literature. The present work has considered a tractable way to combine individual vehicles and a fleet of vehicles which are centrally managed.

Such a model can raise questions that are mathematically difficult. As shown by Silva et al. (2017), a combination of small and large vehicles can lead to some non-equilibrium in the bottleneck problem. This situation cannot be easily avoided and is very unfortunate. The stochastic approach is more realistic and flexible, and it is expected to provide with an equilibrium solution even in such a case.

The relationship between stochastic and deterministic models is an important but difficult topic; see, e.g., Ethier and Kurtz (2005). In the present context, two relevant questions are the links between the local extrema of (stochastic) stationary distributions and the (deterministic) equilibrium states, and the existence of a solution in a stochastic model while there is no solution in the deterministic case.

The modeling framework presented in this chapter will be tested in the future by means of simulation experiments to explore these processes and stationarity properties. We hope that this work will be a good extension of the work we started with Pavlos.

Acknowledgements We thank the referee and the editor for very helpful and constructive comments. The research of André de Palma was supported by the *ANR Project Elitisme* from the French Agence Nationale de la Recherche. The research of Claude Lefèvre was supported by the *ARC Project IAPAS* from the Fédération Wallonie-Bruxelles. This work has benefited from many discussions with Moshe Ben-Akiva and Pavlos Kanaroglou during several visits to the Massachusetts Institute of Technology (MIT) in the past. Additional comments were made by Théo Roudil-Valentin.

References

Anderson, S. P., de Palma, A., & Thisse, J.-F. (1992). *Discrete choice theory of product differentiation*. Cambridge: MIT Press.

Arnott, R., de Palma, A., & Lindsey, R. (1993). A structural model of peak-period congestion: A traffic bottleneck with elastic demand. *American Economic Review, 83*, 161–179.

Ben-Akiva, M., de Palma, A., & Kanaroglou, P. (1986a). Dynamic model of peak period traffic congestion with elastic arrival rates. *Transportation Science, 20*, 164–181.

Ben-Akiva, M., de Palma, A., & Kanaroglou, P. (1986b). Effects of capacity constraints on peak period congestion. *Transportation Research Record, 1085*, 16–26.

Ben-Akiva, M., de Palma, A., & Kanaroglou, P. (1987). Dynamic network equilibrium: Some comments. *European Journal of Operational Research, 30*, 318–320.

de Palma, A., & Lefèvre, C. (1983). Individual decision-making in dynamic collective systems. *Journal of Mathematical Sociology, 9*, 103–124.

de Palma, A., & Marchal, F. (1996). METROPOLIS: Un outil de simulation du trafic urbain. *Revue Transport, 378*, 304–315.
de Palma, A., & Fosgerau, M. (2011). Dynamic traffic modeling. In A. de Palma, R. Lindsey, E. Quinet, & R. Vickerman (Eds.), *A handbook of transport economics* (pp. 188–212). Cheltenham: Edward Elgar.
de Palma, A., Ben-Akiva, M., Lefèvre, C., & Litinas, N. (1983). Stochastic equilibrium model of peak period traffic congestion. *Transportation Science, 17*, 430–453.
de Palma, A., Lefèvre, C., & Ben-Akiva, M. (1987). A dynamic model of peak period flows and delays in a corridor. *Computers and Mathematics with Applications, 14*, 201–223.
de Palma, A., Marchal, F., & Nesterov, Y. (1997). METROPOLIS: Modular system for dynamic traffic simulation. *Transportation Research Record, 1607*, 178–184.
Dixit, A. K., & Stiglitz, J. E. (1977). Monopolistic competition and optimum product diversity. *American Economic Review, 67*, 297–308.
Ethier, S. N., & Kurtz, T. G. (2005). *Markov processes: Characterization and convergence*. New York: Wiley.
McFadden, D. (2001). Economic choices. *American Economic Review, 91*, 351–378.
Nesterov, Y., & de Palma, A. (2003). Stationary dynamic solutions in congested transportation networks: Summary and perspectives. *Networks and Spatial Economics, 3*, 371–395.
Ross, S. M. (1996). *Stochastic Processes* (2nd ed.). New York: Wiley.
Sheffi, Y. (1985). *Urban transportation networks*. Englewood Cliffs: Prentice-Hall.
Silva, H. E., Lindsey, R., de Palma, A., & van den Berg, V. A. C. (2017). On the existence and uniqueness of equilibrium in the bottleneck model with atomic users. *Transportation Science, 51*, 863–881.
Small, K. A. (1982). The scheduling of consumer activities: Work trips. *American Economic Review, 72*, 467–479.
Vickrey, W. S. (1969). Congestion theory and transport investment. *American Economic Review, 59*, 251–260.
Xiao, L.-L., Huang, H.-J., & Liu, R. (2015). Congestion behavior and tolls in a bottleneck model with stochastic capacity. *Transportation Science, 49*, 46–65.
Xiao, Y., Coulombel, N., & de Palma, A. (2017). The valuation of travel time reliability: Does congestion matter? *Transportation Research Part B: Methodological, 97*, 113–141.
Yessenalina, A., & Cardie, C. (2011). Compositional matrix-space models for sentiment analysis. In *Proceedings of the Conference on Empirical Methods in Natural Language Processing* (pp. 172–182). Stroudsburg: Association for Computational Linguistics.
Zeifman, A. I. (1995). Upper and lower bounds on the rate of convergence for nonhomogeneous birth and death processes. *Stochastic Processes and their Applications, 59*, 157–173.

Factors Influencing Journey-to-Work by Public Transit in Mega Canadian Cities

Rahaf Husein, Hanna Maoh, and Dimitris Potoglou

Abstract In this chapter, a modeling exercise is carried out to analyze the factors influencing the journey-to-work by public transit in the largest two Canadian urban centers: Toronto, Ontario and Montreal, Quebec. Transit trip generation at the census tract level for the year 2011 is regressed against urban form and network design variables after controlling for several socio-economic and demographic variables. The simultaneous auto-regressive (SAR) modeling approach is employed to account for the presence of spatial dependencies in the modeled data. These spatial effects are usually neglected in travel demand models, leading to bias in the estimates obtained. While the estimated socio-economic and demographic parameters reinforce previous findings from the literature, the inclusion of urban form and network design variables offers new insights regarding the role that urban planning can have on transit trip generation. The results reported in this chapter can, thus, help promote sustainable urban planning policies.

1 Introduction

Over the past six decades, intense highway development has encouraged suburbanization in many Canadian cities. The suburbs were perceived as the ideal location to live due to the trade-offs between commuting and housing costs. That is, the additional commuting cost from the suburbs was balanced out by the low cost of land and housing. According to Anderson et al. (1996), suburbanization can be categorized as the outcome of two concurrent spatial trends: (1) the increase of population and employment in urban areas, and (2) the dispersion of land use

R. Husein · H. Maoh (✉)
Department of Civil and Environmental Engineering, University of Windsor, Windsor, ON, Canada
e-mail: huseinr@uwindsor.ca; maohhf@uwindsor.ca

D. Potoglou
School of Geography and Planning, Cardiff University, Cardiff, Wales, UK
e-mail: potogloud@cardiff.ac.uk

H. Briassoulis et al. (eds.), *The Practice of Spatial Analysis*,
https://doi.org/10.1007/978-3-319-89806-3_8

activities within these urban areas. This results in converting the form of cities from compact to sprawled over time which in turn gives rise to increased commute times. The later has been fueled by the ease of owning motorized vehicles.

Urban sprawl along with the widespread use of motorized vehicles has downplayed the role of public transit in Canada. As such, the use of public transit in Canadian megacities is not as pronounced as in the case of other places such as London, England where, according to Transport for London (2016), the modal share for public transit was 43% in 2011. Obviously, lower usage of transit and high auto-dependency are not sustainable in the long run especially with the continuous increase in travel demand over time. Over the past few decades, academics, planners, and community organizations, among other stakeholders, have been working on various land use and transportation strategies under the "smart growth" paradigm. The latter is focused on promoting sustainable growth outcomes that minimize negative environmental and social impacts while insuring the maximization of economic benefits. Examples of smart growth strategies include, but are not limited to, Urban Residential Intensification or compact development (Behan et al. 2008) and/or polycentrism (Maoh and Tang 2012). The objective in both cases is to decrease the number of motorized trips and promote greater transit usage in urban areas (Handy 2005). To promote public transit, a major thrust of the research conducted to date has been to explore the impacts that socio-economic and demographic factors have on public transit usage (Wiley et al. 2011; Chakraborty and Mishra 2013). However, little research has been undertaken to determine the impacts of land use configuration and road network design characteristics on public transit ridership.

The work by Chakraborty and Mishra (2013) is an exception in which the relationship between land use types and transit ridership was explored. This chapter takes the analysis one step further by investigating the relationship between road network design characteristics and transit ridership after controlling for other known factors. More specifically, the research question being addressed is the following: other things equal, do urban form and road network design significantly influence transit ridership in Canadian megacities? To date, the majority of existing studies identify population density as a key factor when it comes to the role of land use on transit ridership (Handy 2005; Filion and McSpurren 2007; Chakraborty and Mishra 2013). This is not surprising since public transit must serve dense areas within the metropolitan region. However, while population density is an important factor, it is argued here that the nature of land use itself could also explain, and eventually promote, public transit usage. More specifically, areas that enjoy a higher level of land use mixing or density mixing are expected to be more public transit oriented. Furthermore, areas with curvilinear road network design seem to be less public transit-oriented since these areas represent enclaved auto-oriented and suburban neighborhoods.

The present study tries to address these hypotheses by focusing on public transit ridership in the two largest Canadian metropolitan areas: Toronto, Ontario and Montreal, Quebec. Using journey-to-work data from the 2011 Canadian census, the analysis sets out to evaluate the degree in which urban design characteristics, represented by the structure of the road network and land use configuration, affect public transit ridership while controlling for typical socio-economic and demographic variables. The analysis also investigates whether these factors are systematic across the two studied metropolitan areas. The statistical analysis uses the simultaneous auto-regressive (SAR) modeling technique given the spatial nature of the problem in hand. The findings are expected to help planners to better understand the relationship between urban form and public transit ridership in large metropolitan areas and to contribute to the existing body of knowledge on the subject.

This chapter consists of five sections. The next section discusses urban form and its relation to transportation, the nature of the road network and its expected effect on transit ridership, and the factors that have been used in previous studies to explain public transit ridership. Section 3 outlines the data sources, briefly describes the study areas and presents the modeling approach. The results and the contribution of the analysis conducted are discussed in the fourth section. The conclusion of this Chapter is provided in section five.

2 Factors Influencing Public Transit Ridership

2.1 Urban Form

Anderson et al. (1996) defined urban form as the spatial configuration of different land use activities (i.e., buildings), which is the outcome of the land development process. Urban form in Canada emerged from the traditional monocentric to the more commonly sprawled or dispersed pattern that is observed in many cities nowadays (Maoh et al. 2010). Arguably, changes in urban form can influence the nature of intra-urban trips but this is not the only factor that determines trip flows. Monocentric urban form belongs to the so-called compact form. The latter is characterized by high density, mixed land use (apartments, commercial space, and offices) and strong presence of public transportation systems (OECD 2012, p. 31).

Historically, metropolitan areas were monocentric (i.e., had a single core) because all types of social and economic activities were located in one center (OECD 2012, p. 31). However, this is not the case anymore in most urban areas. Several metropolitan areas evolved to have multiple central locations of high densities that are components of a wider spatial functional entity. This gave rise to polycentricism. In a polycentric metropolitan area, the central city is not widely dispersed but it is typically linked to satellite centers via a well-connected public transportation system (OECD 2012, p. 32). Hence, it can be concluded that a monocentric urban form is more affiliated with mixed density and less auto-dependency, whereas a polycentric urban form is more affiliated with public transit.

By comparison, a sprawled city is typically an area with an extensive spatial growth pattern resulting in inefficiencies due to the strong presence of low-density and leapfrog developments scattered in outer suburbs (Maoh et al. 2010). This form portrays high level of segregation in land uses and lack of centralized planning (Nam et al. 2011; Behan et al. 2008). The segregation of land uses has been blamed for increasing commuting and energy consumption due to auto-dependency. The negative sentiment against urban sprawl development encouraged urban planning initiatives that have focused on promoting "smart growth" principles such as mixed land use development and greater reliance on public transportation for journey-to-work (Handy 2005). According to Behan et al. (2008: XX), smart growth "*seeks to revitalize established urban areas and provide for a greater range of housing, employment and social activities.*" Smart growth strategies aspire to achieve urban sustainability outcomes; that is, to promote growth that will minimize negative environmental and social impacts while maximizing economic benefits. An example of such strategies, as shown in Behan et al. (2008), is adoption of urban residential intensification (URI) in the central part of urban areas. URI is another synonym for compact development. OECD (2012, p. 45) stresses that compact urban form can make a more sustainable usage of space in cities. For instance, it is noted that compact development plays a significant role in reducing CO_2 emissions from both the transportation and building sectors, lowering the dependency on automobiles, and reducing traveled distances and fuel consumption. In addition to compact development, another smart growth strategy is to promote polycentrism. According to Maoh and Tang (2012: XX), "*smart growth via mixed land use could indeed help to curb auto-dependency*." The authors further note that polycentrism could help reduce commute over long distances in Canadian cities. Eventually, an effective polycentric urban form is one that provides reliable transit connections between the various nuclei and the main center.

Quantifying the type of urban form is an important step towards understanding the nature of travel in a given city. Tsai (2005) showed that the *Global Moran's I statistic* can be used to quantify the type of urban form based on zonal population and employment. This was possible by simulating various urban form patterns and calculating the *Moran's I* value that is associated with each urban form type. More specifically, Tsai (2005) found that values close to one indicate autocorrelation and clustering in space, thus representing a monocentric type of urban form. On the other hand, values close to zero indicate random scattering or a polycentric urban form. Lastly, values close to minus one represent a "chessboard" development pattern indicating a sprawled urban form.

2.2 Other Factors Influencing Public Transit Ridership

According to the literature, land use and socio-demographic factors can influence public transit ridership. Different measures have been proposed in the literature to operationalize land use: mixed density index (MDI) (Behan et al. 2008), entropy index (EI) (Cervero and Kockelman 1997; Frank and Pivo 1994), accessibility index (Cervero et al. 1999), and land use types (Chakraborty and Mishra 2013). The MDI computes the level of mixing of population and employment density within a given geographic area. Generally, high MDI values are linked with higher job-housing balance. In other words, higher level of population and employing mixing could be affiliated with shorter commuting distances and greater public transit ridership (Behan et al. 2008). EI is a descriptive statistic that measures land use heterogeneity to determine the evenness of the distribution of land use types, and it is calculated to range between 0 and 1 (Cervero and Kockelman 1997; Frank and Pivo 1994). Potoglou and Kanaroglou (2008) noted that an EI equal to zero implies a single type of land use, while a value of one indicates the presence of mixed land uses (i.e., heterogeneity) in a zone. Thus, the closer EI is to 1, the easier the access to active modes of transportation, namely, transit, walking, and/or biking. A third measure which is based on the gravity formulation could be used to determine the potential accessibility of a given zone from all other zones. This measure will depend on the size of land use activities (i.e., employment or housing) in all the zones comprising the study area but it is typically scaled by the amount of travel time it takes to reach a zone from all other zones (Cervero et al. 1999). More recently, Chakraborty and Mishra (2013) examined the relationship between land use types and public transit ridership. The authors introduced a number of land use-related measures that represent the square footage of various land use types (e.g., dining, health care, housing, industry, office, recreation, shopping, and warehouses) in their transit ridership model. The analysis suggested that, other things being equal, public transit ridership was negatively correlated with health care and recreational square footage in the study area. Besides land use, the impacts of the nature of the road network on transit trips could be assessed. According to the literature, gridded networks have more transit routes and greater efficiency leading to higher transit modal shares (Marshall and Garrick 2010; Messenger and Ewing 1996).

The literature has considered a host of socio-economic and demographic factors also, with the following being the most prevalent: age, gender, household size, income, visible minorities, population density, and employment density. With respect to the age factor, the evidence from the literature suggests that older populations (e.g., 55 years or older) are more likely to use public transport for their daily travel (Cervero 2002; Chee and Fernandez 2013; Patterson et al. 2005). With respect to gender, studies have found that females are less likely to use public transit compared to males (Cervero 2002; Chen et al. 2007; Chee and Fernandez 2013). On the other hand, the literature suggests that household size has a negative impact on public transit in that larger households are less prone to rely on transit (Sultana and Weber 2007). The latter could be attributed to higher auto-ownership level among larger families, which tend to reside in single family houses in suburban areas. Likewise, people with high income are more likely to own a vehicle and, thus, tend to be less likely dependent on public transit (Cervero 2002; Potoglou and Kanaroglou 2008; Chakraborty and Mishra 2013). As for the visible minorities, the current evidence from the literature concluded that these population groups are more prone to use public transit since they usually travel shorter distances to work (Sultana and Weber 2007). Population density was found to have a positive effect on public transit ridership (Handy 2005; Filion and McSpurren 2007; Wiley et al. 2011; Chakraborty and Mishra 2013). Finally, Chakraborty and Mishra (2013) suggested that public transit ridership is positively correlated with employment density in the State of Maryland, USA.

3 Modeling Approach

3.1 Study Areas

This chapter considers the two largest Census Metropolitan Areas (CMAs) in Canada; Toronto and Montreal. Toronto is the provincial capital of Ontario, Canada. It is located in Southern Ontario on the northwest shore of Lake Ontario. According to the 2011 Canadian Census, Toronto had a population of 5,583,064 living in 2,079,459 dwellings. Also, the regional district has a land area of approximately 5,906 km^2 and a population density of roughly 945 per km^2. Montreal is the second largest metropolitan area in Canada. It is located in the Southwest part of the province of Quebec about 542 km away from Toronto. In 2011, Montreal had a population of 3,824,221 living in 1,696,210 dwellings. The regional district has a land area of approximately 4,258 km^2 and a population density of around 898 per km^2 (Statistics Canada 2011).

The two CMAs had the highest transit market shares all across Canada in 2011. As shown in Fig. 1, they also have a disproportionally higher transit share compared to the national average. Toronto is in the lead with a transit share of 23% for work trips in 2011. This is followed by Montreal with a share of 22% (Statistics Canada 2011). Compared to the nationwide modal share distributions, public transit shares are 10–11% higher in the two CMAs. These higher transit shares can be attributed to the well-established transit systems in these urban centers. Furthermore, both Toronto and Montreal have subway systems that provide high speed travel by transit. For more information about the Toronto Subway system, see the Toronto Transit Commission's website at www.ttc.ca. For information about the Montreal Subway system, see the Société de transport de Montréal website at http://www.stm.info/en.

3.2 Multivariate Explanatory Model

The multivariate analysis is performed at the zonal level using the delineations of the census tracts in each CMA. Given the spatial nature of the data (i.e., area data), the simultaneous auto-regressive (SAR) model is employed to study the factors explaining work trip generation by public transit at the census tract level. When the observations representing the dependent variable (i.e., the area data) are spatially autocorrelated, the SAR model is more suitable than the ordinary least square (OLS) regression model (Bailey and Gatrell 1995). Failing to account for spatial autocorrelation could lead to bias in the estimated parameters. The SAR model can be formulated as follows:

$$\boldsymbol{T} = \boldsymbol{X\beta} + \boldsymbol{\rho TW} + \boldsymbol{\varepsilon}$$

where $\boldsymbol{T}$ is the dependent variable representing the number of public transit work trips originating in a given census tract in the metropolitan area; $\boldsymbol{\beta}$ is a vector

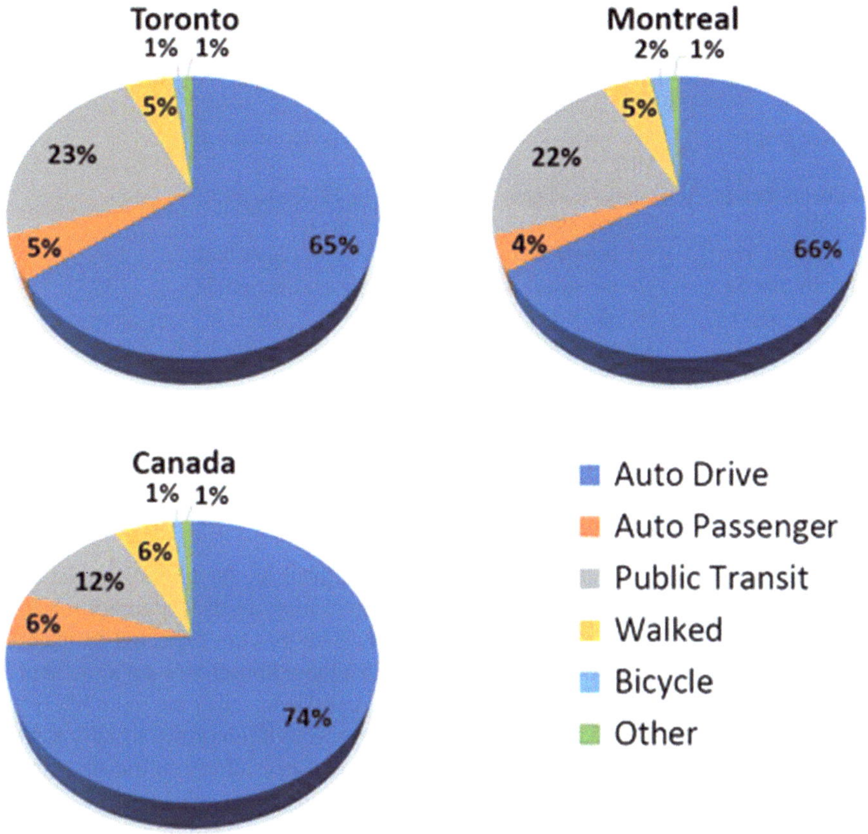

Fig. 1 Travel mode shares for journey-to-work in 2011

representing the parameters pertaining to covariates $\boldsymbol{X}$; ρ is a spatial lag parameter that accounts for the presence of spatial autocorrelation while $\boldsymbol{W}$ is a weight matrix representing the spatial affinity between the census tracts. A first-order rook structure was used to design weight matrix $\boldsymbol{W}$ in the model. Finally $\boldsymbol{\varepsilon}$ is a set of independent error terms.

Several datasets were processed and organized in order to conduct the analysis. The 2011 Journey-to-Work Canadian Census dataset was used to obtain the dependent variable $\boldsymbol{T}$. The Canadian Census asks about the modes of travel (i.e., auto-drive, auto-passenger, transit, walk, bike, other) used by workers to commute to their ordinary place of work. The census data were also used to extract the socio-economic variables used in the specification of the model. It was hypothesized that older people are more prone to use transit when compared to other population groups. Therefore, variable "Population 55+" was used to represent the number of travelers in the age group 55 years or older in the census tract. The rationale for using this age group is inspired by the idea that young workers who grew accustomed to

Table 1 List of variables used analysis

Variable	Description	Expectation
Entropy (EI)[a]	The entropy index calculated per census tract: $EI = -(1/L)\sum_c(p_c \, ln\,(p_c))$	+
Mixed Density (MDI)[b]	The mixed density index calculated per census tract: MDI = (PD × ED)/(PD + ED)	+
Network Design (NDI)	1 if the road network in gridded in the census tract; 0 otherwise	+
Population 55+	The population of 55 years or older in a census tract	+
Family Size 5+	The number of families with five or more persons in a census tract	−
Minorities	The percentage of minorities in the total population per census tract	+
Population Density	The population density per square kilometer per census tract	+

[a]L is total number of land use classes in the census tract; p_c is proportion of land use type c in the tract
[b]PD = Population Density; ED = Employment Density

using the subway system in Toronto and Montreal back in the 1970s adopted this mode of travel as a life-style choice for their commuting needs. Large households with five or more family members (i.e., Family Size 5+) are expected to have a negative impact on public transit. The reason is that these households are more likely to live in suburban areas and own more than one car. Hence, they are more likely to drive to work. On the contrary, the literature suggests that minorities usually travel less distance to work and are more likely to have a positive effect on transit as they are more prone to live in central locations that are usually accessible by transit. In the present model, this group represents low income population and immigrants who are recent movers to Canada. As it was mentioned earlier, population density (population per km^2) is expected to have a positive impact on public transit ridership. Therefore, the density in each census tract was utilized in the specification of the model.

Land use and road network geographic information system (GIS) files for the year 2011 were extracted from the DMTI spatial geo-database and its route logistics package. Socio-economic and demographic variables were also retrieved from the 2011 Canadian census. These data were processed in the ArcMap 10.1 GIS platform. Table 1 lists the variables used in the statistical model following the literature. In order to examine the impact of land use configuration and road network design on the transit trips generated, the following variables were introduced: entropy index (EI), mixed density index (MDI), and network design index (NDI).

The entropy index, EI, is expected to have a positive influence on public transit. That is, higher entropy values (closer to 1), where mixed land uses are more pronounced, are expected to represent areas with better access to public transit. Likewise, the mixed density index, MDI, is expected to have positive effect on public transit trip production since a higher MDI value suggests the presence of higher population and employment densities in a zone. That could be associated with

Fig. 2 Road network designs within Toronto CMA: (**a**) Gridded and (**b**) Curvilinear

shorter commuting distances (Behan et al. 2008), making public transit more appealing to travelers.

As discussed before, gridded road networks are expected to facilitate public transit service. Hence, census tracts dominated by gridded networks (as opposed to curvilinear networks) are more likely to increase transit trip production, other things being equal. An example of the gridded and curvilinear network designs for the CMA of Toronto is shown in Fig. 2. To introduce the effect of network road design in the model, the categorical variable NDI was created. The variable is set to 1 if the majority of roads in the census tracts are gridded; otherwise, it is set to 0.

4 Results and Discussion

4.1 Land Use and Network Design Characteristics

The spatial distribution of population and employment in the two Canadian CMAs is explored following the approach proposed in Tsai (2005). The estimated *Moran's I statistic* for population and employment are presented in Table 2. According to the results, population is clustered in the two CMAs. However, Montreal's population is highly clustered compared to Toronto based on the magnitude of the achieved *Moran's I* values. With respect to jobs, the two CMAs demonstrate similar and high levels of clustering.

To understand the nature of clustering, population and employment density maps were generated (Figs. 3 and 4). The population map for Toronto indicates that density is highest in the urban core and its surrounding areas. However, population density appears to be more localized compared to Montreal. By comparison, Montreal's population density is also at its peak in the core and its surrounding areas but more central tracts have higher density. These findings are in line with the

Table 2 *Moran's I* statistic[a] for population and employment densities

	Toronto		Montreal	
	Population	Employment	Population	Employment
Moran's I	0.118	0.289	0.310	0.252
z-score	6.166	16.023	14.295	14.008

[a]All *Moran's I statistic* values were calculated using ArcGIS 10.1

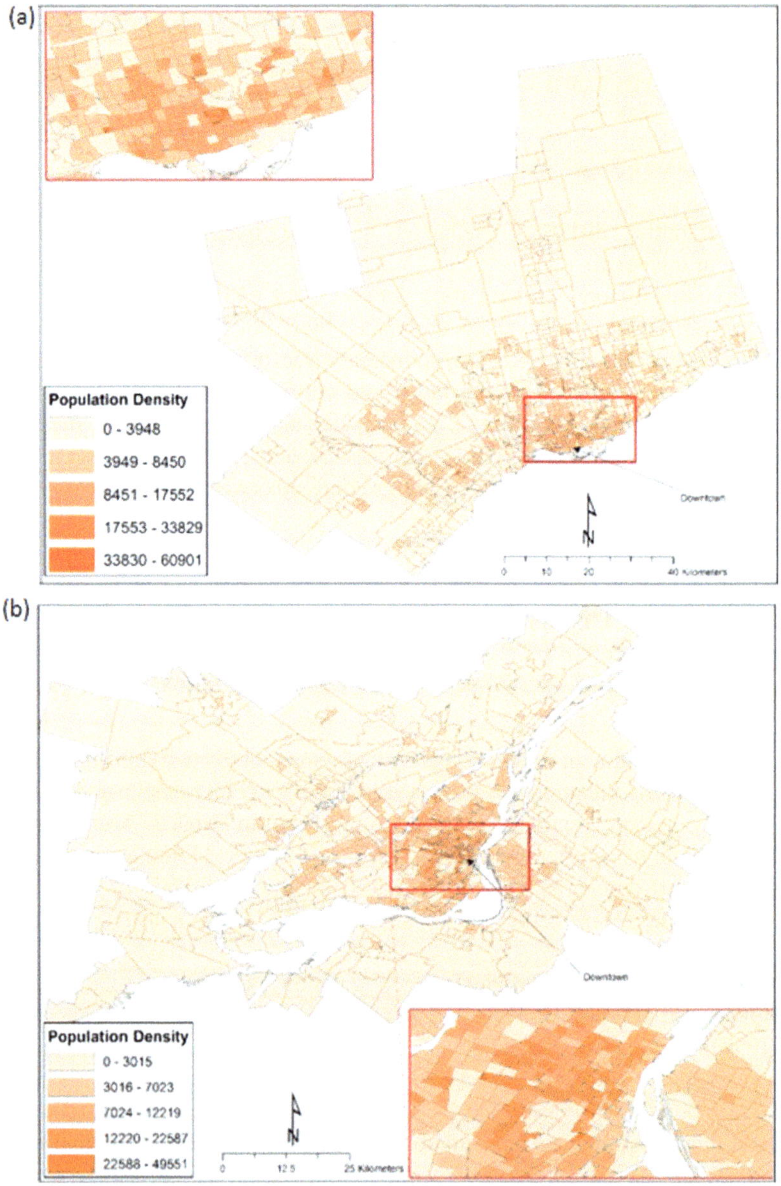

Fig. 3 Population density in the (**a**) Toronto CMA, and (**b**) Montreal CMA, in 2011

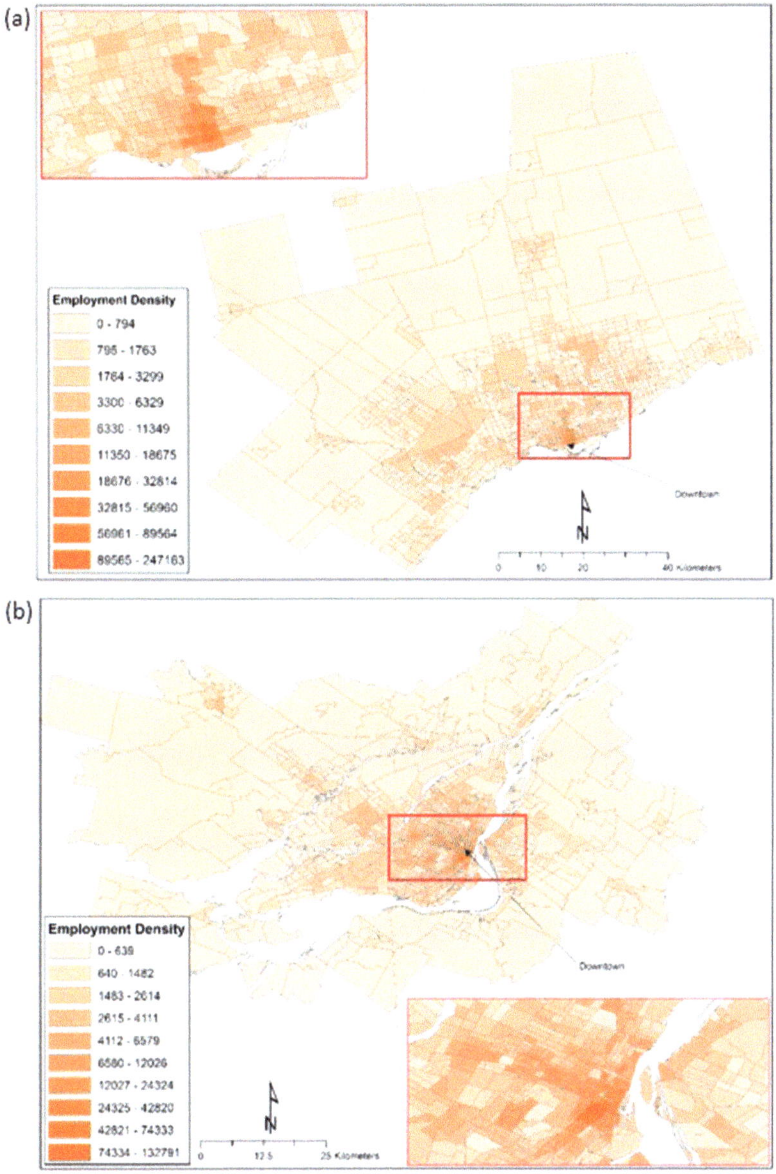

Fig. 4 Employment density in the (**a**) Toronto CMA, and (**b**) Montreal CMA, in 2011

Moran's I values reported in Table 2. Likewise, the patterns depicted in the employment density maps are in line with the *Moran's I Statistic values* reported in Table 2. As demonstrated in Fig. 4, employment density is at its peak in the urban core (i.e., downtown) in the two CMAs, albeit population density in Toronto is more localized.

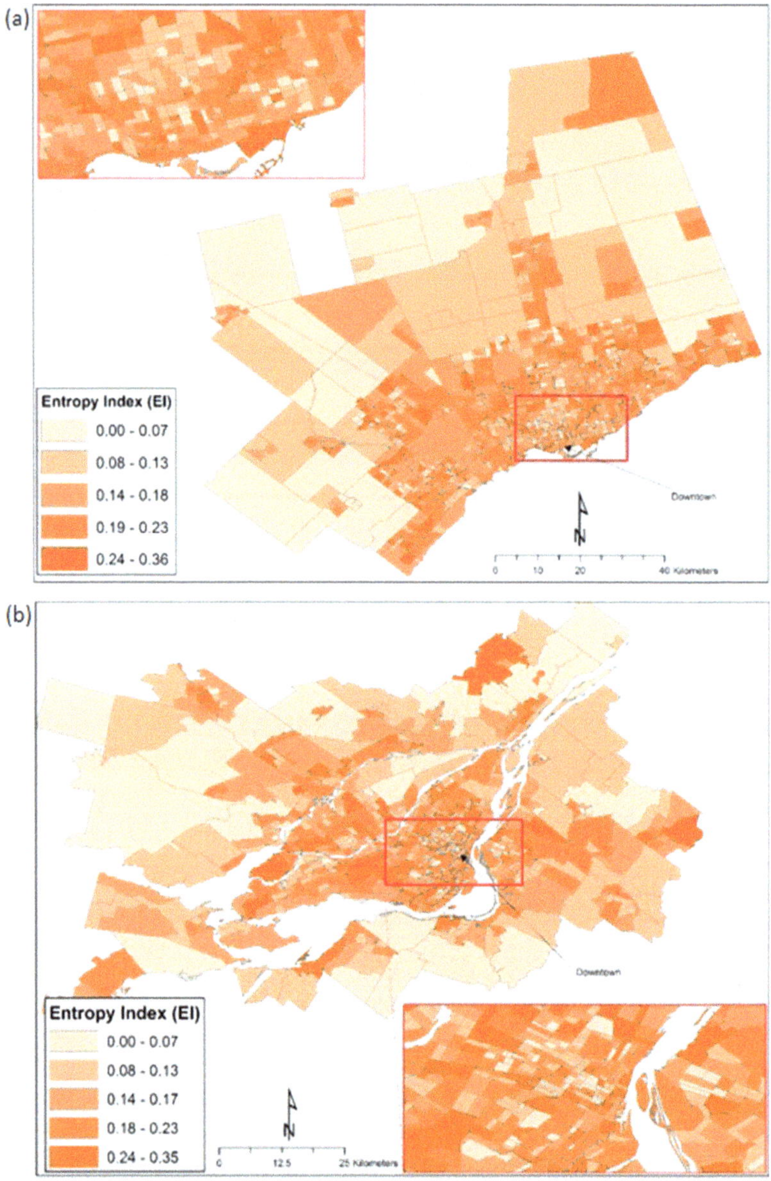

Fig. 5 Entropy Index (EI) in the (**a**) Toronto CMA, and (**b**) Montreal CMA, in 2011

Figures 5 and 6 present the spatial distribution of the calculated entropy (EI) and mixed density (MDI) values at the census tract level. The EI maps suggest a similar pattern in the two CMAs where land use mixing is relatively low. The maximum EI index in the two CMAs is very close (0.36 and 0.35) and so is the average (0.16 and 0.15), as shown in Table 3. Interestingly, the *Moran's I statistic* for the EI values in

Fig. 6 Mixed Density Index (MDI) in the (**a**) Toronto CMA, and (**b**) Montreal CMA, in 2011

the two CMAs is positive, and it has relatively the same magnitude. As such, the entropy values appear to exhibit some clustering over space.

The MDI maps shown in Fig. 6 suggest that the central areas in the two CMAs enjoy the highest level of job-house mixing. However, Montreal appears to have a

Table 3 MDI and EI Statistics

		Toronto ($n = 1088$)		Montreal ($n = 921$)	
		EI	MDI	EI	MDI
Descriptive statistics	Mean	0.16	1030	0.15	1266
	Minimum	0	0	0	0
	Maximum	0.36	18,936	0.35	14,270
	S.D.	0.06	1742	0.07	1531
Clustering statistics	*Moran's I*	0.1383	0.6206	0.1375	0.6836
	z-score	7.153	32.500	6.360	31.612

slightly higher mixed density as discerned from the average MDI values shown in Table 3. Furthermore, MDI is highly clustered as depicted by the *Moran's I* values of 0.62 and 0.68 for Toronto and Montreal, respectively (see Table 3). Figure 7 provides a visual depiction of the network design index (NDI) devised for the two CMAs based on the nature of the road networks in the two regions. The maps suggest that central census tracts are more affiliated with gridded network structure, whereas peripheral tracts in the inner and outer suburbs are more associated with curvilinear networks. The spatial distribution of the transit trips is presented in Fig. 8. The observed pattern exhibits some similarities with the population density distribution, albeit there are some visible differences. More specifically, transit trips appear to be more dispersed compared to population density. The maps in Fig. 7 also tend to overlap with the spatial distribution of transit trips in the two CMAs. However, as in the case of the population density maps, the patterns are not exact.

To obtain a better understanding of the strength of the relationship between transit trips and the variables listed in Table 1, the Pearson correlation coefficient (r) is calculated and reported in Table 4. Toronto appears to have a stronger relation between its population density and transit work trips compared to Montreal. Not surprisingly, transit is expected to have some affiliation with density as was observed in the population density maps. On the other hand, employment density has no relationship with the observed public transit work trips produced from the tracts. This is especially true in the case of Montreal with a correlation coefficient of 0.01. The network design index (NDI) is positively correlated to the number of generated public transit work trips in the two CMAs. Also, population 55 years of age or older is correlated to public transit trips. The strength of the relationship is somehow similar in the two CMAs. Likewise, minorities are also affiliated with public transit trips especially in Montreal, where this variable exhibits the strongest correlation among all factors (i.e., $r = 0.66$). Not surprisingly, the mixed density index (MDI) is positively correlated to the public transit trips although the relationship is stronger in the case of Toronto. Finally, the land use entropy index (EI) bears no significant impact on the generated public transit work trips as indicated by the very low correlation for the two CMAs (i.e., 0.01 and 0.07).

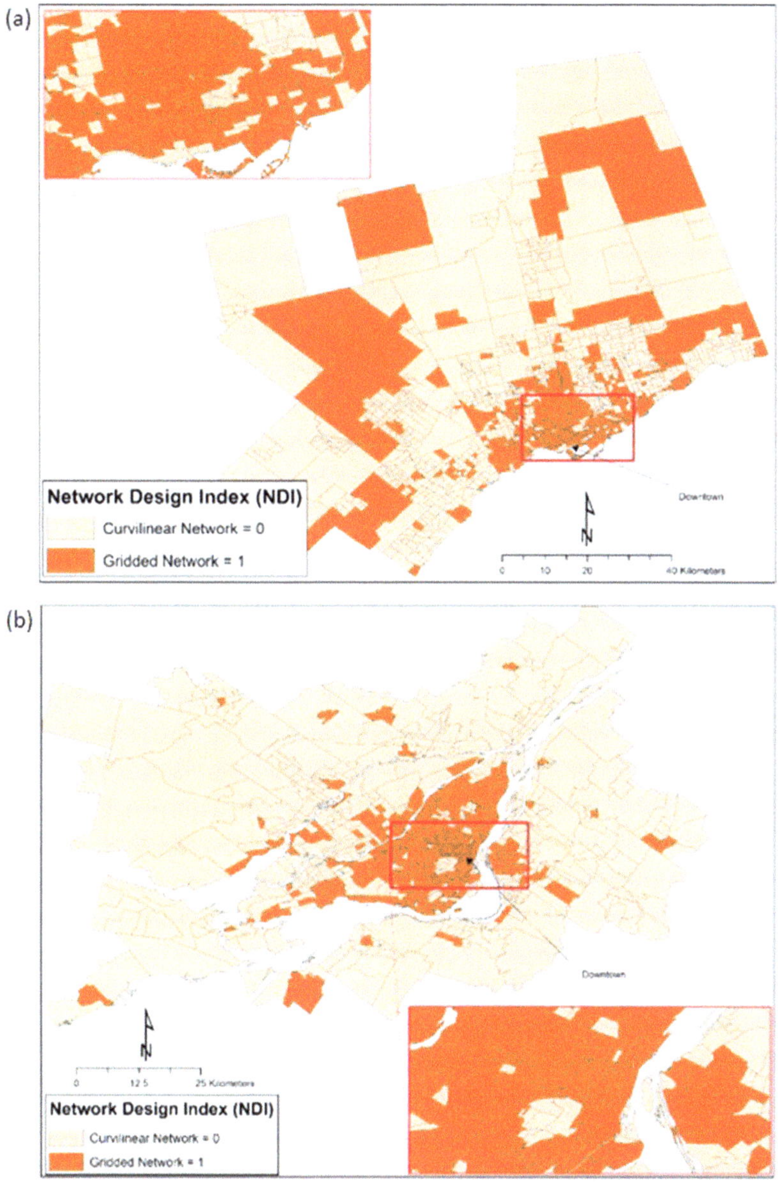

Fig. 7 Network Design Index (NDI) in the (**a**) Toronto CMA, and (**b**) Montreal CMA, in 2011

4.2 Statistical Modeling Results

Given the spatial nature of the utilized public transit data (i.e., area data), the global *Moran's I statistic* was applied to examine if the observed transit trips exhibit any spatial autocorrelation. The results, shown in Table 5, suggest that public transit trips

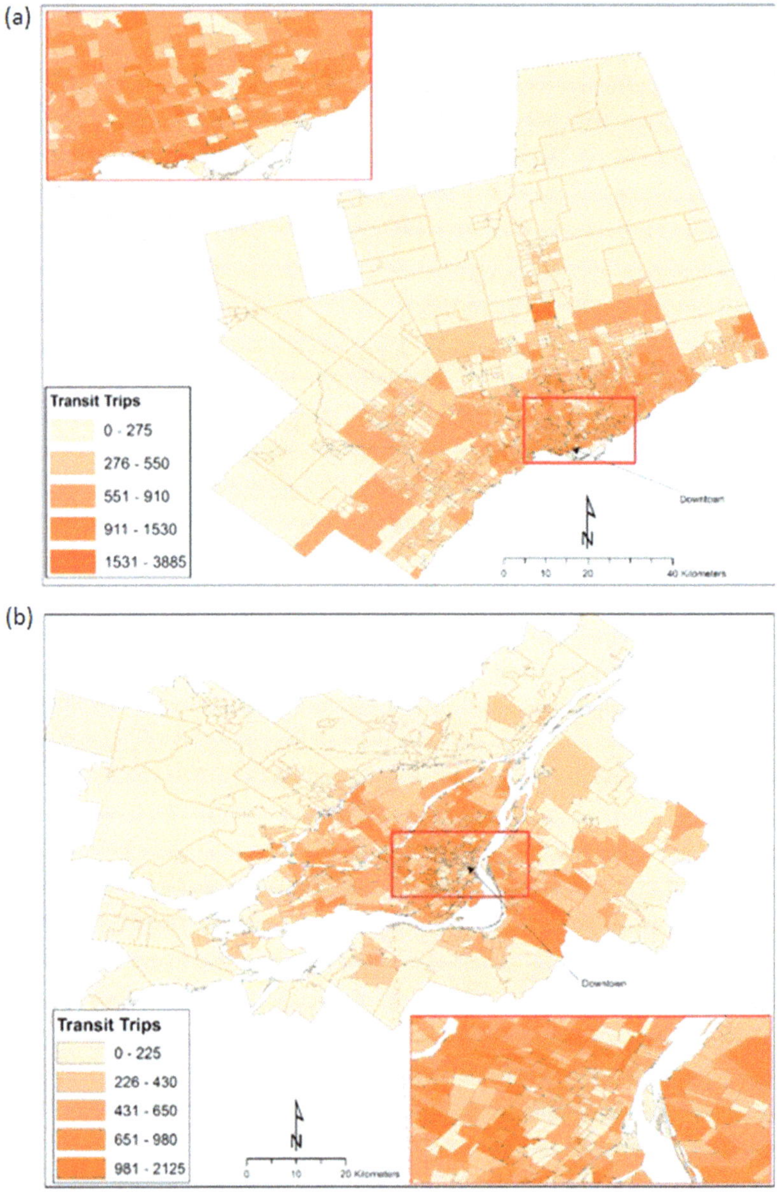

Fig. 8 Spatial distribution of work trips by transit in the (**a**) Toronto CMA, and (**b**) Montreal CMA, in 2011

in the two CMAs are spatially autocorrelated. More specifically, *Moran's I statistic* in each case is significant; thus, the null hypothesis of no spatial association among the analyzed transit trips is rejected. Consequently, the application of an SAR model in lieu of an OLS model would be more suitable for both CMAs.

Table 4 Pearson correlation coefficient between transit trips and other variables

Variable	Toronto	Montreal
Entropy (EI)	0.01	0.07
Mixed Density (MDI)	0.41	0.29
Population Density	0.54	0.45
Employment Density	0.11	0.01
Network Design (NDI)	0.46	0.42
Population 55+	0.32	0.25
Family Size 5+	−0.02	0.11
Minorities	0.33	0.66

Table 5 Global *Moran's I Statistics* for 2011 transit trips

	Moran's *I*	*E*[*I*]	Variance	*z*-score	*p*-value
Toronto	0.490	−0.00092	0.0004	25.30	<0.0000
Montreal	0.479	−0.00108	0.0005	22.04	<0.0000

Table 6 Estimation results of the Toronto models

	OLS Model 1		SAR Model 1		OLS Model 2		SAR Model 2	
Variables	Beta	*t*-stats	Beta	Beta	*t*-stats	*t*-stats	Beta	*t*-stats
Spatial Lag (ρ)	–	–	0.44	15.43	–	–	0.50	17.76
Constant	−74.16	−2.18	−269.78	−8.37	76.78	2.21	−186.10	−5.72
Entropy (EI)	307.85	2.36	225.80	1.95	−41.22	−0.30	−25.48	−0.21
Mixed Density (MDI)	–	–	–	–	0.05	8.62	0.03	5.47
Population Density	0.03	16.38	0.02	12.93	–	–	–	–
Network Design (NDI)	280.74	15.80	167.20	10.26	298.40	15.35	169.56	9.95
Population 55+	0.24	13.75	0.24	15.78	0.22	11.56	0.23	14.20
Family Size 5+	−0.86	−6.96	−0.40	−3.49	−1.19	−8.91	−0.62	−5.09
Minorities	0.07	10.99	0.05	9.45	0.09	14.93	0.07	12.19
Number of Obs.	1088		1088		1088		1088	
Pseudo-R^2	0.56		0.66		0.49		0.62	
AIC Statistic	15250		15042		15419		15166	

The results obtained from the multivariate modeling exercise are presented in Tables 6 and 7 for Toronto and Montreal, respectively. While the SAR model is favored over the OLS model, both were estimated using the GeoDa ver. 1.8.16.4 1 Spatial Statistics Analysis Software (Anselin et al. 2006). For each CMA, two specifications are tested; population density is used in the first and mixed density is used in the second specification. Population density and MDI are not included in the model simultaneously to avoid multicollinearity since the correlation between them is 0.65 and 0.69 for Toronto and Montreal, respectively. The estimation results for the two CMAs suggest well-behaved models. Also, the SAR models in all cases are better than the OLS models. The spatial lag parameter (ρ) is positive and significant

Table 7 Estimation results of the Montreal models

	OLS Model 1		SAR Model 1		OLS Model 2		SAR Model 2	
Variables	Beta	*t*-stats	Beta	Beta	*t*-stats	*t*-stats	Beta	*t*-stats
Spatial Lag (ρ)	–	–	0.40	13.73	–	–	0.43	14.69
Constant	29.38	1.15	−103.44	−4.30	95.31	3.97	−62.47	−2.76
Entropy (EI)	154.39	1.61	117.13	1.36	−52.01	−0.55	−73.71	−0.88
Mixed Density (MDI)	–	–	–	–	0.02	4.10	0.03	6.02
Population Density	0.01	7.82	0.01	8.12	–	–	–	–
Network Design (NDI)	112.64	7.33	61.16	4.34	140.96	9.31	78.14	5.70
Population 55+	0.12	10.33	0.12	11.72	0.11	9.43	0.11	11.04
Family Size 5+	−0.39	−2.91	−0.02	−0.19	−0.55	−3.96	−0.07	−0.57
Minorities	0.17	17.92	0.13	12.81	0.19	21.09	0.14	14.65
Number of Obs.	1088		1088		1088		1088	
Pseudo-R^2	0.58		0.66		0.56		0.66	
AIC Statistic	12,223		12,061		12,266		12,089	

in the estimated SAR models, which suggests the presence of spatial autocorrelation in the dependent variable. This result is in line with the exploratory analysis provided in Table 5.

In the case of Toronto, the specification including the population density variable provides a better fit and explanatory power. However, the specification using the mixed density (SAR Model 2) suggests that areas with mixed density have a positive influence on public transit trips, other things being equal. The same could be said about the effect of mixed density on public transit trips in Montreal although the two specifications are similar in terms of the overall fit of the model in this CMA. According to the estimation results, all the estimated parameters of the socio-economic and demographic variables are significant and exhibit intuitive signs. The older population variable has a positive and significant parameter which suggests that this population group is more reliant on public transit in the two CMAs. Also, the negative effect reported elsewhere in the literature of household size with respect to public transit ridership is confirmed in the present models. However, the *Family Size 5+* parameter is not significant in the Montreal models although it retains a negative sign. In line with the literature, the presence of minority populations, as depicted by the *Minorities* parameter, has a positive and significant impact on public transit trips in the two CMAs. This is not surprising since low income population and new immigrants are less likely to afford a vehicle, which leads them to rely heavily on public transit for their work mobility needs. On the land use side, the entropy index parameter *EI* meets the a priori expectation with a positive and significant affinity to public transit ridership in SAR Model 1 where population density is a covariate in the model. Other things being equal, areas with heterogeneous land uses are more likely to increase the number of public transit trips in Toronto. However, the entropy variable is not significant in Montreal although it retains a positive sign.

The models for the two CMAs verified that people living in areas with predominant gridded network design are more likely to utilize public transit compared to people living in areas with curvilinear network design.

5 Conclusion

This study examined the degree to which the built environment has affected public transit use for journey-to-work in 2011 while controlling for socio-economic and demographic factors in the two largest Canadian metropolitan areas. The analysis utilized the simultaneous auto-regressive modeling technique since spatial autocorrelation exists in the public transit trips generated at the census tract level. The results pertaining to the socio-economic and demographic variables reinforce findings found in previous studies. In addition, for the most part, the results showed systematic effects between the two CMAs and emphasized the important role that land use (namely, mixed land uses and network design) has on public transit ridership after controlling for population density. The latter is found to be an important factor that explains part of the variability in the observed transit trips. As such, this research makes a meaningful contribution to the existing body of transportation literature.

The work presented in this chapter has some meaningful policy implications. The findings pertaining to the land use variables emphasize the nature of the relationship between land use configuration and public transit ridership in large metropolitan areas. The same could be said about the nature of the network design structure on transit ridership. In general, the findings in this study lead us to conclude that the promotion of smart growth urban planning (e.g., encouraging compact, diverse, and pedestrian-oriented urban forms) could lead to increased public transit ridership. Finally, the findings that higher land use heterogeneity and/or job-house mixing are associated with higher levels of public transit ridership suggest that the promotion of mixed and heterogeneous land uses could lead to higher levels of public transit ridership in large Canadian metropolitan areas.

References

Anderson, W., Kanaroglou, P., & Miller, E. (1996). Urban form, energy and the environment: A review of issues, evidence and policy. *Urban Studies, 33*(1), 7–36.

Anselin, L., Syabri, J., & Kho, Y. (2006). Geoda: An introduction to spatial analysis. *Geographical Analysis, 38*, 5–22.

Bailey, T., & Gatrell, A. (1995). *Interactive spatial data analysis*. Essex: Prentice Hall.

Behan, K., Maoh, H., & Kanaroglou, P. (2008). Smart growth strategies, transportation and urban sprawl: Simulated futures for hamilton, Ontario. *Canadian Geographer, 52*(3), 291–308.

Cervero, R. (2002). Built environments and mode choice: Toward a normative framework. *Transportation Research Part D: Transport and Environment, 7*(4), 265–284.

Cervero, R., & Kockelman, K. (1997). Travel demand and the 3Ds: Density, diversity, and design. *Transportation Research Part D: Transport and Environment, 2*(3), 199–219.

Cervero, R., Rood, T., & Appleyard, B. (1999). Tracking accessibility: Employment and housing opportunities in the San Francisco Bay Area. *Environment and Planning A, 31*(7), 1259–1278.

Chakraborty, A., & Mishra, S. (2013). Land use and transit ridership connections: Implications for state-level planning agencies. *Land Use Policy, 30*, 458–469.

Chee, W. L., & Fernandez, J. L. (2013). Factors that influence the choice of mode of transport in Penang: A preliminary analysis. *Procedia - Social and Behavioral Sciences, 91*, 120–127.

Chen, C., Gong, H., & Paaswell, R. (2007). Role of the built environment on mode choice decisions: Additional evidence on the impact of density. *Transportation, 35*(3), 285–299.

Filion, P., & McSpurren, K. (2007). Smart growth and development reality: The difficult co-ordination of land use and transport objectives. *Urban Studies, 44*(3), 501–523.

Frank, L. D., & Pivo, G. (1994). Impacts of mixed use and density on utilization of three modes of travel: Single-occupant vehicle, transit, and walking. *Transportation Research Record, 1466*, 44–52.

Handy, S. (2005). Smart growth and the transportation-land use connection: What does the research tell us? *International Regional Science Review, 28*(2), 146–167.

Maoh, H., Koronios, M.; Kanaroglou, P. (2010). Exploring the Land Development Process and its Impact on Urban Form in Hamilton, Ontario. *The Canadian Geographer, 54*(1), 68–86.

Maoh, H., & Tang, Z. (2012). Determinants of normal and extreme commute distance in a sprawled midsize Canadian city: Evidence from Windsor, Canada. *Journal of Transport Geography, 25*, 50–57.

Marshall, W. E., & Garrick, N. W. (2010). Street network types and road safety: A study of 24 California cities. *Urban Design International, 15*(3), 133–147.

Messenger, T., & Ewing, R. (1996). Transit-oriented development in the sun belt. *Transportation Research Record, 1552*, 145–153.

Nam, K., Lim, U., & Kim, B. H. (2011). 'Compact' or 'sprawl' for sustainable urban form? Measuring the effect on travel behavior in Korea. *The Annals of Regional Science, 49*(1), 157–173.

OECD. (2012). *OECD green growth studies compact city policies a comparative assessment: A comparative assessment*. Paris: OECD.

Patterson, Z., Ewing, G., & Haider, M. (2005). Gender-based analysis of work trip mode choice of commuters in suburban Montreal, Canada, with stated preference data. *Transportation Research Record, 1924*, 85–93.

Potoglou, D., & Kanaroglou, P. S. (2008). Modelling car ownership in urban areas: A case study of Hamilton, Canada. *Journal of Transport Geography, 16*(1), 42–54.

Statistics Canada. (2011). *2011 Census Ottawa*. Ontario: Government of Canada.

Sultana, S., & Weber, J. (2007). Journey-to-work patterns in the age of sprawl: Evidence from Two Midsize Southern Metropolitan Areas. *The Professional Geographer, 59*(2), 193–208.

Transport for London. (2016) Travel in London—Report 9. Retrieved September 21, 2017, from http://content.tfl.gov.uk/travel-in-london-report-9.pdf.

Tsai, Y. (2005). Quantifying urban form: Compactness versus 'sprawl'. *Urban Studies, 42*(1), 141–161.

Wiley, K., Maoh, H., Kanaroglou, P. (2011). Exploring and modeling the level of service of urban public transit: The case of the Greater Toronto and Hamilton Area, Canada. *Transportation Letters: The International Journal of Transportation Research, 3*(2), 77–89.

Freight Transportation: The Evolving Landscape of Matters and Models

Clarence Woudsma

Abstract The 1980s marked the beginning of tumultuous changes in transportation with the forces of deregulation and privatization undertaken by states across the globe. It also marks a time where computing power began its dispersion and ascension, contributing to significant developments in the models applied to transportation challenges. This chapter provides a critical synopsis of the major matters that have influenced freight transportation over the past 30 years and the methods and models applied to those matters. Through this synopsis, the research contributions of Dr. Kanaroglou to the field of freight transportation are highlighted. In the 1980s, studying freight transportation in the intercity and metropolitan context was not a high priority and nor was it politically important. "Freight matters"—in one meaning is a call for recognition, and this chapter argues the call has been answered, concluding with a discussion of exciting prospects for the future.

1 Introduction

The inspiration for this chapter developed through reflection on the career scholarly work of Dr. Pavlos Kanarolgou and our relationship. Pavlos and I first met in the Fall of 1987, when I began pursuit of my Master's degree and he was in his first academic position as an assistant professor in Geography. Like a typical new graduate student, I had many topic ideas in my head, with most relating to transportation and the trucking industry. Pavlos was very patient in hearing me out, and while admitting to not knowing much about the trucking industry at that time, he was confident in being able to help me develop the best methodology—grounded theoretically with a thoughtfully appropriate quantitative method, rigorously developed, analyzed, and implemented. He was a master of spatial quantitative methods, and this collaborative approach became one of the hallmarks of his scholarly life and made him a sought after research partner. Reviewing his CV, one is struck by the breadth of subject

---

C. Woudsma (✉)
School of Planning, University of Waterloo, Waterloo, ON, Canada
e-mail: cwoudsma@uwaterloo.ca

H. Briassoulis et al. (eds.), *The Practice of Spatial Analysis*,
https://doi.org/10.1007/978-3-319-89806-3_9

matter where his expertise made important contributions: traffic engineering, population migration and aging, facility location, policy evaluation, urban planning, highway infrastructure, water, energy, public health, air quality, travel behavior, public transit, and a range of quantitative methods. In each of these areas, Pavlos was responding to prominent issues of the day, tackling major societal challenges with his quantitative skills and understanding.

We did not publish any works directly out of that first research project together; however, it did provide the impetus for a publication by Pavlos many years later (Kanaroglou 1996) on spurious correlation. The particular model we developed for that first research project together was hotly contested by one of my thesis reviewers and the quantitative argument between them carried on well past the conclusion (successful) of my defense. Pavlos patiently explained the crux of the modeling issue to me and promised on that day to publish an article on it. I share this anecdote because it further illustrates his tenacity, expertise, and desire to share the insights learned through his research work with the broader community. As our respective careers evolved, Pavlos and I had a number of occasions to work together again following our established collaborative approach.

The structure of this chapter draws from our working relationship, emphasizing the parallel streams of major freight transportation issues and the quantitative models applied to them. The threads of Dr. Kanarolgou's work are woven through this structure, with the primary intent of framing his work in the context of major transportation issues and disciplinary foci. The chapter is not intended to be a thoroughly systematic review of transportation modeling efforts nor an exhaustive treatment of issues spanning the three decades of his scholarship: from the mid-1980s through to the mid-2010s. Examples of such systematic model reviews include important texts like "Modeling Transport" (Ortúzar and Willumsen 2011), "The Handbook of Transport Geography and Spatial Systems" (Hensher et al. 2004), or more recent focused texts: "Modeling Freight Transport" (Tavasszy and de Jong 2014) and "Freight Transport Modeling" (Ben-Akiva et al. 2013). On the issue side, examples include broader examination of freight issues (Regan et al. 2000) as well as more topic-specific reviews dealing with issues such as logistics and sustainability (Marchet et al. 2014).

Following this introduction, Sect. 2 establishes the starting point for this review chapter, outlining the traditional emphases of freight research and modeling as they looked in the mid-1980s. Section 3 employs the major issue themes of Dr. Kanaroglou's work as mirror of a number of the major issues or "matters" related to freight transportation: (1) Deregulation, (2) Sustainability, and (3) Globalization and the rise of logistics. The final section of this chapter reflects back to the traditional emphases outlined in Sect. 2, the impact of Dr. Kanaroglou's work, and offers closing thoughts on the trends impacting freight and modeling today.

2 Foundations: Traditions in Freight Research and Modeling

The 1970s was likely a lonely time to be a freight modeler. The aggregate level models that were developed during that time were constrained by data availability and main frame computing limitations. The fields of Operations Research and Transportation Engineering were certainly dealing with freight matters: estimation of intercity demands for example. The economists were also very much interested in rates, prices, trade, and flows (Harker 1985). Freight models, as today, varied in their scope from point-to-point estimation of flows, mode splits, or optimal routing to larger scale model frameworks seeking to understand interzonal flows at the city, region, or national scale (De Jong et al. 2004). It is this latter class of modeling effort that is primarily addressed in this chapter—a narrow scope of freight modeling but undertaken to parallel the work of Dr. Kanaroglou.

The modeling activity that was at the center of research in the 1970s focused on passenger demand, with the emergence of the now well known and established "Classic Transport Model" (Ortúzar and Willumsen 2011). Known also as the Urban Transportation Planning Model or "four-step model," it replicates the key individual decisions of (1) to make a trip, (2) where to take that trip, (3) by what mode, and (4) by what route, employing aggregation to a zonal structure and system network. The primary issue or motivation for the development of this classic modeling structure was supporting the need for more effective infrastructure planning and development in the booming post-war western economy of the 1960s and 1970s. The consideration of freight was absent from those early modeling efforts yet would emerge in subsequent years, adapted for freight with the steps then being freight amount generation in zones (in tonnes or dollars of goods), distributing those flows in and out of zones, making a mode choice (rail, truck marine, etc.), and finally assigning those modal flows to vehicle units on specific links (De Jong et al. 2004).

2.1 *The Place of Freight in Transportation Research*

The adoption of the passenger-focused four-step model to freight activity in the 1970s suggests that transportation research was more passenger focused. Indeed, a basic analysis using the search terms "freight and modeling" versus "passenger and modeling" in transport publications presented in Fig. 1 would support that argument. Based on analysis from Web of Science and Scopus, a significant gap between passenger and freight publication volumes emerges in the early 1970s and persists in the 15 years that follow.

The reasons for this gap include the dominance of passenger flows, a lack of data on freight activity (private firms not providing data), a lack of political capital (freight does not vote like people do), and the greater complexity of freight movements (Ambrosini and Routhier 2004; Dablanc 2007; Woudsma 2001). Figure 2

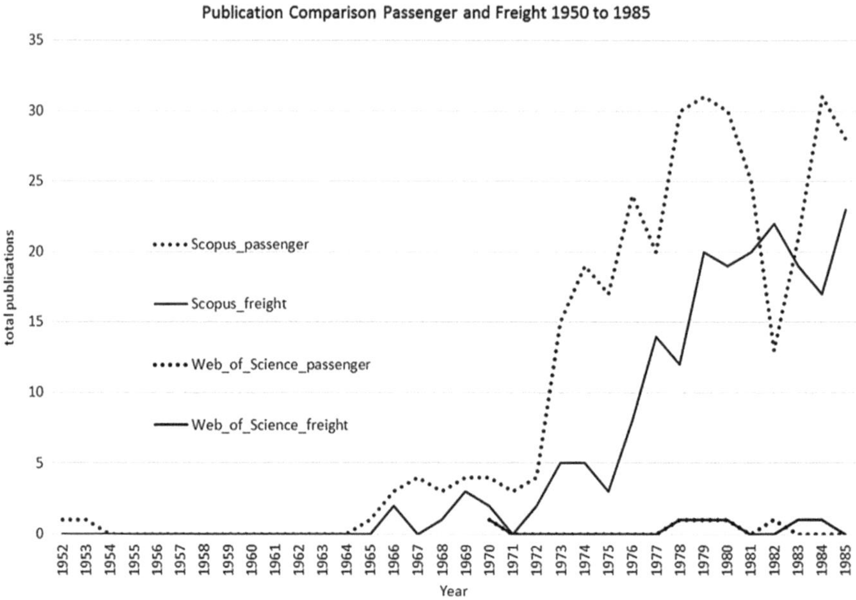

Fig. 1 Publications in transport modeling 1950 to 1985

Fig. 2 Framework for the study of freight (Woudsma 2001)

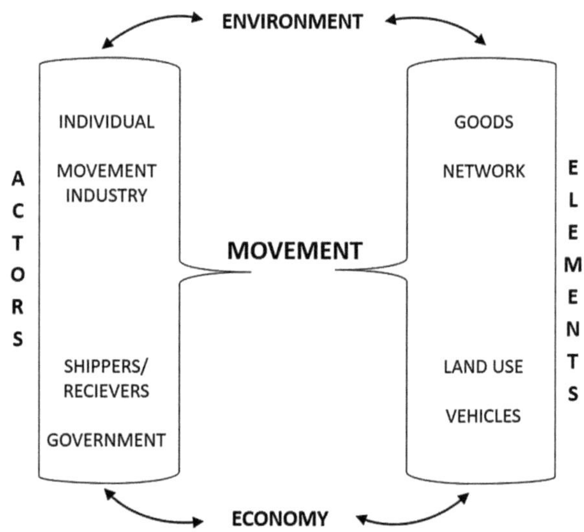

reflects that complexity, highlighting the diverse set of actors and elements that interact within the economic and environmental drivers that shape freight activity.

It is worth noting that the overall volume of publications in Fig. 1 is relatively low, perhaps reflecting the aforementioned data and computing limitations. Further, the only year in which Freight modeling publications exceeded that of passenger coincides with the time immediately following deregulation of both trucking and rail in the USA (laws came into effect in 1980).

3 Major Issues in Freight Transportation Since the 1980s

Moving beyond 1985, attention now turns to the major themes impacting freight and modeling scholarship, which have arguably advanced considerably since their early stages described previously. It also provides the framework for presenting the scholarly contributions of Dr. Pavlos Kanaroglou.

3.1 Deregulation and Privatization

One of the key matters from the 1980s was the deregulation of surface freight transportation (Harker 1985). The US deregulated in 1980, with other jurisdictions following on in subsequent years; for example, Canada in 1987. The arguments for the need to regulate surface freight hinged on market competition—too little in rail and too much in trucking with the overall objective being to provide safe and efficient services to all communities. There were regulatory provisions around an "obligation to serve," fair pricing, and an implied cross-subsidization of rates for services, with rate setting by regulators common. The arguments against economic regulation focused on firms earning excessive profits, inefficiencies in operations, and welfare gains to labor.

The debate on the merits of regulation had a particularly spatial dimension, given the implied cross-subsidization of rates and the license requirements of an obligation to serve. In the Canadian case, small isolated communities were concerned that they would see their transport services disappear, arguing that firms provided services at a loss but were able to make excessive profits serving markets where competition (via entry control) was limited. Woudsma and Kanaroglou (1994) present the arguments and evidence related to the rate differences based on markets before and after deregulation. An econometric model of trucking industry revenues (revenues = rates × tonne/km) was developed with variables introduced that reflect community size and proximity to major markets (Woudsma and Kanaroglou 1996).

The econometric modeling was developed from fundamental principles which described the hypothesized influence of important factors like distance and weight on shipment revenues, but also accessibility variables characterizing the size and place of origin/destination communities. On the topic of computing power, the calculations for accessibility took weeks to complete on then powerful Unix workstations. The major conclusions from this work were that deregulation had little impact on the revenues or provision of services in Ontario communities, with the suggestion that economic regulation as poorly enforced previously, had little impact to begin with.

In retrospect, however, deregulation of transportation and other markets, alongside the privatization of government assets (both transport providers and facilities) are identified as major drivers in the subsequent globalization of modern commerce (Madar 2000). In the post-fordist era, trade liberalization and restructuring of global commerce was supported by the rise of containerization and concomitant improvements in transportation services underpinned by developments like deregulation of surface freight services in many countries. Deregulation drove efficiency improvements, contributed to industry consolidation and greater flexibility in service offerings, and reshaped the logistics landscape in the 2000s and beyond. Thus, it is argued that deregulation is a key matter in the study of freight systems and one of the areas in which Dr. Kanaroglou made early contributions.

3.2 *Sustainability*

The broad topic of sustainable transportation encompasses a number of important matters which have been tackled by freight transportation researchers. In defining sustainable transport, there is agreement that it equally weights consideration of environment, economy, and equity (society), with the aim of mitigating the negative externalities associated with transportation including GHG emissions, air pollution, injuries and fatalities, congestion, noise, and addressing equality (Zhou 2012). The energy crises of the 1970s sparked the emergence of "energy supply and security" concerns and related policy developments to improve fuel efficiency—early efforts to address sustainability. California's concern with poor air quality (California Air Resources Board—CARB) can be viewed as a driver of cleaner vehicles in the North American context and beyond. The oil glut of the 1980s contributed to a shift in focus from energy conservation to environmental preservation (Anderson et al. 1996), paradoxically also contributing to continued urban sprawl—with fuel efficiency gains of the 1970s countered by increasing levels of VMT. In the 1990s, major policy efforts were directed at mitigating GHG emissions in conjunction with the signing of the United Nations Framework Convention on Climate Change (UNFCCC) in 1992 and the subsequent Kyoto Protocol. And today, there is little doubt about the importance of tackling not only GHG emission reductions, but investigating how transportation will be adapting to the changing climate (Meyer 2010). As an important matter for freight transportation scholarship, sustainability-related research has been very prominent, and not surprisingly, as will be reviewed in this section, central in the works of Pavlos Kanaroglou.

Urban Form and Air Quality

The seminal book "Sustainability and Cities" (Newman and Kenworthy 1999) highlights the central argument of the important relationship between urban form—the spatial configuration of the built environment and associated networked infrastructure—and transportation systems performance. Specifically, it presents an aggregate analysis of urban density and fuel consumption, comparing a range of global cities and concluding that higher density is associated with lower fuel consumption per capita. As referenced in Anderson et al. (1996), their particular analysis (originally published in 1990) included calls for "reurbanization," arguing that cities may follow policy paths to increase density and thereby enhance the potential for decreasing energy intensity or more generally, reduced auto dependency. Rather than advocating for policy instruments in the technology (zero emission vehicles), or behavioral realm (mode switching or reduced trip making), their argument advocates for changing the land use—the physical form of the city in order to enhance transport sustainability (Anderson et al. 1996). They also issue a call for further efforts in the realm of "integrated land use/transport planning" model development.

Integrated land use/transport models, or transport/land use models, have their origins in the quantitative revolution of the 1960s and 1970s. At their core, they contain land use market and transport market components, the latter based on the classic four-step transportation demand model. Well-known versions include the Lowry, ITLUP, and MEPLAN (Rodrigue 2017), with their value as planning models being their ability to evaluate policies like urban growth management, congestion pricing, and a wide range of environmental regulations.

Kanaroglou and South (1999) explore such questions as they introduce an integrated urban simulation model that forms the basis of many subsequent research efforts undertaken by Dr. Kanaroglou. IMULATE—Integrated Model of Urban Land use and Transportation for Environmental Analysis is constructed of four separate submodules that respectively represent (1) residential mobility and the housing market, (2) interzonal trip demands, (3) network allocation of trip demands, and (4) travel time estimation, with the latter two modules working iteratively until network equilibrium is achieved. Developed for Hamilton Census Metropolitan Area (the community home to McMaster University), the initial use of IMULATE was to explore future growth scenarios, linking the core model to two other modules to estimate emissions (air quality) and fuel consumption (energy and GHG) (Kanaroglou and South 1999).

Of relevance to the theme of this chapter, IMULATE has also been employed in one of the few efforts to address urban freight in an integrated urban model (Kanaroglou and Buliung 2008). The approach involved generating a matrix of urban freight vehicle flows, based on a local vehicle survey and traditional intersection and cordon counts. Passenger car equivalents were employed to address freight vehicle impacts on traffic operations, and average freight vehicle types were used to represent the freight vehicle fleet. Their goal was to evaluate the contribution of truck traffic to mobile source emissions. A key conclusion based on their analysis was the disproportionate contribution of freight vehicle activity to particulate matter pollution with a caveat concerning the challenging nature of modeling freight activity. Not surprisingly, they point to "a lack of data" as one of the major limitations, with no mention of computational issues. They capture the essence of their modeling effort in the following: "Policymakers face a clear challenge with respect to addressing trade-offs between the essential role of trucking in the economy and our daily lives, and negative externalities associated with on-road goods movement" (Kanaroglou and Buliung 2008, p.274).

Building on this initial freight-related modeling effort and picking up on the theme of "tradeoffs," Ferguson et al. (2012a) presented an ambitious effort to combine passenger and freight activity within the concept of an urban gateway. In this effort, they introduced a MRIO (Multiregional input-output) model at the regional level to explore the impacts of economic growth scenarios and then link those impacts (GDP impacts and job changes) to the local level through IMULATE to explore impacts on travel demand, traffic flow, and emissions. At the local level, they investigated how these impacts vary depending on growth scenarios of sprawl, compact, and compact plus the introduction of light rail transit (LRT). The motivation for this particular modeling effort was to demonstrate the effects of balancing the desire to enhance the city's role as a gateway (economic activity) with maintenance of quality of life—or put another way that economic prosperity can be achieved in a sustainable manner. Their unique approach of mixing regional and local modeling is also novel in the emphasis on goods movement impacts. Key in their findings is the optimization of commercial movement under the compact-plus-LRT scenario and high economic growth. There is an increase in jobs and consequent freight movements in the core of the city but the LRT provides a modal option that benefits the freight activity.

Climate Change

The 1990s witnessed considerable efforts in transport research related to the mitigation of GHG emissions, in particular from autos, road freight, and aviation (Chapman 2007). These efforts explored the potential for mitigation from either technological (e.g., vehicle or fuel systems) or behavioral solutions (modal shift). Despite progress, for example, improving heavy truck fuel efficiencies, evidence suggests that the increasing demands for trucking, have seen freight-related emissions continue to rise in many countries (Kamakate and Schipper 2009).

The potential impacts of a changing climate have also been heavily researched and widely accepted through the assessment reports of the Intergovernmental Panel on Climate Change—beginning with the first assessment in 1990. Specific transportation impacts from sea level rise to extreme temperature events have been researched extensively across the globe (Koetse and Rietveld 2009). Contributing to this body of work, Maoh et al. (2008) addressed the common under-representation of freight transportation with a modeling project to explore climate change impacts on rail and trucking services in Canada and on the broader Canadian economy.

Employing a novel modeling approach, they constructed a simulation model (CLIMATE-C) that links changes in weather events to transportation system performance and interregional trade flows. In the Canadian context, the large distances and diverse physiography and weather conditions translate into significant impacts on regional economies and transportation systems. We know how severe weather events and episodes of poor weather conditions cause transport system delays—snow storms, rainfall events, strong winds, extreme heat and cold (especially rail track). A gap in the climate change impact research is translating changes in the frequency of these events into broader economy impacts. For example, major snow events in the Canadian Rocky Mountains can see the closure of mainline railways for days at a time due to a combination of avalanches and narrow passages. Under a changed climate, if those events increase in frequency, how would mode choice change and how would trade flows change?

The linkages between transportation and the economy are handled in this model through a random utility-based multiregional input–output model (RUBMRIO), which predicts interregional trade flows by truck and rail among the 76 economic regions of Canada for 43 commodities. The influence of weather on transportation is handled through speed adjustment factors that account for the reduction in travel speeds because of changes in the frequency of various weather events. Therefore, changes in the frequency of weather events translate into travel delays, which in turn influence trade flows between regions. Sensitivity analysis with the implemented model illustrated its ability to assess the impact of climate change on transportation and the economy in Canada.

The results illustrated that the increase in the frequency of severe weather conditions can degrade the performance of the transportation system, which in turn will impact trade flow by truck and train. The system was applied to four weather scenarios that included an optimal scenario and three weather base scenarios for 2001, 2020, and 2050 to assess how climate conditions impact trade flows. The optimal scenario reflects a hypothetical situation where the transportation system performance is optimal and is not influenced by any weather conditions. The weather scenarios, on the other hand, reflect the 2001 weather patterns and an extrapolation of historical weather data to reflect the weather conditions in 2020 and 2050. The analysis suggested that the simulated weather conditions have minimal impacts on the transport system performance and trade flows at the system-wide level. Those findings are in line with arguments that the transportation system can more easily adapt and that climate change will be a relatively minor impact on the performance of the transportation system (Woudsma et al. 2007).

3.3 Globalization and the Rise of Logistics

The third major matter of interest in this chapter is the transformation of freight transportation through globalization, e-commerce, and the widespread adoption of logistics and supply management. Rodrigue (2017) outlines the role of containerization, ICT, liberalized trade agreements, and deregulation/privatization in driving the volumes of goods moving across the globe to unprecedented levels. Further, increasing rates of urbanization, consumerism, and heightened concerns for transport sustainability (congestion) have contributed to the development of "city logistics" as a research priority realm in the 2000s.

A number of excellent reviews provide a window into modeling in the area of city logistics and urban freight (Ambrosini et al. 2008; Anand et al. 2012; Browne et al.

2007; Crainic et al. 2009; Gonzalez-Feliu and Routhier 2012; Tavasszy et al. 2012). The range of reviews speaks not only to the rise in importance but the amount of scholarship in this area. Each review offers a categorization of modeling activity, for example; "that two main model families were used: operational models primarily directed towards the improvement of flow management, and systemic models to evaluate the impact of urban logistics modifications on the flows generated" (Ambrosini et al. 2008, p. 3).

Kanaroglou's research work in this area included addressing one of the data limitations of the IMULATE commercial vehicle study (Kanaroglou and Buliung 2008) through the development of a microsimulation model for estimating urban commercial vehicle movements (Ferguson et al. 2012b). His work also included a non-modeling effort to tackle the challenge of developing a standardized typology for logistics centers (Higgins et al. 2012). One particularly novel research effort is noted for its focus on the transport/land use connection from a logistics land use perspective (Ambrosini et al. 2008).

Woudsma et al. (2008) tackled the complex issue of the relationship between logistics-land use location and the influence of congestion. All else being equal, if a logistics firm chose between two available land parcels to build a new distribution center, how would current congestion patterns influence their decision, if at all? Their modeling effort dealt with the inherent spatial variability, spatial dependence, and temporal variability, constructing a spatial autoregressive model with key variables related to network accessibility reflected in congested travel times. In order to isolate the influence of congestion from distance, they employed the percentage change between free flow travel time and congested link travel time, doing so in 5-year increments over a database that spanned 40 years of actual traffic data for the City of Calgary. Anecdotally, that traffic data was made available in paper map form, with the digitization of the changes in the road network and related traffic volumes into the GIS framework one of the major efforts in this project—interesting to reflect back on this in the age of "big data."

The data for the land use side was captured through a digital parcel level database that identified the use of the parcel (logistics) as well as the year of construction. Importantly, this allowed for modeling of spatial and temporal lags, with one of the key findings being the significant congestion lag of 5–10 years. Intuitively, this suggests that the choice of location (land use) is influenced by the historic congestion in the area (10 years prior), rather than the immediate system performance (Woudsma et al. 2008).

4 Issues Worth Investigating? The Future for Freight and Models

Section 3 serves to place the scholarly works of Pavlos Kanaroglou in the framework of the major issues in freight scholarship since his career began in the early 1980s. One can appreciate the breadth of his work, as well as the originality of a number of

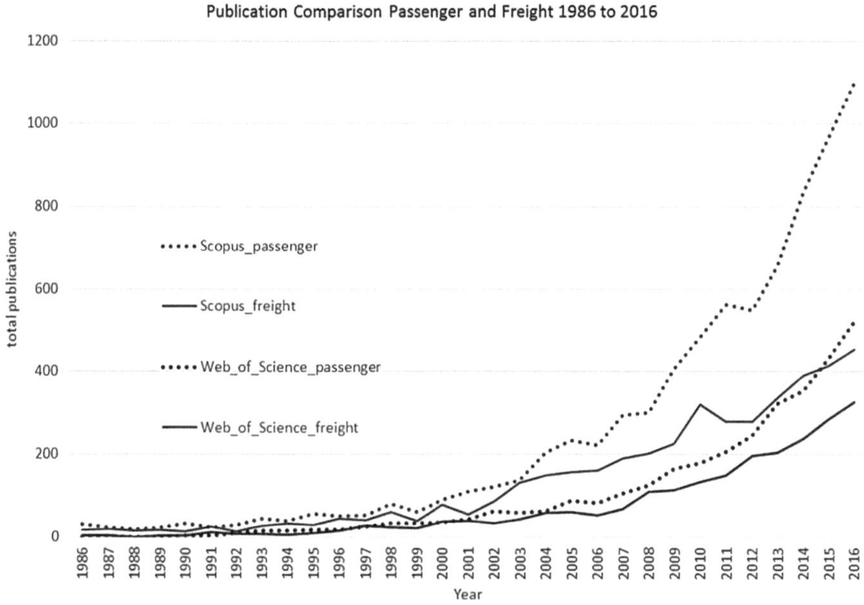

Fig. 3 Publications in transport modeling 1986 to 2016

the examples reviewed. His spatial analytical expertise and impressive output were part of a wave of scholarship devoted to providing solutions to the many pressing challenges facing our society.

Returning to the literature analysis introduced in Sect. 2, Fig. 3 provides the publication trajectory on the search terms "freight" and "passenger" "modeling" from the 1985 to 2016 period. Rather than employing a more sophisticated literature search of other works (Anand et al. 2012; Wolpert and Reuter 2012), the method used here was a basic keyword search within Scopus and the Web of Science. Picking up on the idea of a "wave of scholarship," an obvious change is the scale of the graph which is in increments of 200, rather than 5 as in Fig. 1. Naturally, there is easier reporting and indexing of scholarship since the advent of personal computing and the rise of the web in the 1990s and beyond, but there is no disputing the increasing levels of transport modeling scholarship shown in Fig. 3.

As for the gap between passenger and freight, the early 2000s appear to be a period where there was an approaching balance in terms of volume of output—equal publications in both. However, that has since changed, particularly the output in Scopus which indicates almost twice as much passenger modeling output compared to freight modeling output.

Building on the explanations for this gap presented earlier, continued emphasis for reducing congestion and automobility, rising interest in active transportation, smart growth, transit, livability, and health are a few of the drivers that may account for the significant gap, especially on the social science side. Given that an iPhone-4 has more processing power than a Cray-2 super computer of the 1980s, computing power is not relevant, nor would it really influence that gap—however, a lack of data may still persist as a detriment to freight modeling scholarship. Despite this speculation, the mainstream adoption of the classic four-step transportation model in freight has witnessed a proliferation of modeling efforts at the regional and national scale.

The good news of increasing freight modeling work is tempered by the realization that the increases on the passenger side have been even more significant. However, perhaps this should not be of prime concern, considering this quote drawn from work in 1967: "And yet while the problems of passenger transport have become subjects of great concern, those of freight transport have received relatively scant attention from either government or the research community. They should receive more. It is not logical to concentrate nearly all research attention on the passenger transportation subsystem when freight and passenger transportation together compose the transportation system for the nation and for individual urban areas." (Goeller 1971, p. 149). It is apparent that this issue has not changed much over the course of 50 years.

What of the future? What will the next 30 years of freight modeling scholarship look like? Of the major matters discussed in this chapter, sustainability and the rise of logistics present complex societal challenges that will no doubt fuel research and policy agendas for years to come. Experts in a major review conclude that (1) urban freight is associated with many externalities that affect the quality and livability of urban areas, (2) the spatial distribution of freight supply and demand needs better understanding, and (3) available data are lacking (Giuliano et al. 2013). Within these three areas, further topics of research relate to road pricing, information technology, e-commerce and home delivery, new logistics management approaches, new manufacturing technologies, alternative fuels, electric vehicles, and co-modality to name a few (Regan et al. 2000; Taniguchi et al. 2016). The disruptions in retailing supply chains are very much changing the face of cities and the material flows within—both in terms of how demands are changing and how the supply of transportation is evolving. Smartphone applications and cloud computing are especially impacting last mile deliveries in ways we could not have imagined a decade ago. Add into this mix recent launches of heavy duty electrified trucks and the buzz surrounding the potential for autonomous vehicles (Milakis et al. 2017), with significant implications for freight and logistics (Bergvall and Gustavson 2017). Smart infrastructure and connected vehicles, with enough impetus from sustainability mandates, may contribute to the long bemoaned lack of freight activity data that would be critical to advancing the freight modeling research agenda. There is excitement around the potential for new data sources related to GPS, cloud-based data, and vehicle informatics that are already providing new kinds of modeling opportunities moving beyond our traditional approaches.

In addition to his role as a scholar, Dr. Kanaroglou was a visionary leader and one of the founders of the McMaster Institute for Transportation and Logistics. It remains vibrantly active in freight and modeling research and in his spirit, very committed to working with the community. Their list of current projects reflects the major themes explored in this chapter—sustainability, goods movement, freight villages, and much more. The matters affecting freight will no doubt continue to evolve in the future and the work of Dr. Pavlos Kanaroglou will also continue to provide insights and inspiration for the next generation of scholars.

References

Ambrosini, C., & Routhier, J.-L. (2004). Objectives, methods and results of surveys carried out in the field of urban freight transport: An international comparison. *Transport Reviews, 24*(1), 57–77.

Ambrosini, C., Routhier, J. L., Sonntag, H., & Meimbresse, B. (2008). *Urban freight modelling: A review. Innovations in city logistics* (pp. 197–211). Nova Science: New York.

Anand, N., Quaka, H., van Duina, R., & Tavasszya, L. (2012). City logistics modeling efforts: Trends and gaps—A review. *Procedia: Social & Behavioral Sciences, 39*, 101–115.

Anderson, W. P., Kanaroglou, P. S., & Miller, E. J. (1996). Urban form, energy and the environment: A review of issues, evidence and policy. *Urban Studies, 33*(1), 7–35.

Ben-Akiva, M., Meersman, H., & Van de Voorde, E. (2013). *Freight transport modelling*. Bingley: Emerald.

Bergvall, J., & Gustavson, C. (2017). *The economic impact of autonomous vehicles in the logistics industry*. Masters thesis, Jönköping International Business School.

Browne, M., Piotrowska, B., Woodburn, A., & Allan, J. (2007). Literature review WM9: Part I—Urban Freight Transport, Transport Studies Group, University of Westminster.

Chapman, L. (2007). Transport and climate change: A review. *Journal of Transport Geography, 15* (5), 354–367.

Crainic, T. G., Ricciardi, N., & Storchi, G. (2009). Models for evaluating and planning city logistics systems. *Transportation Science, 43*(4), 432–454.

Dablanc, L. (2007). Goods transport in large European cities: Difficult to organize, difficult to modernize. *Transportation Research Part A: Policy and Practice, 41*(3), 280–285.

De Jong, G., Gunn, H., & Walker, W. (2004). National and international freight transport models: An overview and ideas for future development. *Transport Reviews, 24*(1), 103–124.

Ferguson, M., Maoh, H., & Kanaroglou, P. S. (2012a). Simulating sustainable urban gateway development: Illustration from Hamilton, Ontario, Canada. *Transportation Research Record: Journal of the Transportation Research Board, 2269*, 135–144.

Ferguson, M., Maoh, H., Ryan, J., Kanaroglou, P., & Rashidi, T. H. (2012b). Transferability and enhancement of a microsimulation model for estimating urban commercial vehicle movements. *Journal of Transport Geography, 24*, 358–369.

Giuliano, G., O'Brien, T., Dablanc, L., & Holliday, K. (2013). *NCHRP report 23: Synthesis of freight research in urban transportation planning*. Washington, DC: Transportation Research Board.

Goeller, B. F. (1971). *Freight transport in urban areas: Issues for research and action* (Urban Commodity Flow, Special Report, 120). Rand Corporation.

Gonzalez-Feliu, J., & Routhier, J. L. (2012). Modeling urban goods movement: How to be oriented with so many approaches? *Procedia-Social and Behavioral Sciences, 39*, 89–100.

Harker, P. T. (1985). The state of the art in the predictive analysis of freight transport systems. *Transport Reviews, 5*(2), 143–164.

Hensher, D., Button, K. J., Haynes, K. E. & Stopher, P. R. (2004). *Handbook of transport geography and spatial systems* (1st ed., Handbooks in transport, 5). Amsterdam: Elsevier.

Higgins, C. D., Ferguson, M. R., & Kanaroglou, P. S. (2012). Varieties of logistics centres: Towards a standardized typology and hierarchy. *Transportation Research Record: Journal of the Transportation Research Board, 2288*, 9–18.

Kamakaté, F., & Schipper, L. (2009). Trends in truck freight energy use and carbon emissions in selected OECD countries from 1973 to 2005. *Energy Policy, 37*(10), 3743–3751.

Kanaroglou, P. S. (1996). On spurious correlation in geographical problems. *The Canadian Geographer, 40*(3), 194–202.

Kanaroglou, P. S., & Buliung, R. N. (2008). Estimating the contribution of commercial vehicle movement to mobile emissions in urban areas. *Transportation Research Part E: Logistics and Transportation Review, 44*(2), 260–276.

Kanaroglou, P. S., & South, R. (1999). Can urban form affect transportation energy use and emissions? An analysis of potential growth patterns for the Hamilton Census Metropolitan Area. *Energy Studies Review, 9*(2), 22–40.

Koetse, M. J., & Rietveld, P. (2009). The impact of climate change and weather on transport: An overview of empirical findings. *Transportation Research Part D: Transport and Environment, 14*(3), 205–221.

Madar, D. (2000). *Heavy traffic: Deregulation, trade, and transformation in north American trucking*. Vancouver: UBC Press.

Maoh, H., Kanaroglou, P. S., & Woudsma, C. (2008). Simulation model for assessing the impact of climate change on transportation and the economy in Canada. *Transportation Research Record: Journal of the Transportation Research Board, 2067*, 84–92.

Marchet, G., Melacini, M., & Perotti, S. (2014). Environmental sustainability in logistics and freight transportation: A literature review and research agenda. *Journal of Manufacturing Technology Management, 25*(6), 775–811.

Meyer, M. D. (2010). Greenhouse gas and climate change assessment: Framing a transportation research agenda. *Journal of the American Planning Association, 76*(4), 402–412.

Milakis, D., van Arem, B., & van Wee, B. (2017). Policy and society related implications of automated driving: A review of literature and directions for future research. *Journal of Intelligent Transportation Systems, 21*(4), 1–25. https://doi.org/10.1080/15472450.2017.1291351.

Newman, P., & Kenworthy, J. (1999). *Sustainability and cities: Overcoming automobile dependence*. Washington, DC: Island Press.

Ortúzar, J., & Willumsen, L. G. (2011). *Modelling transport* (4th ed.). Chichester: John Wiley & Sons.

Regan, A., Holgun-Veras, J., Chow, G., & Sonstegaard, M. H. (2000). *Freight transportation planning and logistics, transportation in the new millennium*. Washington, DC: Transportation Research Board.

Rodrigue, J. P. (2017). *The geography of transport systems* (4th ed.). New York: Routledge.

Taniguchi, E., Thompson, R. G., & Yamada, T. (2016). New opportunities and challenges for city logistics. *Transportation Research Procedia, 12*, 5–13.

Tavasszy, L., & de Jong, G. (2014). *Modeling freight transport* (1st ed.). Amsterdam: Elsevier.

Tavasszy, L. A., Ruijgrok, K., & Davydenko, I. (2012). Incorporating logistics in freight transport demand models: State-of-the-art and research opportunities. *Transport Reviews, 32*(2), 203–219.

Wolpert, S., & Reuter, C. (2012). Status quo of city logistics in scientific literature: Systematic review. *Transportation Research Record: Journal of the Transportation Research Board, 2269*, 110–116.

Woudsma, C. (2001). Understanding the movement of goods, not people: Issues, evidence and potential. *Urban Studies, 38*, 2439–2455.

Woudsma, C. G., & Kanaroglou, P. S. (1994). Deregulation of the motor carrier industry: A Canadian example. *Environment and Planning A, 26*, 343–360.

Woudsma, C. G., & Kanaroglou, P. S. (1996). The impacts of trucking deregulation in Ontario: A market-specific analysis. *Canadian Public Policy/Analyse de Politiques, 22*(4), 356–377.

Woudsma, C., Kanaroglou, P., Maoh, H., & Marshall,S. (2007). Climate change and Canadian road transport: Assessing impacts and adaptations. Final Report submitted to Natural Resources Canada, Climate Change Impacts and Adaptations Directorate, Ottawa.

Woudsma, C., Jensen, J. F., Kanaroglou, P., & Maoh, H. (2008). Logistics land use and the city: A spatial-temporal modeling approach. *Transportation Research Part E: Logistics and Transportation Review, 44*(2), 277–297.

Zhou, J. (2012). Sustainable transportation in the US: A review of proposals, policies, and programs since 2000. *Frontiers of Architectural Research, 1*(2), 150–165.

A Structural Equation Model of Commercial Vehicle Ownership

Toka S. Mostafa and Matthew J. Roorda

Abstract Behavioral freight transportation modeling has recently emerged as an approach to enhance the quality of freight and logistics-related decisions. Commercial vehicle fleet ownership and composition is one important decision a business establishment faces that has not been adequately addressed in the literature. This research investigates vehicle ownership of small/medium-sized business establishments in the Greater Toronto and Hamilton Area (GTHA) using a structural equation modeling approach. The model reveals a complementary relationship between the number of cars and trucks owned, indicating that, on average, for each owned car there is a 17% increase in the number of owned trucks. This model can be used for estimating the number of owned cars, pickups/vans, and trucks for establishments located in the GTHA according to their location, industrial sector, freight demand, and some establishment characteristics.

1 Introduction

Recently, there has been an increased interest in behavioral freight transportation modeling as an approach to enhance the quality of decisions related to freight and logistics. Business establishments (BEs) are the key agents acting within freight systems. BEs, including shippers, receivers, and logistics service providers, interact to develop the complex system of goods and service truck movements. Decisions of BEs resulting in shipment generation and goods movement within and between urban areas have been identified and modeled (Roorda et al. 2010; Hunt and Stefan 2007; Fischer et al. 2005; Wisetjindawat et al. 2007; de Jong and Ben-Akiva 2007; Liedtke 2006; Boerkamps et al. 2000).

One of the essential decisions a BE needs to make is whether to own a private vehicle fleet or outsource logistics to other companies (Abdur Razzaque and Chen Sheng 1998; Deepen 2007). BEs can either perform all their shipment operations

T. S. Mostafa (✉) · M. J. Roorda
Department of Civil Engineering, University of Toronto, Toronto, ON, Canada
e-mail: Toka.muhammad@mail.utoronto.ca; roordam@ecf.utoronto.ca

H. Briassoulis et al. (eds.), *The Practice of Spatial Analysis*,
https://doi.org/10.1007/978-3-319-89806-3_10

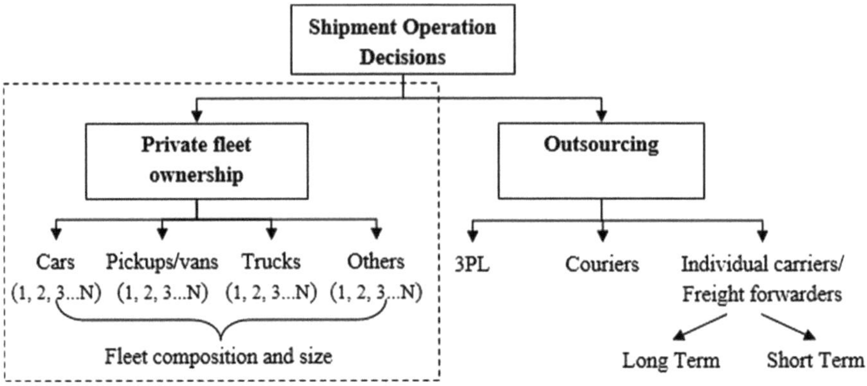

Fig. 1 Shipment operation decision hierarchy schematic

using their own fleet (whether the fleet is owned by the same BE or by other owned subsidiaries) or outsource all or part of their operations to other BEs (Mostafa and Roorda 2017). If a BE decides to perform part/all of their own shipment operations, the decision to own a private fleet follows and the question of how many and what type of vehicles arises. Here, it is hypothesized that fleet composition and fleet size are interrelated. For instance, owning one truck might be more beneficial than owning two pickups.

The decisions of vehicle ownership and outsourcing of shipment transportation operations are usually made at early stages of a BE's life cycle. Such decisions may be revisited in later stages depending on the growth of the BE and their employed long/short-term strategies, and/or changes in the economy (Mostafa and Roorda 2015). Once the decision to own a private fleet is made, two subsequent decisions follow: determining fleet size and composition. Figure 1 shows a schematic of shipment operation decisions hierarchy.

In this chapter, it is hypothesized that the ownership decisions of different vehicle fleet types are reciprocal, i.e., the ownership of one vehicle type influences the decision of owning other vehicle types either positively or negatively. In other words, a simultaneous relationship between the ownership of different vehicle types is hypothesized, either directly measured by observed variables or indirectly measured by other unobserved (latent) variables. For instance, the geographic suitability of a BE's location, proximities to resources/shippers/customers, and accessibility to the transportation network are expected to influence vehicle ownership decisions. Location suitability cannot be measured directly by itself; instead, it can be estimated using other variables such as land use types, transportation network road densities, and proximities to highways and other business locations. Such assumptions can be investigated and statistically evaluated using a structural equation modeling (SEM) approach.

This chapter presents a structural equation model (SEM) for fleet size and composition of commercial vehicle ownership decisions. The decisions of freight operation outsourcing have been covered by Mostafa and Roorda (2017). The term "vehicle ownership" in this chapter is used to include both owning and leasing of vehicles. The proposed model considers decisions to own/lease three vehicle types: cars, pickups/vans, and trucks for BEs located in the GTHA, Canada. The model investigates the relationships between the ownership of cars, pickups/vans, and trucks, and whether they are substitution or complementary ones. The model also evaluates the effect of some BE characteristics (e.g., employment size, age, and area), shipment demand, and some BE location characteristics (e.g., land use and geographic proximities) on vehicle fleet ownership decisions. The SEM presented helps to understand the overall commercial vehicle ownership that is useful for freight planning and policy assessment purposes. This model could potentially be extended and used as part of the large-scale freight microsimulation introduced by Mostafa and Roorda (2015).

The chapter is organized as follows. First, a review of previous studies of vehicle ownership modeling is offered. The SEM method is then explained followed by a data description section. The SEM structure and estimation results are then provided. Finally, conclusions and recommendations are summarized.

2 Literature Review: Freight Vehicle Fleet Ownership Modeling

Business establishment fleet ownership can be treated using either a normative or a behavioral approach (Rashidi and Roorda 2017). The normative approach deals with vehicle ownership as an optimization problem that aims to minimize an establishment's operations cost by treating vehicle ownership as an outcome of the vehicle routing problem (Li and Tao 2009; Arunapuram et al. 2003; Klincewics et al. 1990; Schroeter 1983). The behavioral approach attempts to replicate behavioral patterns of vehicle ownership given information about the BE attributes. This approach has been widely applied to the case of household vehicle ownership at both the aggregate and the disaggregate level (de Jong and Kitamura 2009, de Jong et al. 2004; Choo and Mokhtarian 2004; Hensher and Ton 2002). However, most behavioral models of freight fleet size have intended to estimate the total commercial vehicle stock at the aggregate level for policy assessment purposes (Huo and Wang 2012; Hao et al. 2011; TREMOVE 2007; Belhaj 2002; Wasserfallen and Güntensperger 1988; Sorrell 1992; Greene et al. 1986).

For transport emission and energy estimation and for fuel consumption studies, vehicle stock ownership is estimated using classical scrap and sales models (Huo and Wang 2012; TREMOVE 2007; Wasserfallen and Güntensperger 1988; Sorrell 1992; Greene et al. 1986). Scrap and sales models are aggregate approaches. Scrapped vehicles are estimated based on scrap rates. Total vehicle sales by mode are then determined by subtracting the total vehicle stock required to fulfill transport demand from the remaining stock. Disaggregation of the total vehicle stock sales by mode according to vehicle type and technology is then conducted to identify fleet composition.

TREMOVE (2007), a policy assessment model that studies the effects of transportation and environmental policies on the transportation sector, has a "vehicle stock turnover module" that is responsible for producing aggregate vehicle demand forecasts for each mode of transport for a specific year. The vehicle stock turnover first estimates previous year's surviving vehicle stock based on a scrappage module with survival functions. Then the new vehicle demand is calculated based on the needed vehicle-kilometers in the simulated year. This average mileage is converted into number of vehicles needed to fulfill this demand based on historical fleet volume statistics of different vehicle modes. The estimated number of vehicles by mode is further disaggregated into vehicle types using a logit discrete purchase choice model that estimates the market shares of different vehicle types based on the total lifetime price of the vehicle, vehicle characteristics, and consumer preferences. TREMOVE treats vehicle stock ownership in an aggregate fashion and accounts for generic production factors such as estimates of production quantities and prices only. The behavior and attributes of BEs as well as the geographical attributes of establishment locations are ignored.

Palencia et al. (2012) proposed a model that describes the evolution of the light-duty vehicle fleet between the years of 2010 and 2050 for CO_2 emission estimation. The model describes the main stages of the vehicle life cycle using three models: vehicle production, usage, and disposal models. Given the information of vehicle model, survival profiles, vehicle sales, and base year vehicle stock, the vehicle stock for a specific year is forecasted.

Belhaj (2002) introduced a structural equation model (SEM) to estimate the vehicle demand and fuel consumption in Morocco. The model assumed a correlation between vehicle demand and fuel consumption. The structural model considered gross domestic product, average road tax, car prices, and deflated fuel prices as explanatory variables.

Another case regarding freight vehicle fleet ownership concerns the need to model the strategic decision of vehicle replacements either for vehicle failure incidents or for potential failure prevention (Redmer 2009). Strategies have been modeled to determine the optimal life time for the vehicle fleet and suggested replacement intervals by considering vehicle deterioration conditions, fleet age, and the replacement policies of fleet operators (Redmer 2009; Elion et al. 1966; Simms et al. 1984; Suzuki and Pautsch 2005). Furthermore, other models have used cross-sectional and time series analysis to estimate trends of ownership growth (Hao et al. 2011).

To the authors' knowledge, no previous research has captured the behavioral aspect of business establishments' fleet vehicle ownership, except for Rashidi and Roorda (2017), who introduced a combined multinomial logit and trivariate probit model for BE vehicle ownership. They based their model on data from a commercial travel survey of 500 BEs located in the Region of Peel in Canada (McCabe et al. 2013). This chapter builds upon the efforts of Rashidi and Roorda (2017) to describe commercial fleet size and composition decisions using multivariate regression analysis in the form of a structural equation model. The approach uses a sample of approximately 2,000 BEs in the GTHA, Canada.

3 Structural Equation Modeling (SEM)

SEM is a statistical method that is used to represent complex relationships between a set of independent variables and a set of dependent variables. SEM is also a means for multivariate regression analysis, which defines error terms for the derived system of equations as well as the correlation between latent (unobserved) variables. SEM differs from multivariate regression in that it allows the quantification of the effect of latent variables that are believed to have a direct relationship with the dependent variables and are influenced by measured (exogenous) variables. It also allows to define the relationship between more than one dependent variable and measure their potential simultaneous relationship (Ullman and Bentler 2003).

Three basic steps are followed in the development of a SEM: (1) hypothesis formation, a model hypothesis is a suggested relationship between some observed and unobserved variables to explain a specific behavior, (2) structural model development, and (3) measurement model specification (Bollen and Long 1992). The purpose of a SEM is to statistically test a hypothesis with a corresponding model structure. One hypothesis can be represented and evaluated using different structural and measurement models.

The structural model allows the evaluation of potential relationships between the dependent and the independent variables. The model can be visualized in the form of a path diagram in which arrows represent the causal relationships between variables and arrow heads point to the dependent variables. Observed variables are represented in rectangles while latent variables and error terms are represented in circles (Acock 2013). A measurement model is the mathematical interpretation of the structural model in which the observed and latent variables, error terms, variances, and covariance are identified.

For example, in the commercial vehicle ownership model shown in Fig. 2, one model hypothesis is that the ownership of cars, pickups/vans, and trucks are independent (i.e., no complementary or substitution relationships exist). Another hypothesis is that geographic suitability of BE locations (a latent variable), that reflects transportation network accessibility and BE access to resources, has a direct influence on vehicle ownership decisions. However, this variable cannot be measured; instead, it is hypothesized that this latent variable can be quantified using a set of observed variables such as proximity to major roads and land use characteristics of census tracts of the BE locations. The model structure also suggests that there is a direct influence of demand-related variables (such as values of inbound and outbound shipments, value of provided services, and number of outbound shipments) and BE attributes on the ownership of the modeled vehicle types.

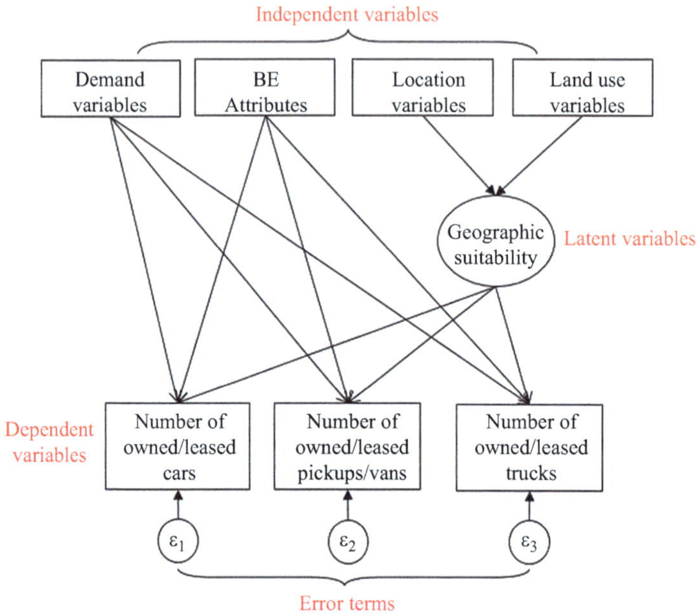

Fig. 2 Schematic example of a model structure with a latent variable. In this example, geographic suitability is hypothesized to be a latent variable that is affected by a set of observed variables including land use and road proximity variables, while BE attributes and demand variables are considered as direct dependent variables

This model structure can be interpreted with the measurement model shown in Eq. 1:

$$\begin{aligned} y_1 &= \alpha_1 + \beta_1 X + \gamma_1 S + \varepsilon_1 \\ y_2 &= \alpha_2 + \beta_2 X + \gamma_2 S + \varepsilon_2 \\ y_3 &= \alpha_3 + \beta_3 X + \gamma_3 S + \varepsilon_3 \\ S &= \alpha_4 + \beta_4 Z \end{aligned} \tag{1}$$

where.

y_1, y_2, and y_3 are the dependent variables of number of cars, pickups/vans, and trucks owned/leased by the establishment, respectively,

X is the vector of independent variables representing demand-related variables and BE attributes,

S is the latent variable of the geographic suitability of BE locations,

Z is the vector of the independent variables of location and land use,

$\alpha_{(n)}$ is the vector of model constants,

$\beta_{(n)}$ and $\gamma_{(n)}$ are vectors of model coefficients, and

$\varepsilon_{(n)}$ are the error terms.

The SEM allows the statistical evaluation of model hypotheses and structures both by evaluating the overall goodness-of-fit of the model and the statistical significance of each of the variables (with or without latent variable structures) using statistical measures such as coefficient of determination R^2 and *t-statistics* for each parameter (Ullman and Bentler 2003). In addition, SEM has simultaneous equation modeling capabilities which quantify relationships between dependent variables (with or without the consideration of latent effects). For instance, the truck ownership-dependent variable can have a substitution or complementary relationship with the ownership of pickups/vans and/or cars. Equation 2 models three dependent variables y_1, y_2, and y_3 using a vector of explanatory variables X. SEM allows for simultaneous consideration of the effect of other dependent variables, represented in the vector Y, and the quantification of the error terms (ε) associated with each equation.

$$\begin{aligned} y_1 &= \alpha_1 + \beta_1 X + \gamma_1 Y + \varepsilon_1 \\ y_2 &= \alpha_2 + \beta_2 X + \gamma_2 Y + \varepsilon_2 \\ y_3 &= \alpha_3 + \beta_3 X + \gamma_3 Y + \varepsilon_3 \end{aligned} \tag{2}$$

4 Data Sources

Two different sets of variables are considered: BE attributes and geographic characteristics of census tracts in which the BEs are located. Attributes of approximately 2000 small/medium-sized establishments (less than 100 employees), located in the GTHA Canada (Fig. 3), were collected from three surveys that used the same survey instrument: the Region of Peel Commercial Travel Survey (2006/2007), the Region of Durham Commercial Travel Survey (2010) and the Greater Toronto and Hamilton Area Commercial Travel Survey (2012). 2,114 BEs belonging to different industry classes (excluding transportation and logistics service providers) responded with information about their daily shipment activities. The data collected from the three surveys include information only for BEs that shipped goods or provided services

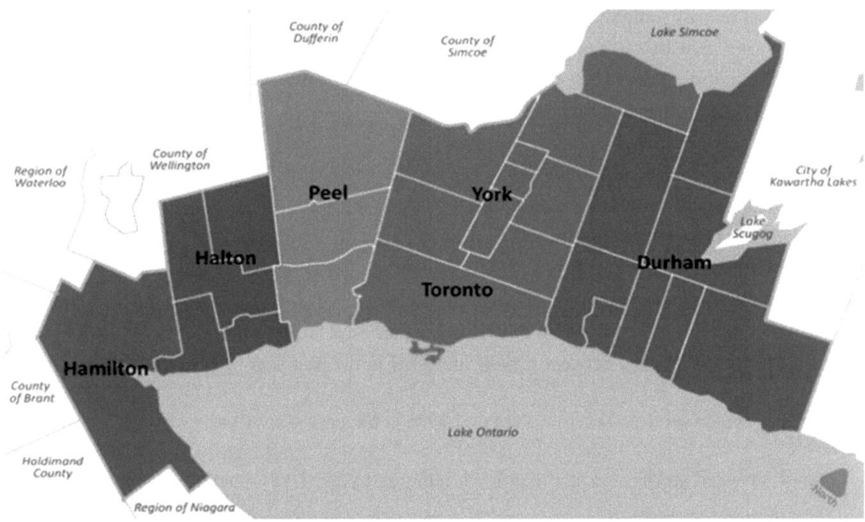

Fig. 3 Study area

that required transportation at least once during the week before the survey. The survey data include attributes of the establishments such as number of employees, number of years in business and industry class. The data set also includes estimates of BE goods and service demand such as previous year's inbound and outbound shipment and service values. Vehicle ownership information of different types of vehicles is also included.

Census tract attributes were prepared using a geographic information system (GIS) analysis. BE locations were geocoded using ArcGIS to identify the census tracts. GIS data (DMTI Spatial Inc. 2012) included geographical attributes of the census tracts such as land use information and road characteristics. The final combined data set was then filtered to exclude large-sized BEs (i.e., BEs with more than 100 employees) and to include records for establishments that own or lease at least one commercial vehicle only. A final sample size of 1,676 complete BE records out of 2,114 is considered for analysis and modeling (see Table 1 for the list of variables used in the model).

Table 1 Data set variable list

Variable description	Mean	Std. Dev.
Dependent variables		
Number of cars owned/eased by the BE	1.54	3.87
Number of pickups owned/eased by the BE	1.45	2.64
Number of trucks and tractors owned/eased by the BE	0.80	3.85
Explanatory variables		
Attributes of business establishments		
Number of employees	13.55	17.14
Number of years in business	15.41	14.61
Square footage	14,317	52,215
Employment density (employee/100 sqft)	0.42	1.61
Manufacturing industry class (1 yes, 0 no)	0.18	–
Construction industry class (1 yes, 0 no)	0.12	–
Retail/wholesale trade industry class (1 yes, 0 no)	0.21	–
Offices and services industry class (1 yes, 0 no)	0.22	–
Utilities industry class (1 yes, 0 no)	0.10	–
Professional/scientific services industry class (1 yes, 0 no)	0.10	–
Other industry class (1 yes, 0 no)	0.07	–
Demand variables		
Demand varies over the year (1 yes, 0 no)	0.51	0.50
Value of provided services less than $500K in the past year (1 yes, 0 no)	0.45	–
Value of provided services between $500K and $1M in the past year (1 yes, 0 no)	0.20	–
Value of provided services between $1M and $5M in the past year (1 yes, 0 no)	0.26	–
Value of provided services greater than $5M in the past year (1 yes, 0 no)	0.08	–

(continued)

Table 1 (continued)

Variable description	Mean	Std. Dev.
Value of received shipments less than $500K in the past year (1 yes, 0 no)	0.55	–
Value of received shipments $500K and $1M in the past year (1 yes, 0 no)	0.17	–
Value of received shipments between $1M and $5M in the past year (1 yes, 0 no)	0.19	–
Value of received shipments greater than $5M in the past year (1 yes, 0 no)	0.08	–
Value of outbound shipments less than $500K in the past year (1 yes, 0 no)	0.51	–
Value of outbound shipments between $500K and $1M in the past year (1 yes, 0 no)	0.13	–
Value of outbound shipments between $1M and $5M in the past year (1 yes, 0 no)	0.24	–
Value of outbound shipments greater than $5M in the past year (1 yes, 0 no)	0.11	–
Number of outbound shipments less than 100 in the past year (1 yes, 0 no)	0.35	–
Number of outbound shipments between 100 and 500 in the past year (1 yes, 0 no)	0.30	–
Number of outbound shipments greater than 500 in the past year (1 yes, 0 no)	0.35	–
Location variables		
BE located within the central business district of Toronto (1 yes, 0 no)	0.03	–
BE located in the City of Toronto (1 yes, 0 no)	0.19	–
BE located in the Region of Peel (1 yes, 0 no)	0.31	–
BE located in Durham Region (1 yes, 0 no)	0.23	–
Proximity to the nearest car pool location (km)	10.32	6.25
Proximity to the nearest major road (km)	0.28	0.39
Proximity to the nearest highway (km)	1.81	2.16
Proximity to the nearest railway (km)	1.44	1.82
Proximity to the nearest expressway (km)	2.84	5.16
Land use variables		
Proportion of commercial land use in the census tract in which the BE is located	0.09	0.51
Proportion of industrial land use in the census tract in which the BE is located	0.62	0.67
Proportion of government land use in the census tract in which the BE is located	0.09	0.25
Proportion of residential land use in the census tract in which the BE is located	0.54	0.67
Density of the highway network of the census tract in which the BE is located (km/km^2)	1.07	1.79
Density of the expressway network of the census tract in which the BE is located (km/km^2)	2.19	3.86
Density of the major road network of the census tract in which the BE is located (km/km^2)	3.37	2.68
Density of carpool locations within the census tract in which the BE is located ($location/km^2$)	0.01	0.13
Density of BEs within the census tract in which the BE is located (BE/km^2)	4.65	9.80

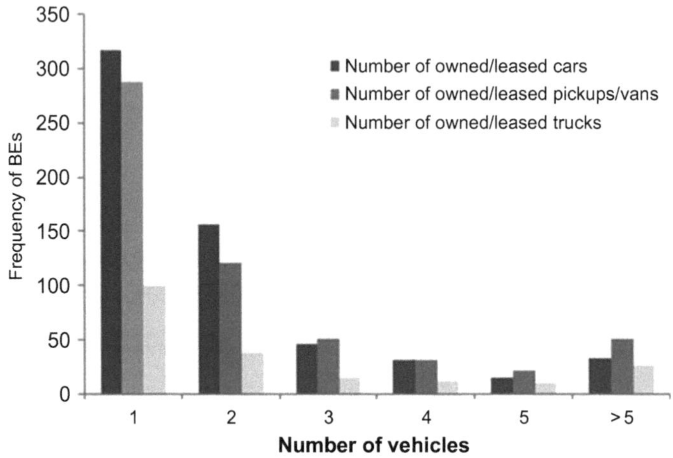

Fig. 4 BE vehicle ownership distribution

5 Data Analysis

Three types of vehicles were considered in this modeling exercise: cars, pickups/vans, and trucks. The database also includes information about other vehicle types such as trailers and containers. However, the number of establishments owning such vehicle types is too small to be considered within this modeling exercise (40 out of 1,676 BEs); hence, these types are excluded. Therefore, only BEs that at least own one unit of any of the three vehicle types are considered and BEs with zero ownership are excluded. Distributions of the three modeled vehicle type variables with regard to the other independent variables are analyzed. Figure 4 presents the distribution of BEs by vehicle types. Figure 5 presents the sample distribution of vehicle ownership for some industry classes.

6 SEM for Commercial Vehicle Ownership: Model Hypothesis and Structure

Various model structures for vehicle ownership decisions have been tested. Some of them considered latent variables (such as those shown in Fig. 2) and others considered only direct relations between independent and dependent variables. The selected model structure, which is the subject of this chapter, is a multivariate regression structure where the three dependent variables (number of cars, pickups/vans, and trucks) are directly affected by the set of independent variables, with simultaneous effects between the dependent variables, and no latent effects are found to be statistically significant (Fig. 6). The measurement model was presented previously in Eq. 2.

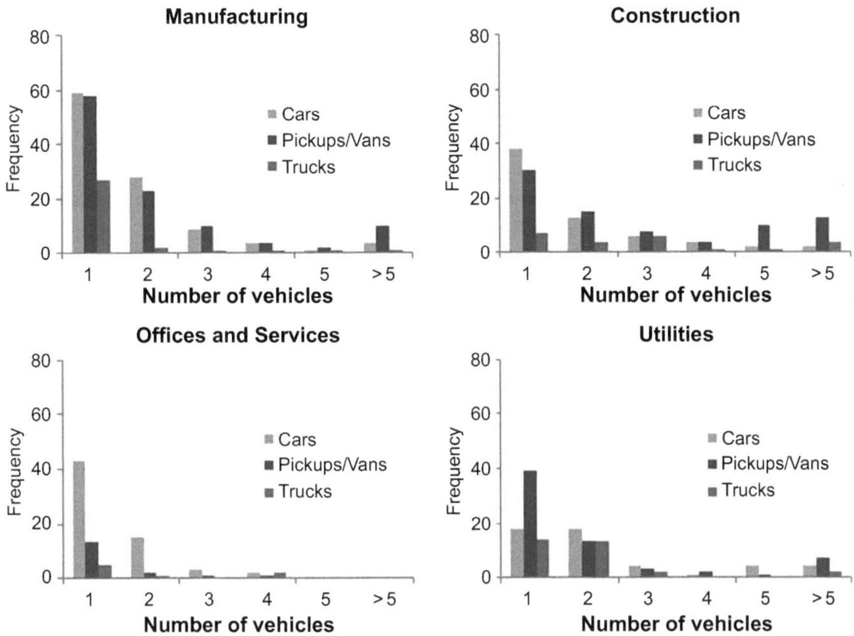

Fig. 5 BE vehicle ownership distribution for selected industry classes. Both manufacturers and construction industry classes have more cars and pickups/vans than trucks. BEs that belong to the offices and services industrial class own more cars, while utility BEs own more pickups/vans

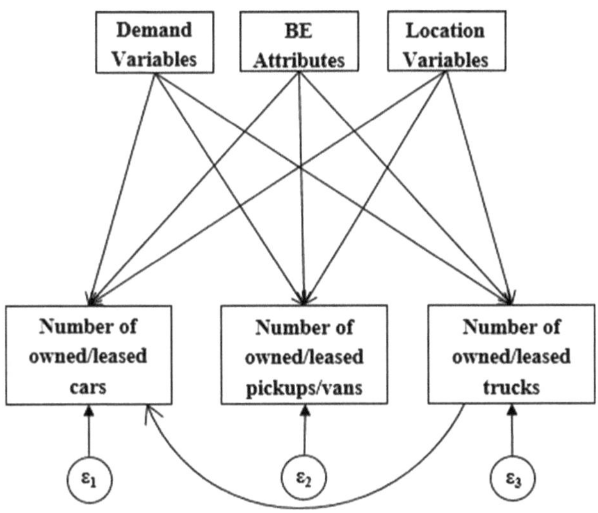

Fig. 6 Final model structure. The model reveals a simultaneous relationship between the ownership of trucks and the ownership of cars

7 Model Results and Interpretations

Table 2 summarizes the parameter estimates for the final model. All the reported variables are statistically significant with at least an 80% confidence level.

Simultaneous relationships: It can be inferred from the model that there is a complementary relationship between car and truck ownership levels indicated by the positive sign of the number of trucks parameter coefficient. The model results indicate that for each owned car there is, on average, a 17% increase in the number of owned trucks. This may indicate that owning more cars implies more demand on provided services/logistics, and, hence, owning trucks would become complementary to accommodate the increase in service demand.

BE characteristics: There is a positive relationship between the age of BEs and the number of owned trucks. Another positive relationship is found between the ownership of cars and employment density; the increase in employment density by 1 employee/100 sq.ft. increases the number of owned cars by 0.76. In other words, the growth of BEs is multidimensional in terms of employment, space, and car ownership level.

Industry classification: BEs belonging to construction industry class are inclined to own more pickups/vans than cars and trucks conforming to the results found by Rashidi and Roorda (2017). This can be explained by the nature of the construction business in which the use of pickups/vans is favored to facilitate visits to construction sites. On the other hand, the model suggests that BEs providing professional and technical services tend to own more trucks.

Demand variables: Establishments that provide services with value of more than \$5M own fewer trucks. Conversely, when the value of the outbound shipments increases, the ownership of trucks increases, while BEs that receive more shipments tend to own fewer trucks. This makes sense; the more a BE ships, the better it is to own a larger fleet size, while outsourcing might be a better option in case a BE receives more shipments.

Location characteristics: Most geographical characteristics, such as proximity to different roadways and density of land use types within census tracts were not found to have statistically significant effects on vehicle ownership level at the 80% confidence level. However, the region in which a BE is located has a significant and intuitive effect. For example, the Region of Peel is a commercial/transportation/warehousing hub in the Toronto area producing a lot of logistics operations, hence, the model identifies that BEs located in this region own more trucks. BEs located in Durham Region own fewer cars, and BEs located in the City of Toronto own fewer pickups or vans.

Table 2 Final model results

	Passenger car			Pickup/van			Trucks		
Variable name	Coef.	SE	Z	Coef.	SE	Z	Coef.	SE	Z
Constant	−0.08	0.46	−0.18	0.55	0.16	3.46	−0.67	0.31	−2.17
Number of years in business/10							0.18	0.11	1.72
Number of employment/10	0.45	0.08	5.40	0.42	0.06	7.50	0.45	0.10	4.47
Employment density (emp./100 sqft)	0.76	0.07	10.71						
Industry class									
Manufacturing							−0.87	0.39	−2.23
Construction				1.49	0.29	5.08			
Office and service	1.20	0.33	3.58						
Utilities	−0.71	0.43	−1.66				−1.98	0.59	−3.33
Professional and tech. Services							2.13	0.53	4.05
Others				2.43	0.38	6.44	1.53	0.62	2.48
No. of cars									
No. of pickups/vans									
No. of trucks/tractors	0.17	0.03	5.15						
Value of provided services									
Greater than $5M							−1.45	0.72	−2.02
Value of inbound shipments									
Less than $500K	0.67	0.44	1.53						
Between $500K and $999K	0.80	0.48	1.65	0.80	0.25	3.13	−0.82	0.47	−1.74
Between $1000K and $5M	0.65	0.46	1.42	0.77	0.24	3.17	−1.18	0.47	−2.52
Value of outbound shipments									
Between $500K and $999K							0.93	0.53	1.75
Between $1M and $5M							1.41	0.47	2.98
Greater than $5M							3.65	0.62	5.88
Census tract characteristics									
In Durham Region (yes/no)	−0.64	0.30	−2.10						
In Peel Region (yes/no)							1.05	0.38	2.73
In the City of Toronto (yes/no)				−0.76	0.26	−2.90			
Variances									
e1	9.06	0.52							
e2	5.21	0.30							
e3	12.88	0.75							

Table 3 Coefficient of determination R^2

Dependent variables	R-squared
Car ownership	0.30
Pickup/van ownership	0.21
Truck ownership	0.21
Overall	0.53

7.1 Goodness-of-Fit

The coefficient of determination (R^2) is used to evaluate the model as shown in Table 3. Four R^2 values are reported; three are for the individual evaluation of the models for each vehicle type, and one is for the combined performance of the three models. The R^2 value indicates a better fit in the case of the number of cars owned (0.3), than for pickups (0.21) and trucks (0.21). Also, the error term variances (Table 2) confirm the reported R^2 values; the model reports a relatively higher error term value for the truck ownership model and fairly low values for the cars and pickups/vans models. The overall R^2 value (0.53) indicates a good model fit for the performance of the three models.

8 Conclusions and Recommendations

In this chapter, a structural equation modeling approach has been used to model the commercial vehicle ownership level for three vehicle types. The model considers the simultaneous effects between dependent variables. This model could potentially be extended and used to estimate vehicle ownership and composition in a large-scale behavioral freight microsimulation introduced by Mostafa and Roorda (2015) that could be used for planning and policy assessment purposes. The results have revealed one complementary relationship between the number of cars and trucks owned. The model represents differences in commercial vehicle ownership by BEs in different industries, different regions, and different scales of operation, as measured by employment, inbound and outbound goods shipments and services.

The overall goodness-of-fit of the model structure is good for this kind of modeling. The individual goodness-of-fit of the car ownership model gives a moderately better prediction than for pickup/van and truck ownership. There are several improvements that may be considered in the future to improve the predictive ability of the model though most would require additional data. The effect of lagged endogenous variables could be investigated to see whether or not the ownership of commercial vehicles is dependent on the already owned fleet composition. This would require acquisition of longitudinal data for the same BEs. In addition, an enriched data set that includes more BE attributes that reflect the size and nature of BE operations (e.g., produced commodities, location of customers, etc.) and information about owned vehicle technology, fleet age, and costs associated with the fleet operation might enhance the quality of the model.

Commercial vehicle ownership could be heterogenous across industries. Estimating industry-specific models is another future step that could help specify ownership patterns across industries. Furthermore, other model structures (e.g., multivariate ordered models) could be explored to enhance the predictive performance of estimating vehicle fleet size and composition. Finally, future research might compare the performance of the presented structural equation model to other previously estimated model structures (i.e., Rashidi and Roorda 2017).

References

Abdur Razzaque, M., & Chen Sheng, C. (1998). Outsourcing of logistics functions: A literature survey. *International Journal of Physical Distribution and Logistics Management*, 89–107.

Acock, A. C. (2013). *Discovering structural equation modeling using Stata*. College Station, TX: Stata Press books.

Arunapuram, S., Mathur, K., & Solow, D. (2003). Vehicle routing and scheduling with full truckloads. *Transportation Science, 37*(2), 170–182.

Belhaj, M. (2002). Vehicle and fuel demand in Morocco. *Energy Policy, 30*(13), 1163–1171.

Boerkamps, J. H. K., van Binsbergen, A. J., & Bovy, P. H. L. (2000). Modeling behavioral aspects of urban freight movement in supply chains. *Transportation Research Record: Journal of the Transportation Research Board, 1725*, 17–25.

Bollen, K. A., & Long, J. S. (1992). Tests for structural equation models: Introduction. *Sociological Methods & Research, 21*(2), 123–131.

Choo, S., & Mokhtarian, P. L. (2004). What type of vehicle do people drive? The role of attitude and lifestyle in influencing vehicle type choice. *Transportation Research Part A, 38*(3), 201–222.

de Jong, G., & Ben-Akiva, M. (2007). A micro-simulation model of shipment size and transport chain choice. *Transportation Research Part B: Methodological, 41*(9), 950–965.

de Jong, G. C., & Kitamura, R. (2009). A review of household dynamic vehicle ownership models: Holdings models versus transactions models. *Transportation*, 733–743.

de Jong, G. C., Fox, J., Pieters, M., Daly, A., & Smit, R. (2004). A comparison of car ownership models. *Transport Reviews, 24*(4), 379–408.

Deepen, J. M. (2007). *Logistics outsourcing relationships: Measurement, antecedents, and effects of logistics outsourcing performance*. New York: Springer Science & Business Media.

DMTI Spatial Incorporation. (2012). *CanMap RouteLogistics Toronto (CMA) subset*. Toronto: Maps and Data Library, University of Toronto.

Elion, S., King, J., & Hutchinson, D. (1966). A study in equipment replacement. *Operational Research Quarterly, 17*(1), 59–71.

Fischer, M. J., Outwater, M. L., Cheng, L. L., Ahanotu, D. N., & Calix, R. (2005). An innovative framework for modeling freight transportation in Los Angeles County. *Transportation Research Record: Journal of the Transportation Research Board, 1906*, 105–112.

Greene, D. L., Meddeb, N., & Liu, J. (1986). Vehicle stock modelling of highway energy use: Tunisian and US applications. *Energy Policy, 14*(5), 437–446.

Hao, H., Wang, H., & Yi, R. (2011). Hybrid modeling of China's vehicle ownership and projection through 2050. *Energy, 36*(2), 1351–1361.

Hensher, D. A., & Ton, T. (2002). TRESIS: A transportation, land use and environmental strategy impact simulator for urban areas. *Transportation, 29*(4), 439–457.

Hunt, J. D., & Stefan, K. J. (2007). Tour-based microsimulation of urban commercial movements. *Transportation Research Part B: Methodological, 41*(9), 981–1013.
Huo, H., & Wang, M. (2012). Modeling future vehicle sales and stock in China. *Energy Policy, 43*, 17–29.
Klincewics, J. G., Luss, H., & Pilcher, M. G. (1990). Fleet size planning when outside carrier services are available. *Transportation Science, 24*(3), 169–182.
Li, Z., & Tao, F. (2009). On determining optimal fleet size and vehicle transfer policy for a car rental BE. *Computers and Operations Research, 37*(2), 341–350.
Liedtke, G. (2006). *An actor-based approach to commodity transport modelling*. Baden Germany: Nomos Verlagsgesellschaft.
McCabe, S., Kwan, H., & Roorda, M. J. (2013). Comparing GPS and non-GPS survey methods for collecting urban goods and service movements. *International Journal of Transport Economics, 40*(2), 183–206.
Mostafa, T. S., & Roorda, M. J. (2015). A Conceptual framework for modelling firmography. In *Canadian Transportation Research Forum 50th Annual Conference-Another 50 Years: Where to From Here?//Un autre 50 ans: qu'en est-il à partir de maintenant?* Montreal, Quebec, May 24–26, 2015.
Mostafa, T., & Roorda, M. J. (2017). Discrete choice modeling of freight outsourcing decisions of Canadian manufacturers. *Transportation Research Record: Journal of the Transportation Research Board, 2610*, 76–86.
Palencia, J. C. G., Furubayashi, T., & Nakata, T. (2012). Energy use and CO_2 emissions reduction potential in passenger car fleet using zero emission vehicles and lightweight materials. *Energy, 48*(1), 548–565.
Rashidi, T. H., & Roorda, M. J. (2017). A business establishment fleet ownership and composition model. *Transportation*, 1–17.
Redmer, A. (2009). Optimisation of the exploitation period of individual vehicles in freight transportation companies. *Transportation Research Part E, 45*(6), 978–987.
Roorda, M. J., Cavalcante, R., McCabe, S., & Kwan, H. (2010). A conceptual framework for agent-based modelling of logistics services. *Transportation Research Part E, 46*(1), 18–31.
Schroeter, J. R. (1983). A model of taxi service under fare structure and fleet size regulation. *The Bell Journal of Economics, 14*(1), 81–96.
Simms, B., Lamarre, B., Jardine, A., & Boudreau, A. (1984). Optimal buy, operate and sell policies for fleets of vehicles. *European Journal of Operational Research, 15*, 183–195.
Sorrell, S. (1992). Fuel efficiency in the UK vehicle stock. *Energy Policy, 20*(8), 766–780.
Suzuki, Y., & Pautsch, G. (2005). A vehicle replacement policy for motor carriers in an unsteady economy. *Transportation Research Part A, 39*(5), 463–480.
Transport and Mobility Leuven, TREMOVE "Service contract for the further development and application of the transport and environmental TREMOVE model Lot 1 (Improvement of the data set and model structure)". December 2007, online: http://www.tremove.org/documentation/TREMOVE_Short_Description.pdf
Ullman, J. B., & Bentler, P. M. (2003). *Structural equation modeling*. New York: John Wiley & Sons, Inc.
Wasserfallen, W., & Güntensperger, H. (1988). Gasoline consumption and the stock of motor vehicles: An empirical analysis for the Swiss economy. *Energy Economics, 10*(4), 276–282.
Wisetjindawat, W., Sano, K., Matsumoto, S., Raothanachonkun, P., 2007. Micro-simulation model for modeling freight agents interactions in urban freight movement. In: *CD Proceedings, 86th Annual Meeting of the Transportation Research Board, Washington DC*, 21–25 January 2007.

Applying Behavior Change Theory to Predict Travel Behavior of University Commuters

Panagiotis Papaioannou and Ioannis Politis

Abstract Selecting the right alternatives to an existing travel mode for commuting to a major trip generator at the center of a metropolitan area is often a complex project. In this chapter, the Aristotle University of Thessaloniki, Greece is the major trip generator in a study that compares the predictions of two approaches, both of which use discrete choice modeling, regarding the number of commuters who are willing to change travel mode if the right alterative is offered to them. The first is the conventional approach that, using travel cost and travel time as the main travel mode choice determinants, is applied to a selected (aggregate) sample of the commuter population. The second approach uses the same determinants but stratifies the sample into four strata according to the behavioral change stage to which the commuters belong. The model results indicate that the predicted travel mode choice behavior differs significantly among the persons in the sample depending on which behavioral change stage they are in. This study suggests that transport policy makers might better devise targeted interventions to maximize the positive socioeconomic and environmental impacts of travel mode alternatives.

1 Introduction

Transport research on travel behavior has greatly benefited from behavior change theories and models developed in the Social and the Health Sciences. Research in these two domains has preceded pertinent research in the transport sector. Despite differences between the relevant theories and models of these two domains, they have many common elements (TAPESTRY 2007; MAX 2007a, b; 2009). Travel behavior change can be attained much better if the drivers, and the pertaining circumstances, that can trigger and/or accelerate that change are known. Social psychology suggests that it is necessary to consider people's endogenous behavioral characteristics which measure an individual's internal process for changing

P. Papaioannou (✉) · I. Politis
Department of Civil Engineering, Aristotle University of Thessaloniki, Thessaloniki, Greece
e-mail: papa@civil.auth.gr; pol@civil.auth.gr

H. Briassoulis et al. (eds.), *The Practice of Spatial Analysis*,
https://doi.org/10.1007/978-3-319-89806-3_11

something that is perceived as wrong or as a bad habit by the broader community (Prochaska and Norcross 2001). In the transport sector, excessive and aimless usage of private cars is considered a "bad" habit/behavior nowadays and changing into public transit usage or to other environmentally friendly modes, such as biking or walking, is considered as right travel behavior.

Most theories and models about individual's behavior change in the Social Sciences and, consequently, in their application in the transport sector accept that change stages are discrete intervals which can be used to stratify the targeted population in an ordinal way. This applies to all behavior stages, which number from 4 to 7 and even to 11 stages depending on the model structure. Extensive travel behavior research indicates that when latent endogenous factors which characterize trip makers are considered in designing interventions to change people's everyday travel habits, the outcomes are much better compared to those that rely on traditional transportation planning processes.

This chapter examines the impact that endogenous characteristics of commuters to a large urban trip generator have on their travel mode choice when transport alternatives with specific attributes are provided. This is attempted by combining a specific behavioral change model in conjunction with discrete choice modeling. The major trip generator considered is the Aristotle University of Thessaloniki in Greece, the largest university in the country and one of the largest universities in the greater Balkan area with more than 80,000 active students and 5000 staff members (academic and administrative). The university is located at the city center of Thessaloniki, and it is one of the main contributors to the daily traffic congestion observed in the road network.

The chapter is structured into six sections. The second section reviews the literature on behavior change in traffic psychology theories and models as well as on mobility management issues at university campuses. The case study and the main statistics of the data and information obtained are described in the third section. The fourth section presents the classification of commuters into different behavioral change stages according to the change model employed. The fifth section presents the logit models derived and the respective predictions for mode shift. The model results and the main conclusions of this study are discussed in the last section.

2 Literature Review

The present research is based on two main pillars: (a) traffic psychology and interpretation of human travel behavior and (b) mobility plans and demand management schemes for universities. This section summarizes past research and key findings related to these two pillars.

2.1 *Traffic Psychology and Behavioral Stage for Change*

According to Schlag and Schade (2010), there are two main research areas where Social Psychology can be applied to the transportation sector. The first area is known as *"Traffic Psychology"* and investigates the user behavior as well as the psychological processes that produce this behavior. In addition, traffic psychology examines the relation between travel behavior and road accidents. The second area, known as *"Mobility Psychology,"* investigates issues related to mobility itself, examines individual and social parameters for people and commodity travel and it is highly related to Travel Demand Management applications.

Traditional transportation planning theories and methodologies focus primarily on the examination of trip characteristics such as travel time and travel cost (for current or planned transport modes). However, the new trend highlights the need for an integrated approach where human behavior and its characteristics are in the core of the analysis; some people may be more skeptical to change their current travel behavior, some other may need some pull or push measures to adopt new behavior, while still some others may pay more attention on comfort, safety, or reliability. It is clear that mobility management for universities sounds a suitable test bed for this integrated approach where both Traffic and Mobility Psychology research is applicable.

For the interpretation of human behavior, a number of explanatory psychology models have been developed. These models are trying to measure a number of human endogenous factors such as perceptual and attention behavior, cognitive, motivational, and emotional parameters. (Balcetis and Dunning 2007; Golledge and Garling 2007). One of the most interesting parts of interpreting human behavior is the examination of the behavioral change process. During the last decades, behavioral change theories have been implemented in health (Chen and Land 1986; NIH 2003), education (Nutbeam 2000; Elton 2003), and criminology (Akers et al. 1989; Maguire et al. 2007). Gradually, these theories and models are also being implemented for the interpretation of travel behavior.

The most well-known models are the *Need Opportunity Ability (NOA)* model (Shiftan 2008), the *Health Belief Model (HBM)* (Rosenstock et al. 1994), the *Protection Motivation Theory (PMT)* (Boer and Seydel 1996), the Theory of Planned Behavior (TPB) (Ajzen 1985), the Norm Activation Model (NAM) (Klockner and Matthies 2004), the Social Cognitive Theory (SCT) (Bandura 1977), the Transtheoretical Model (TTM) (Prochaska and DiClemente 1983), and the MaxSem Model (MAX 2009).

Most of these models use the concept of behavior stage for change; each individual is allocated into a (behavioral) stage which identifies his willingness to change a norm, a habit, a behavior, etc. The higher the behavioral stage is, the highest is individual support to adopt (or maintain) a new behavior.

The behavioral model that was used in this study was developed within the framework of an EU funded project called "MAX" and it is entitled "MaxSEM." It consists of four stages (Fig. 1): the pre-contemplation stage, the contemplation

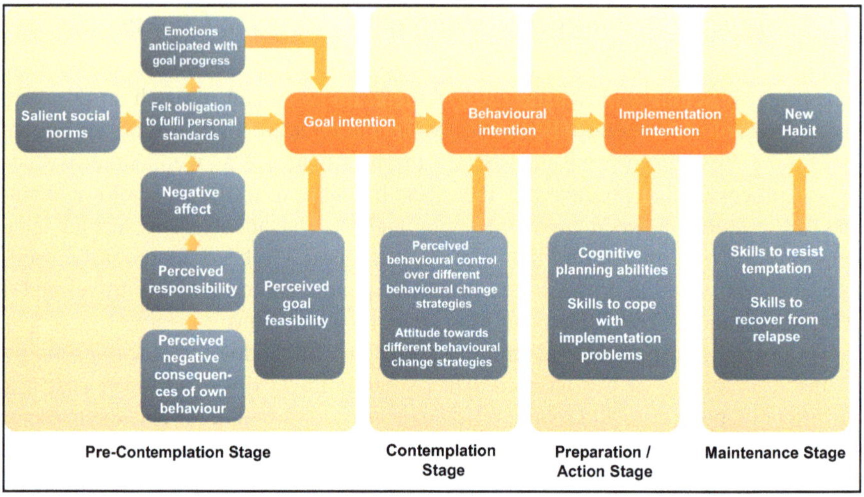

Fig. 1 The MaxSEM behavioral model

stage, the preparation/action stage, and the maintenance stage (MAX 2009). A short description of these behavioral stages follows.

Stage 1: Pre-contemplation stage. Persons in this stage are habitual car drivers who have no intention to reduce their current car use. The aim here is to make this group think of possible change.

Stage 2: Contemplation stage. Persons in this stage are thinking about reducing their current car use, i.e., they have formed a personal car reduction goal. The aim here is to present to this group attractive options for changing their behavior.

Stage 3: Preparation/action stage. Individuals in this stage have selected a concrete behavioral strategy for reaching their car reduction goal (e.g., using the bike instead of the car to go to work) or they have already occasionally tried the new behavior. The aim here is to have the group actually try out the new behavior and to facilitate its maintenance.

Stage 4: Maintenance stage. Individuals in this stage have adopted the new behavior and have formed a new habit. The aim here is to reward the new habit and to prevent relapse to the old behavior.

The individuals' stage position can be measured through a revealed preference questionnaire survey; the individual is being asked to answer six simplified questions (called diagnostic questions) regarding his/her current travel behavior and how ready or not he/she is to adopt a new one (MAX 2009).

The usage of behavioral stage for change is a relatively new approach for transportation analyses. It is based on the idea that the targeted population may not support (is not ready) a particular intervention, or may strongly support (is ready) another intervention. People assigned to each stage have completely different needs, tastes and, in turn, possibilities to adopt a new behavior or to support a planned action or measure. In this respect, aggregating this mixed population is not correct and disaggregating it according to behavioral stage should be taken into account by policy makers in the decision-making process.

The examination of behavioral stage in transportation analyses, and more specifically in analyses of commuting trips to universities and colleges, is limited in the international literature. Shannon et al. (2006) divided staff and student population at the University of Western Australia into three behavioral stages and found substantial differences among the stages on issues related to current travel behavior as well the willingness to adopt new alternative modes for commuting to the university and traveling in general. In addition, Politis et al. (2012) found that behavioral stage for change is a key consideration for interpreting the behavior of university commuters in the case of introducing flexible transport modes.

Taking into consideration the above, the aim of this chapter is to offer additional knowledge about the interpretation of travel behavior by utilizing the concept of behavioral stage for change. In this context, different population segments are defined for commuting trips to the Aristotle University of Thessaloniki, the largest university in Greece, and various scenarios are examined regarding the mode of commuting. The main research question is whether or not the disaggregation of the population into discrete behavioral stages may provoke different policy outcomes based on alternative travel mode scenarios for commuting to the university.

2.2 *Mobility Management for University Campuses*

Universities and colleges are traditionally considered as huge trip generators; students and working staff visit the campus on a daily basis, sometimes more than once per day. Very often, campuses are located at the city center or in densely populated areas and their premises consist of buildings, stadiums, parks, parking spaces, etc. The growing demand for visiting the campus, both by staff and students, generate the need for intervention schemes and accessibility plans. These schemes and plans usually focus on both physical measures and awareness campaigns, aiming to promote sustainable mobility from/to or within the university campus.

The objective of sustainable mobility is too difficult to be fulfilled in the case of universities since the problem is multidimensional. One main dimension is the targeted population and the accomplishment of their needs and priorities: teaching staff, administrative employees, undergraduate and postgraduate students are different population segments with mixed socioeconomic backgrounds, different trip characteristics and needs but with a common trip destination. Another dimension is the temporal features of travel demand: with the exception of the administrative staff, all other population segments visit university at different periods during the weekdays. Finally, another main dimension is the authority for interventions: local schemes within a campus are exclusively coordinated by the university authorities. However, the trip chain extends beyond this area and, therefore, any supporting intervention outside the campus presupposes coordination with the local municipal authorities and stakeholders, such as the transit operators, something that is not

Table 1 Indicative list of university mobility plans

University	Reference period	Staff/ Students	Measures		Evaluation	Reference
			Soft	Hard		
Technical University of Cracow	2006–2008	N.A/ 18,000	v	v	Yes	McDonald M. et al. (2010)
University of Limerick	2012–2015	420/ 17000	v	x	Not yet	Higgins and Collins (2015)
University of Hawaii	2011–2012	30,000 total	v	x	Yes	Nelson (2012)
University of Malta	2011–2014	4000/ 11,000	v	x	No	Attard et al. (2014)
University of Thessaloniki	2010–2013	4000/ 80,000	v	v	No	Pitsiava-Latinopoulou et al. (2013)
University of Essex	2010–2012	1800/ 12,000	v	v	Yes	Un. of Essex (2015)
University of Massachusetts Lowell	2005–2010	1400/ 15000	v	v	Yes	Black et al. (2011)
North Carolina State University	2012–2022	7000/ 35000	v	x	Yes	HDR Engineering (2012)
University of Warwick	2005–2018	1400/ 23,000	v	v	Yes	Un. Of Warwick (2007)
Liverpool John Moores University	2004–2015	1600/ 24000	v	x	Yes	Un. Of Liverpool John Moores (2014)

always secured. The multidimensionality of the concept of sustainable mobility in the case of universities is highlighted by Longo (Longo et al. 2015) where an Analytic Hierarchy Process (AHP) was applied to prioritize mobility interventions in the University of Trieste, Italy.

University Mobility Plans (UMoPs), also referred as Access Plans, Transportation Demand Management Plans, Green Travel Plans, etc., are key policy instruments to deal with the issues mentioned previously and overcome barriers and obstacles. Nowadays, more and more universities are developing these plans either as separate localized plans or as components of an integrated transport master plan of the city or the metropolitan area. UMoPs can be planning documents, describing qualitative and quantitative targets, or even detailed operational tools with specific measures, budgets, time plans, etc. (Silva and Ferreira 2008). The international literature provides plenty of such documents. Table 1 presents an indicative list. Toor and Havlick (2004) also have presented mobility problems, examples, and solutions at American Universities in a structured way.

The impacts of UMoPs have been extensively analyzed and evaluated. Balsas (2003), for example, examined eight pedestrian and bicycle friendly campuses in the USA. He found that the modal share for the manpower modes at these universities where higher compared to the national trend. Black et al. (1999) observed substantial improvement in transit operation as well as in road traffic nearby the University of

South Wales in Australia, after the implementation of an integrated Transport Demand Management scheme. Similar evaluations were performed for mobility plans at the University of Toronto at Mississauga (Conway et al. 2008), the University of Sydney in Australia (Baboulet and Lenzen 2010; Atherton and Giurco 2011), the Monash University in Melbourne (Rose 2008), the University of Florida (Bond and Steiner 2006), etc.

Regarding commuters, the main efforts focus primarily on shifting people from their individual cars to public transport. From a revealed preference questionnaire survey of 23 American Universities, Daggett and Gutkowski (2003) found that transit performance was enhanced dramatically by certain fare policies. In a similar study, Brown et al. (2003) found an increase on transit share together with a simultaneous drop of private car share when an unlimited access program was implemented at the UCLA. Similar outcomes derived from the studies of Daniels and Mulley (2013), MacDonald and MacDonald (2001), and Finlay and Massey (2012).

3 Description Case Study

3.1 The Aristotle University of Thessaloniki, Greece

The Aristotle University of Thessaloniki (A.U.Th.) is the largest University in Greece and one of the largest academic institutions in the Balkan area. The total number of registered students is approximately 110,000 who are distributed in 41 departments. The university is located at the city center and significantly contributes to the traffic and environmental problems that are observed on a daily basis in that area. Politis et al. (2012) have summarized the main factors that contribute to this situation into: (a) the landscape particularities of the greater area, (b) the absence of modern and efficient public transport services since only one bus operator provides services in the Thessaloniki Metropolitan Area of one million inhabitants, and (c) the trip and mobility characteristics; almost 25% of the 2.3 million motorized and non-motorized daily trips have the city center as their origin or destination (or both) point. In addition, the Public Transport Authority of Thessaloniki (PTA) estimates that the car mode share (trips made) in 2010 for the Metropolitan Area increased to 55% compared to 44% 10 years before (PTA 2011).

Regarding commuting trips to the university campus, the current situation is also problematic. Although university employees are almost 5000 people, no travel or mobility plan exists. Furthermore, the large number of available parking spaces inside the campus inevitably makes car an attractive option for commuting. Finally, given that the university is located at the city center, a large number of bus lanes (13 in total) pass close to the campus, but none of them have a stop located in it, increasing the mean walking distance and, ultimately, decreasing the utility of the bus alternative to the car.

3.2 The Survey

For the purposes of this research, a revealed preference survey and a stated preference experiment were conducted on a sample of 450 staff members of the University. The aim of the revealed preference survey was to obtain information about the personal background and trip characteristics of the commuters in the sample; furthermore, by using diagnostic questions, the aim was to assign each respondent to the respective behavioral stage. The aim of the stated preference experiment was to examine the potential for use of alternative modes for commuting. The stated preference experiments conducted are detailed in the following section.

The revealed preference part of the questionnaire was divided into three parts: (a) questions regarding the trip characteristics of the respondent, (b) questions regarding the values of the respondent on issues towards his/her environmental responsibility and the perceptions about mode safety and reliability, and (c) questions about the socioeconomic background of the respondent. From the responses obtained, certain indicator values were calculated, such as cost per kilometer of commuting trip and number of available cars per household member.

The sampling approach used was the proportional stratified random sampling. The population was stratified according to two main criteria (classes):

- The type of employee (teaching or administrative staff); according to the official staff office of the University, 50% of the registered employees are teaching staff.
- The obligation/need of the employee to go/visit another intermediate destination before or after work (accompany children to/from school, visit another location, go for coffee, etc.).

For the sample size estimation, the formula used was that provided by Cohen et al. (2010). For a population of 5000 people, a 95% confidence level, and a 5% sampling error, the appropriate minimum sample size was calculated to be 357 employees. However, in order to overcome any bias in the sample, almost 100 additional employees were included in the survey.

3.3 Descriptive and Inferential Statistical Analysis

Table 2 presents the descriptive statistics of the nominal and ordinal variables used in the study, together with their measurement type (in bold). As expected, the absence of an integrated mobility plan of the university has an impact on the mode choice share; more than 52% of the employees use their private car for commuting to the university, either as drivers or co-drivers. Although this percentage is lower compared to the respective percentages for commuting to universities in the USA (Shannon et al. 2006; Shaheen et al. 2003), it is similar compared to other European Universities (University of Essex 2015). The public transport share for commuting to the university is approximately 21%. The majority (72%) of the

Table 2 Frequencies of the nominal and ordinal variables of the study

Variable	Measurement scale/response	Frequency	Percent	Type
Trip commuting characteristics				
Mode used	**1:** Car (Driver)	233	51.8	Nominal
	2: Car (Co-driver)	13	2.9	
	3: Taxi	9	2.0	
	4:Bus	95	21.1	
	5: Walk	82	18.2	
	6: Other	18	4.0	
Obligation to be at work on certain time	**0:** Yes	323	71.8	Nominal
	1: No	127	28.2	
Obligations before/after work	**0:** Yes	252	56.0	Nominal
	1: No	189	42.0	
Values				
Car usage is the most important factor for air pollution	**1:** Totally disagree	7	1.6	Ordinal
	2: Disagree	62	13.8	
	3: Do not know	100	22.2	
	4: Agree	204	45.3	
	5: Totally agree	76	16.9	
Main reason for car usage reduction is environmental improvement	**1:** Totally disagree	6	1.3	Ordinal
	2: Disagree	38	8.4	
	3: Do not know	85	18.9	
	4: Agree	236	52.4	
	5: Totally agree	85	18.9	
Hybrid and electric cars well reduce the environmental problems in 10 years	**1:** Totally disagree	4	0.9	Ordinal
	2: Disagree	43	9.6	
	3: Do not know	76	16.9	
	4: Agree	255	56.7	
	5: Totally agree	72	16.0	
It is important for me to use a comfort mean of transport	**1:** Totally disagree	3	0.7	Ordinal
	2: Disagree	21	4.7	
	3: Do not know	45	10.0	
	4: Agree	210	46.7	
	5: Totally agree	171	38.0	
It is important for me to use a travel mode that relaxes me	**1:** Totally disagree	2	0.4	Ordinal
	2: Disagree	15	3.3	
	3: Do not know	42	9.3	
	4: Agree	178	39.6	
	5: Totally agree	213	47.3	

(continued)

Table 2 (continued)

Variable	Measurement scale/response	Frequency	Percent	Type
I want to be able to read a book or text during my commuting trip to work	**1:** Totally disagree	68	15.1	Ordinal
	2: Disagree	108	24.0	
	3: Do not know	114	25.3	
	4: Agree	87	19.3	
	5: Totally agree	73	16.2	
It is important for me to use the shortest path to reach my travel destination	**1:** Totally disagree	3	0.7	Ordinal
	2: Disagree	18	4.0	
	3: Do not know	64	14.2	
	4: Agree	179	39.8	
	5: Totally agree	186	41.3	
I want my travel mode to be available any time	**1:** Totally disagree	2	0.4	Ordinal
	2: Disagree	22	4.9	
	3: Do not know	46	10.2	
	4: Agree	192	42.7	
	5: Totally agree	187	41.6	
I want to use the travel mode independently of the weather conditions	**1:** Totally disagree	2	0.4	Ordinal
	2: Disagree	16	3.6	
	3: Do not know	26	5.8	
	4: Agree	176	39.1	
	5: Totally agree	229	50.9	
Socioeconomic background				
Gender	**0:** Man	212	47.1	Nominal
	1: Woman	238	52.9	
Income level (in thousand euros)	**1:** <15	46	10.2	Ordinal
	2: 16–20	56	12.4	
	3: 21–25	49	10.9	
	4: 26–30	45	10.0	
	5: 31–35	38	8.4	
	6: 36–40	71	15.8	
	7: >40	144	32.0	
Teaching or administrative staff	**0:** Teaching	247	54.9	Nominal
	1: Administrative	203	45.1	
Behavioral stage for change	**0:** Pre-contemplation	163	36.2	Ordinal
	1: Contemplation	31	6.9	
	2: Preperation/ action	49	10.9	
	3: Maintenance	207	46.0	

employees stated that they have the obligation to be at work on specific time, whereas almost half of them (56%) stated that they have obligations before or after work (before returning home).

As mentioned in the previous section, the questionnaire contained also questions related to the values/personal beliefs of the respondents. Each one of the three examined values was measured through a set of three questions with the answers given on a 5-point Likert scale (from strongly disagree to strongly agree). Table 2 presents the descriptive statistics for these nine questions (the internal consistency of the answers was also checked for each one of the three values by returning a satisfactory Cronbach alpha of 0.6). From this Table, it can be easily concluded that the only statement that was not strongly supported by the respondents was that of "I want to be able to read a book or text during my commuting trip to work."

From the variables related to the socioeconomic background of the respondents, two of them were used as control variables for the sample selection: gender (it was known that women slightly outnumber men at the university) and the percentages of teaching and administrative staff (almost 50/50).

However, the most important outcome derived from the descriptive statistics is the behavioral stage distribution. Most respondents (46%) were assigned to the "Maintenance" stage. Part of the persons in the sample stated that they use a transport mode other than private car such the bus, bike, or walking, for commuting. A large part of cars users (36% of total sample and 67% among car users) were assigned to the "Pre-contemplation" stage, whereas only 7% (13% among cars users) were assigned to the "Contemplation" stage. The respective percentage for those who were assigned to the "Preparation/Action" stage is 11% (20% among car users). This outcome strengthens the initial assumption that the affected population should not be taken into consideration as a whole since they may not be all ready to accept or choose a new/alternative intervention with the same level of support. People who are assigned to the upper behavioral stages will be more enthusiastic to leave their car and use an alternative mode compared to those assigned to the "Pre-contemplation" stage.

Table 3 presents the mean values for each one of the variables examined for the whole sample ($N = 450$) as well as for each one of the four discrete behavioral change stages. Bivariate inferential tests examined the hypothesis that a given answer is differentiated among the different behavioral stages. From the results obtained it can be concluded that, indeed, people who are assigned to each behavioral stage belong to a certain group with specific characteristics and behavioral profile.

The most important outcomes of the inferential tests are the following:

- The existence of obligations before or after work is differentiated among the behavioral stages; those who have been categorized to the "Maintenance" stage stated primarily that they do not have an obligation before or after work compared to those who were assigned to the other three stages. This obligation may act as barrier to leave their car and shift to another travel mode.

Table 3 Inferential statistics for the total sample and by behavioral change stage

	All ($n = 450$)	Pre-contemplation ($n = 163$)	Contemplation ($n = 31$)	Preparation ($n = 49$)	Maintenance ($n = 207$)	Test	Result
Trip commuting characteristics							
Mode used	2.63	1.07	1.00	1.00	4.49	Kruskal-Wallis	<0.001
Obligation to be at work on certain time	0.28	0.25	0.45	0.33	0.28	Mann-Whitney U	>0.05
Obligations before/after work	0.43	0.36	0.32	0.37	0.52	Chi-square	<0.05
Values							
Car usage is the most important factor for air pollution	3.62	3.40	3.74	3.39	3.84	Wilcoxon	<0.001
Main reason for car usage reduction is environmental improvement	3.79	3.87	3.74	3.65	3.77	Wilcoxon	<0.001
Hybrid and electric cars will reduce the environmental problems in 10 years	3.77	3.79	3.77	3.76	3.76	Wilcoxon	<0.001
It is important for me to use a comfortable mean of transport	4.17	4.19	4.13	4.14	4.16	Wilcoxon	<0.001
It is important for me to use a travel mode that relaxes me	4.30	4.26	4.06	4.22	4.39	Wilcoxon	<0.001
I want to be able to read a book or text during my commuting trip to work	2.98	2.82	3.06	2.76	3.14	Wilcoxon	<0.001
It is important for me to use the shortest path to reach my travel destination	4.17	4.12	4.16	4.06	4.24	Wilcoxon	<0.001
I want my travel mode to be available any time	4.20	4.22	4.23	3.90	4.26	Wilcoxon	<0.001
I want to use the travel mode independently of the weather conditions	4.37	4.39	4.45	4.27	4.36	Wilcoxon	<0.001

Socioeconomic background							
Gender	0.53	0.52	0.58	0.41	0.56	Mann-Whitney U	>0.05
Income level	4.70	5.31	4.90	4.53	4.22	Wilcoxon	<0.001
Teaching or administrative staff	0.45	0.34	0.42	0.49	0.53	Mann-Whitney U	<0.001
Available number of cars per household member	1.71	1.87	1.91	2.06	1.47	One Way ANOVA	<0.001
Trip commuting perception							
Perceived travel length	7.43	10.68	10.05	9.90	3.93	One Way ANOVA	<0.001
Perceived cost per trip length ratio	0.30	0.40	0.56	0.49	0.15	One Way ANOVA	<0.001
Perceived cost per real travel length ratio	1.16	1.14	1.22	1.18	1.15	One Way ANOVA	>0.05

- People in different behavioral stages tend to have different values; all the statements examined were differently answered by the respondents in each of the four behavioral stages. For instance, persons belonging to the "Maintenance" stage are more sensitive to environmental issues, they prefer to travel in a relaxed way, they enjoy reading a document or a book and mode reliability/availability is low in their priorities.
- As expected, important differences were found among people in different stages as regards their socioeconomic background. With the exception of gender, all other parameters (e.g., income level, teaching or administrative staff, available number of cars for commuting per household member) were differentiated among the four behavioral stages.
- Trip length may affect the assignment to a stage; longer trips may constitute a good reason for using the car. The mean perceived trip length for those who were classified in the "Pre-Contemplation" stage, is 10.68 km. whereas for those who were classified in the "Maintenance" stage it is only 3.93 km.
- Apart from other benefits, people classified in the "Maintenance" stage recognize that their commuting trip is less costly, with a mean unit cost per kilometer equal to 0.15 euros/km. The respective value for people in the "Contemplation" stage is 0.40 euros/km.

All these qualitative and quantitative outcomes confirm that people classified into different behavioral stages indeed have different current trip characteristics, perceptions, and habits. However, this finding is not enough to confirm the main hypothesis of the chapter that people in different behavioral stages will be differentiated between each other in the case where alternative travel modes will be available for commuting to the university. The next section makes use of the data derived from the stated preference experiment to confirm or reject this hypothesis. Analytical econometric models of discrete choice methods were used as hypothesis confirmation or rejection tools.

4 Behavioral Change Stage and Demand Estimation

4.1 The Stated Preference Experiment

The stated preference (SP) experiment was carried out for car drivers only, i.e., only for those classified in the first three behavioral stages. The employees were asked to choose between their current travel mode (car) and alternative travel options which are not under operation.

More specifically, four alternative modes to the car were examined; namely collective taxi, carpooling, van pooling, and metro system. The latter is under construction and after its completion in year 2019 will be the second public transport mode in operation together with the buses.

In order to avoid complexity, each employee participated in two SP games; their preferences were tested only for two out of the four alternative options. The distribution of the games among the different types (strata) of employees was the following:

- *SP GAME 1: Car versus Van Pooling:* In this game, the following strata participated: (a) teaching staff *without* any obligation before or after work and (b) administrative staff *without* any obligation before or after work
- *SP GAME 2: Car versus Collective Taxi:* In this game, the following strata participated: (a) teaching staff *with* obligation before or after work and (b) administrative staff *with* obligation before or after work
- *SP GAME 3: Car versus Carpooling:* In this game, the following strata participated: (a) administrative staff *without* any obligation before or after work and (b) administrative staff *with* obligation before or after the work
- *SP GAME 4: Car versus Metro:* In this game, the following strata participated: (a) teaching staff *without* any obligation before or after the work and (b) teaching staff *with* obligation before or after work

Three mode choice attributes were included and examined in the SP games: in-vehicle time, travel cost, and remaining (out of vehicle) time. For each attribute, three levels were defined; namely low, medium, and high. The full factorial design results in 27 possible combinations and for this reason a partial factorial symmetrical design was selected with nine combinations. Table 4 presents the different alternatives, the respective attributes and their levels as shown to the university employees. With the exception of the metro alternative, all other values are percentages or absolute differences compared to the revealed preference statement of the respondent for his/her commuting trip.

4.2 Binary Choice Models and Model Specification

Binary choice models fall into the category of utility maximization models based on micro-economic theories of consumer behavior and psychological choice behavior. In transportation research, choice models assume that the probability of an individual to choose a particular transport mode depends upon the utility to be gained from traveling by that mode as compared to the utility corresponding to another alternative. In other words, the mode with the higher utility will be chosen by each individual (Ben-Akiva and Lerman 1985; Train 2003).

The utility of each mode can be affected by a number of factors related to the mode itself as well as to habits, tastes, and preferences of the individual. In terms of mathematical notation, the above can be written as:

$$\mathrm{U}_{it} = f(X_{it}, S_{it}) \tag{1}$$

Table 4 Attribute levels for each mode of the SP survey

	Van pooling			Collective taxi			Carpooling			Metro		
Attributes	L	M	H	L	M	H	L	M	H	L	M	H
Travel cost[a]	Free	−90	−70	−50	−40	−30	−80	−70	−50	1	1.2	1.7
In-vehicle time[b]	0	+5	+10	−5	+5	+10	−5	+5	+10	5	10	15
Remaining travel time[b]	+5	+7	+10	+2	+5	+10	+2	+5	+10	+7	+10	+15

[a]Cost is reported as percentage difference from the revealed preference cost except for the metro mode in which case absolute prices are given
[b]Vehicle time and remaining travel time are reported as absolute differences from the respective revealed preference times in minutes, except for the "in-vehicle time" of the metro mode in which case specific values are used

where U_{it} is the utility gained by individual (t) for using mode (i), X_{it} are the attribute values for mode (i) with respect to individual (t), such as travel time and cost, and S_{it} are individual-specific attributes of individual (t), such as age, gender, and income.

Due to the weakness of measuring all the elements that compose the utility of each mode for each individual, utility is considered as a random variable and, therefore, it consists of a sum of observed variables (called the *systematic or representative component* of utility—V_{it}) and random components (called *disturbances* of utility—ε_{it}):

$$\mathrm{U}_{it} = \boldsymbol{V}_{it} + \varepsilon_{it} = \beta_t \boldsymbol{x}_{it} + \varepsilon_{it} \tag{2}$$

where β_t is the vector of unknown (estimated) parameters related to variables x_{it} (mode and user specific) (Louviere et al. 2000). Since individual (t) will choose the mode with the higher utility, in the case of a binary choice model the following relationship between the utilities of the two alternative modes i *and* j exists:

$$\mathrm{U}_{it} > \mathrm{U}_{jt} \tag{3}$$

And the probability of mode i to be chosen by individual (t) is written as:

$$\begin{aligned} \boldsymbol{P}_{it} &= \mathrm{Prob}\left(\boldsymbol{U}_{it} > \boldsymbol{U}_{jt}\right) \Rightarrow \boldsymbol{P}_{it} = \mathrm{Prob}\left[\left(\boldsymbol{V}_{it} + \varepsilon_{it}\right) > \left(\boldsymbol{V}_{jt} + \varepsilon_{jt}\right)\right] \\ &\Rightarrow \boldsymbol{P}_{it} = \mathrm{Prob}\left[\left(\boldsymbol{V}_{it} - \boldsymbol{V}_{jt} + \varepsilon_{it}\right) > \varepsilon_{jt}\right] \quad \forall i \neq j \end{aligned} \tag{4}$$

In the general form of a binary choice model, the possibility of choosing mode i is expressed as:

$$\boldsymbol{P}_i = \frac{\boldsymbol{e}^{\left(\beta_0 + \sum_j \beta_j \boldsymbol{x}_j\right)}}{1 + \boldsymbol{e}^{\left(\beta_0 + \sum_j \beta_j \boldsymbol{x}_j\right)}} = \frac{1}{1 + \boldsymbol{e}^{\left(\beta_0 + \sum_j \beta_j \boldsymbol{x}_j\right)}} \tag{5}$$

Regarding the specification of the models examined in this chapter, the idea is the following:

- Each SP game represents a discrete mode choice experiment. This means that each time the car is compared with the alternative mode of the SP game (binary comparison).
- The binary choice models are divided each time into the *simplified* model (model 1) including only the mode-specific variables of in-vehicle-time, out-of-vehicle-time and travel cost, and the *extended* model (model 2) in which some socioeconomic variables such as age, gender, occupation, commitment to be at work on a certain time, and whether or not the employee has obligations before or after work were included. The mathematical notation of the utility for each one of the two models is as follows:

$$\textbf{\textit{Model 1}} V_{alter} - V_{car} = \begin{aligned} & ASC + \beta_1 IVT_{alter} + \beta_2 OVT_{alter} \\ & + \beta_3 COST_{alter} \beta_4 IVT_{car} + \beta_5 OVT_{car} + \beta_6 COST_{car} \end{aligned} \quad (6)$$

$$\textbf{\textit{Model 2}} V_{alter} - V_{car} = \begin{aligned} & ASC + \beta_1 IVT_{alter} + \beta_2 OVT_{alter} \\ & + \beta_3 COST_{alter} \beta_4 IVT_{car} + \beta_5 OVT_{car} + + \beta_6 COST_{car} \\ & + \beta_7 oblig + \beta_8 commit + \beta_9 gender + \beta_{10} occup \\ & + + \beta_{11} Age\_26\_35 + \beta_{12} Age\_36\_45 + \beta_{13} Age\_46\_55 \\ & + + \beta_{14} Age\_56\_67 \end{aligned} \quad (7)$$

where.

"alter" means the alternative mode examined each time at the SP game.
"oblig": obligation to be at work on a certain time.
"commit": obligations before/after work.
"gender": sex of the respondent.
"occup": teaching or administrative staff.
The reference category for the dummy variable of age is the age interval 18–25

- For the comparison of the two models, a likelihood ratio test was being performed each time. The likelihood ratio test is usually used to measure the performance of one model relative to another. The mathematical formula for the two models is written as follows:

$$L^* = -2\left[L_{(\text{model}_1)} - L_{(\text{model}_2)}\right] \quad (8)$$

- Alternative-specific beta coefficients were considered for the variables of travel time and cost. This represents an assumption that a minute and/or a cent has different marginal disutility of the car compared to other alternative travel modes examined in the SP experiment (Ben-Akiva and Lerman 1985).
- For each SP game, the two models (simplified and extended) were developed (a) for the total sample of the SP game (b) for each one of the three behavioral stages. This means that 4 (groups, 1 for all population at the game +3 for each behavioral stage) x 2 (models, simplified and extended) x 4 (SP games) = 32 binary choice models were finally developed within the framework of this study. The results are presented in the next section.

5 Model Results and Discussion

The objective of this section is to interpret and comment on the results of the binary choice models developed. The model parameter values together with their respective p value and t-student statistic (in parenthesis) are presented in Tables 5, 6, 7, and 8. The following subsections present the qualitative and quantitative findings derived from each SP game which represented each time an alternative to the car mode choice scenario.

Table 5 Binary choice model beta estimates for car versus van pool

	All				Pre-contemplator stage				Contemplator stage				Preparation/action stage			
	Model 1		Model 2		Model 1		Model 2		Model 1		Model 2		Model 1		Model 2	
beta	*b*	*p*	*b*	*p*	*b*	*p*	*b*	*p*	*b*	*p*	*b*	*p*	*b*	*p*	*b*	*p*
ASC					−0.495 (−1.89)	0.059	−0.821 (−2.71)	0.007	1.899 (2.76)	0.006	8.640 (3.62)	0	0.962 (3.41)	0.001	1.922 (4.78)	0
β_1									−0.066 (−2.59)	0.01	−0.182 (−3.77)	0	0.074[a] (1.76)	0.078	0.080[a] (1.79)	0.074
β_4	0.026 (3.96)	0	0.044 (4.71)	0	0.024 (2.13)	0.033	0.029 (2.25)	0.025								
β_6	−0.126 (−3.03)	0	−0.163 (−3.44)	0.001												
β_7			−0.592 (−3.21)	0.001			−1.212 (−4.27)	0			−6.061 (−3.56)	0				
β_{10}			1.492 (6.40)	0			1.061 (4.36)	0								
β_{11}			−1.689 (−6.14)	0											−1.710 (−3.91)	0
β_{12}			−1.049 (−3.65)	0												
β_{13}							0.625 (2.56)	0.01								
β_{14}			−0.573 (−2.93)	0.003											−1.394 (−3.08)	0.002
N	644		644		425		425		66		66		153		153	
R^2	0.026		0.157		0.011		0.131		0.156		0.563		0.02		0.136	
Adj R^2	0.034		0.21		0.014		0.174		0.209		0.751		0.028		0.186	

(continued)

Table 5 (continued)

	All				Pre-contemplator stage				Contemplator stage				Preparation/action stage			
	Model 1		Model 2		Model 1		Model 2		Model 1		Model 2		Model 1		Model 2	
beta	*b*	*p*	*b*	*p*	*b*	*p*	*b*	*p*	*b*	*p*	*b*	*p*	*b*	*p*	*b*	*p*
-2LL	875.946		782.68		584.492		529.602		80.031		36.687		196.685		177.542	
L^*	93.266				54.89				43.344				19.143			
LR test	sig. dif. at 99% confidential level				sig. dif. at 99% confidential level				sig. dif. at 99% confidential level				sig. dif. at 99% confidential level			

[a]Generic variable (DTTin = IVTcar-IVTvan)

Table 6 Binary choice model beta estimates for car versus collective taxi

	All				Pre-contemplator stage				Contemplator stage				Preparation /Action Stage			
	Model 1		Model 2		Model 1		Model 2		Model 1		Model 2		Model 1		Model 2	
beta	*b*	*p*	*b*	*p*	*b*	*p*	*b*	*p*	*b*	*p*	*b*	*p*	*b*	*p*	*b*	*p*
ASC	0.374 (1.94)	0	0.567 (4.91)	0	−0.483 (−2.44)	0.015	−0.874 (−4.09)	0					0.785 (2.19)	0.029	1.499 (3.27)	0.001
β_2													0.079 (1.78)	0.075	0.096 (1.95)	0.051
β_4					0.029 (3.21)	0.001	0.034 (3.56)	0	0.024 (3.15)	0.002	0.109 (3.56)	0				
β_5													−0.125 (−1.89)	0.059	−0.203 (−2.59)	0.01
β_6	−0.052 (−3.82)	0.053	−0.057 (−2.05)	0.04	−0.072 (−2.05)	0.041	−0.084 (−2.21)	0.027								
β_7											1.465 (2.39)	0.017				
β_9											−5.066 (−4.59)	0			1.658 (4.08)	0
β_{10}							0.888 (5.87)	0			4.351 (5.75)	0				
β_{11}			0.666 (3.75)	0							−6.486 (−4.72)	0				
β_{12}															−1.764 (−3.57)	0
β_{14}			−0.866 (−6.52)	0											−1.928 (−5.02)	0
N	1216		1216		806		806		157		157		253		253	
R^2	0.003		0.069		0.014		0.056		0.065		0.46		0.016		0.183	

(continued)

Table 6 (continued)

	All				Pre-contemplator stage				Contemplator stage				Preparation /Action Stage			
	Model 1		Model 2		Model 1		Model 2		Model 1		Model 2		Model 1		Model 2	
beta	*b*	*p*	*b*	*p*	*b*	*p*	*b*	*p*	*b*	*p*	*b*	*p*	*b*	*p*	*b*	*p*
Adj R^2	0.004		0.092		0.018		0.075		0.086		0.613		0.024		0.266	
-2LL	1667.152		1584.542		1105.652		1070.134		207.165		120.88		290.363		243.344	
*L**	82.61				35.518				86.285				47.019			
LR test	sig. dif. at 99% confidential level				sig. dif. at 99% confidential level				sig. dif. at 99% confidential level				sig. dif. at 99% confidential level			

Table 7 Binary choice model beta estimates for car versus carpooling

	All				Pre-contemplator stage				Contemplator stage				Preparation/action stage			
	Model 1		Model 2		Model 1		Model 2		Model 1		Model 2		Model 1		Model 2	
beta	*b*	*p*	*b*	*p*	*b*	*p*	*b*	*p*	*b*	*p*	*b*	*p*	*b*	*p*	*b*	*p*
ASC	1.721 (6.59)	0	1.821 (6.43)	0	0.772 (2.38)	0.018	1.219 (3.41)	0.001	1.978 (4.05)	0	2.769 (4.14)	0	1.428 (2.90)	0.004	5.108 (3.58)	0
β_1	−0.026 (−4.01)	0	−0.026 (−3.71)	0												
β_2	−0.067 (−4.04)	0	−0.074 (−4.27)	0												
β_4					−0.047 (−3.48)	0	−0.077 (−4.67)	0	−0.077 (−3.49)	0	−0.062 (−2.41)	0.016	0.137 (4.38)	0	0.724 (2.59)	0.009
β_5					−0.180 (−4.48)	0	−0.247 (−5.11)	0								
β_6					0.355 (5.06)	0	0.586 (6.50)	0					−1.052 (−5.16)	0	−5.291 (−3.06)	0.002
β_7			1.779 (5.13)	0			3.346 (5.30)	0								
β_8															−2.793 (−3.00)	0.003
β_9											−1.344 (−2.41)	0.016				
β_{11}															3.075 (−3.11)	0.002
β_{12}											−1.288 (−2.02)	0.044				
β_{13}			−0.789 (−3.83)	0			−1.491 (−4.61)	0							−8.084 (−2.62)	0.009

(continued)

Table 7 (continued)

	All				Pre-contemplator stage				Contemplator stage				Preparation/action stage			
	Model 1		Model 2		Model 1		Model 2		Model 1		Model 2		Model 1		Model 2	
beta	*b*	*p*	*b*	*p*	*b*	*p*	*b*	*p*	*b*	*p*	*b*	*p*	*b*	*p*	*b*	*p*
β_{14}			−0.973 (−2.24)	0.025			−1.334 (−2.51)	0.012								
N	684		684		386		386		116		116		182		182	
R^2	0.046		0.116		0.156		0.272		0.152		0.309		0.23		0.445	
Adj R^2	0.062		0.156		0.208		0.363		0.205		0.416		0.363		0.7	
-2LL	893.199		841.633		469.608		412.557		138.23		114.498		136.113		76.656	
*L**	51.66				57.01				23.732				59.457			
LR test	sig. dif. at 99% confidential level				sig. dif. at 99% confidential level				sig. dif. at 99% confidential level				sig. dif. at 99% confidential level			

Table 8 Binary choice model beta estimates for car versus metro

	All				Pre-contemplator stage				Contemplator stage				Preparation/action stage			
	Model 1		Model 2		Model 1		Model 2		Model 1		Model 2		Model 1		Model 2	
beta	*b*	*p*	*b*	*p*	*b*	*p*	*b*	*p*	*b*	*p*	*b*	*p*	*b*	*p*	*b*	*p*
ASC	−0.526 (−2.97)	0.003	−0.629 (−3.35)	0.001	−0.509 (−2.50)	0.013	−0.863 (−3.21)	0.001								
β_1									−0.115 (−1.91)	0.056	−0.122 (−1.90)	0.057				
β_2									−0.134 (−1.91)	0.056	−0.150 (−2.01)	0.044				
β_3									−1.398 (−2.04)	0.041	−1.650 (−2.25)	0.025				
β_4	0.035 (4.53)	0	0.036 (4.56)	0	0.043 (4.28)	0	0.055 (5.11)	0					0.030 (4.61)	0	0.042 (5.36)	0
β_5	0.058 (2.60)	0.009	0.048 (2.10)	0.035					1.088 (4.43)	0	1.212 (4.52)	0				
β_6	−0.142 (−4.34)	0	−0.141 (−4.25)	0	−0.235 (−4.77)	0	−0.250 (−4.63)	0	1.075 (3.39)	0.001	1.418 (3.72)	0				
β_8			0.244 (2.06)	0.039			0.332 (2.25)	0.025								
β_9															−0.885 (−3.03)	0.002
β_{11}			0.541 (2.16)	0.031												
β_{12}											−2.102 (−2.90)	0.004				

(continued)

Table 8 (continued)

	All				Pre-contemplator stage				Contemplator stage				Preparation/action stage			
	Model 1		Model 2		Model 1		Model 2		Model 1		Model 2		Model 1		Model 2	
beta	*b*	*p*	*b*	*p*	*b*	*p*	*b*	*p*	*b*	*p*	*b*	*p*	*b*	*p*	*b*	*p*
β_{13}							−0.914 (−4.34)	0								
β_{14}							0.447 (2.51)	0.012								
N	1263		1263		903		903		136		136		224		224	
R^2	0.029		0.036		0.039		0.108		0.494		0.528		0.102		0.139	
Adj R^2	0.038		0.048		0.052		0.145		0.658		0.704		0.136		0.185	
-2LL	1713.202		1703.934		1207.44		1139.617		95.944		86.421		286.496		277.098	
*L**	9.268				67.823				9.523				9.398			
LR test	sig. dif. at 99% confidential level				sig. dif. at 99% confidential level				sig. dif. at 99% confidential level				sig. dif. at 99% confidential level			

5.1 Car Versus Van Pooling

The first direct finding concerns the alternative-specific constant (ASC) (Table 5). ASC reflects the difference in utility of alternative i from that of alternative j, when "all else is equal." In simple words, ASC reflects the consumer priorities when all parameters (such as travel time and travel cost) are the same for all alternatives. In this respect, the results indicate that:

- In the case where the binary choice models were run without segmenting the population by behavioral stage, ASC was not found to be statistically significant. This finding would lead policy makers to the (faulty/incorrect) decision that the working staff of the university does not give any priority to their car or to the alternative mode examined.
- The outcomes are completely different when the segmentation of the population is taken into account. The results indicate that people who have been classified in the "Pre-Contemplation" stage return a negative ASC. A negative ASC indicates a preference for the car. On the other hand, the results indicate that people belonging to the second and third behavioral stage have a strong preference for the alternative option, van pooling.

Regarding the mode-specific variables (beta 1–6), the following results have been obtained:

- Without the segmentation of the population, it was found that only car travel time and travel cost are statistically significant attributes that explain mode choice. The beta estimate for travel time is positive, meaning that an increase in travel time by car returns higher possibility to choose the alternative mode. This finding sounds plausible since the beta coefficients indicate that higher travel time for cars translates into higher utility. Regarding travel cost by car, an unexpected finding came out; the beta coefficient was negative, indicating that the increase of the out-of-pocket cost for commuting by car does not directly translate into higher possibility to choose the alternative mode proposed. This inconsistency with the theoretical expectation was commented by Hess et al. (2004). Following their suggestions, it is concluded that commuters probably related the car cost with the alternative mode travel cost since, for instance, longer trips from home to work will continue to be costly either by cars or the university van. It should be mentioned that this paradox was also observed for the other alternative scenarios with the exception of the carpooling scenario as it is discussed below.
- The findings become plausible when the population is segmented into behavioral stages. The only mode-specific parameter found to have a statistically significant role in the mode choice interpretation is travel time by car for those who are classified in the "Pre-Contemplation" stage and travel time by the van for those classified in the "Contemplation" stage. The differences in the signs indicate that the increase in travel time for one mode will directly increase the potential for the preferring other mode. Finally, travel time was also the one and only mode-specific important parameter for those classified in the "Preparation/Action"

stage. However, this is the only case where generic beta estimates were used, meaning that people of this stage perceive travel time equal either for car or for public transport.

Regarding the user-specific variables, it was found that some of them have an explanatory power. For all scenarios, the likelihood ratio tests indicated an improvement of the initial model (model 1) at the 99% confidence level. To examine the validity of the model, the r-square indices of Cox and Snell (R^2) and Nagelerke (Adj R^2) were used. From the odds ratio calculation, it was found that only 12% of the commuters who do not have obligations before or after work and are classified in the "Pre-Contemplator" stage will choose the van-pooling option. The respective percentage for those located in the "Contemplator" stage is 92%.

5.2 Car Versus Collective TAXI

Following the same approach with the previous scenario, the alternative specific constant (ASC) was first interpreted. The results (Table 6) indicate that the university staff have a preference for the alternative mode (positive ASC). However, this outcome is not valid. The segmentation of the population per behavioral stage shows that only commuters who are classified in the "Preparation/Action" stage actually have this preference; commuters classified in the "Pre-Contemplation" behavioral stage and comprise the majority within the population prefer their car (negative ASC) whereas the "Contemplators" do not have any preference (ASC was not found to be statistically significant).

As far as the mode-specific parameters are concerned, travel cost by car was found to be statistically significant for the whole population and for those in the "Pre-Contemplation" behavioral stage. The beta coefficient is negative; as mentioned and commented in the previous scenario, this finding indicates that commuters tend to relate high cost commuting trips with their preference for the car. This is also related to the proposed alternative itself, the Collective Taxi, since travel cost for this alternative is highly related (in fact it is perceived as similar) to travel cost by car. On the other hand, as expected, the higher the travel time by car, the higher the possibility to choose the alternative mode for those who belong to the "Pre-Contemplation" and "Contemplation" behavioral stage (positive beta 4).

Finally, user-specific variables were found to have an explanatory power in the models. As in the previous scenario, the likelihood ratio tests indicated an improvement of the initial model at the 99% confidence level. The parameters that are statistically significant are those for gender, occupation (teaching or administrative staff), and age (compared to the reference interval category of 18–25). For instance, the odds ratio calculation revealed that 96% of the women classified in the "Preparation" behavioral stage will choose the alternative mode, whereas only 1% of them will make this selection in case they belong to the "Contemplation" behavioral stage. In addition, the odds ratio calculation for those being in the "Pre-Contemplation" behavioral stage showed that the odds for choosing the alternative mode are 2.43 times higher for the administrative staff compared to teaching staff whereas the respective odds for those in the "Contemplation" behavioral stage is increased by 77.56 times.

5.3 Car Versus Carpooling

The results shown in Table 7 indicate that car pooling is considered as the mode with the highest preference against the car since ASC is positive both for the total sample and the different segments of population. Therefore, should a policy be formulated, this should favor a scheme based on car pooling.

Regarding the mode-specific parameters, the beta estimates derived from the total sample indicate that only travel time by the alternative mode (betas 1 and 2) are considered as important in the mode choice process. The negative values indicate that the higher the carpooling travel time is, the lower the possibility to be preferred by the commuters. However, the segmentation of the population by behavioral stage shows that the variables that affect mode choice are only travel time and travel cost by car. Again, different signs in travel cost and travel time were found. For those who are in the "Pre-Contemplation" and "Contemplation" behavioral stage, it is found that they associate the high travel time by car with high travel time with the carpooling mode since the utility of the car increases (negative beta) in this case. In other words, the negative beta value indicates that those who have been classified in these stages do not prefer to travel with others in the car, especially when travel time is long, something which is not the case for those who are in the "Maintenance" stage; positive beta for travel time by car means that longer travel times make people in this stage to give preference to the alternative mode. However, the outcomes are reversed when trip cost is considered.

Finally, in this scenario user-specific variables were found to have an explanatory power. The likelihood ratio tests indicate an improvement of the initial model at the 99% confidence level. A number of parameters are found to be statistically significant such as those for gender, occupation (teaching or administrative staff), and various age levels (compared to the reference interval category of 18–25). From the odds ratio calculation, it was found that people aged between 18 and 25 who are in the "Pre-Contemplation" behavioral stage are 77% likely to choose the alternative travel mode; the respective percentage for this age interval for those in the "Preparation" behavioral stage is 99%.

5.4 Car Versus Metro

The last scenario was that of car versus the metro. This alternative, although it refers to a transport mode with completely different characteristics compared to the other three alternatives (hard measure versus soft measures), was included in the analysis given that it constitutes an alternative that will be available in the future.

The results presented in Table 8 indicate that ASC is negative for the total sample and for models derived from the “Pre-Contemplation” behavioral stage. ASC was not found to be statistically significant for the other two behavioral stages. As already mentioned, the negative ASC means that people give a strong preference for the car when all other factors are equal.

Regarding mode-specific variables, the variables related to the car are those with the higher explanatory interest and statistical significance. Travel time by car was found to have a positive beta coefficient for most models, meaning that those who make long trips are in favor of the metro. Regarding travel cost by car, the beta coefficients are negative for the total sample and for the models of the “Pre-Contemplation” stage, which reveals that the costly trips are in favor of the car. However, the findings for those in the “Contemplation” stage were different since the beta coefficients were found to be positive, meaning that costly trips by car could make “Contemplators” change their commuting mode.

Finally, user-specific variables were found to have an explanatory power, however with no success and power as it was found in the previous scenarios. The likelihood ratio tests indicate an improvement of the initial model at the 99% confidence level with a small likelihood difference (L^*) between models 1 and 2 around the value of 9.

6 Conclusions

Mobility management through soft measures is a growing trend in traditional transportation planning policy. Since public space and road capacity are getting limited over the years, policy makers and stakeholders invent new ways to relieve the transport network and the environment. Alternative approaches to the conventional costly transport infrastructure investments, such as highways, should focus primarily on human behavior change which can be attained with soft measures.

In this chapter, a hypothetical mobility management scheme in a university campus was examined based on the probability of university commuters to change commuting mode. The probability to select an alternative mode out of four proposed was estimated not only with traditional modeling prediction techniques, but also by employing the notion of behavioral change stage from Social Psychology. The MAXSEM model, developed in the context of a European research project, recognizes four successive behavioral change stages and it was used for the purposes of this research. The employment of the behavioral change stage in the binary choice models and the allocation of the sample to different stages based on specific psychometric questions returned quite interesting results which in most cases were different from those obtained when the total sample was used.

The main hypothesis examined when comparing the models using the total sample with those taking into consideration the behavioral change stage was that the majority of the affected population may not always be positive or ready to change existing mode choices and behavior and, therefore, the decision-making process should take into consideration the segmentation of this population by behavioral change stage. The results confirmed this hypothesis. Population segments per behavioral stage do have completely different preferences and priorities related to car use and to alternative travel modes. In all scenarios examined, it was found that, indeed, those in the "Maintenance" behavioral stage have high interest in and potential to change their current habits and travel pattern compared to those being in the "pre-contemplation" behavioral stage who are extremely difficult to abandon the private car for an alternative travel mode.

The present study reveals that transport research can really benefit from other scientific disciplines such as Social Psychology which studies individual attitudes, preferences, and behavior. The research findings of this study demonstrate that the right approach can help decision and policy makers in the transport sector to proceed to targeted interventions and maximize positive impacts. Cost-benefit analysis, as an additional exercise, will demonstrate the additional benefits of the disaggregate approach compared to the conventional aggregate approach in travel mode choice studies.

References

Ajzen, I. (1985). From intentions to actions: A theory of planned behavior. In J. Kuhl & J. Beckmann (Eds.), *Action control: From cognition to behavior*. New York: Springer.

Akers, R., Krohn, M., Lanza-Kaduce, L., & Radosevish, M. (1989). Social learning and deviant behavior: A specific test of a general theory. *American Sociological Review, 44*(4), 636–655.

Atherton, A., & Giurco, D. (2011). Campus sustainability: Climate change, transport and paper reduction. *International Journal of Sustainability in Higher Education, 12*(3), 269–279.

Attard M., Bajada T., & Gauci P. (2014) *Green Transport Plan for the Msida Campus-University of Malta*.

Baboulet, O., & Lenzen, M. (2010). Evaluating the environmental performance of a university. *Journal of Cleaner Production, 18*(12), 1134–1141.

Balcetis, E., & Dunning, D. (2007). Cognitive dissonance and the perception of natural environments. *Psychological Science, 18*(10), 917–921.

Balsas, C. J. (2003). Sustainable transportation planning on college campuses. *Transport Policy, 10* (1), 35–49.

Bandura, A. (1977). *Social learning theory*. New York: General Learning Press.

Ben-Akiva, M. E., & Lerman, S. R. (1985). *Discrete choice analysis: Theory and application to travel demand*. Cambridge, MA: MIT Press.

Black, J., Mason, C., & Stanley, K. (1999). Travel Demand Management: Policy context and an application by the University of New South Wales (UNSW) as a Large Trip Generator. *Transport Engineering in Australia, 5*(2), 1–11.

Black D., Sloan-Rossiter S., Ng A.,Guidoboni E., & Cranshaw W. (2011, December). *Campus Transportation Plan, University of Massachusetts Lowell*. Prepared by Vanase Hangen Brustlin Inc for the University of Massachusetts Lowell.

Boer, H., & Seydel, E. R. (1996). Protection motivation theory. In M. Connor & P. Norman (Eds.), *Predicting health behavior*. Buckingham: Open University Press.

Bond, A., & Steiner, R. (2006). Sustainable campus transportation through transit partnership and transportation demand management: A case study from the University of Florida. *Berkeley Planning Journal, 19*(1), 125–142.
Brown, J., Hess, D. B., & Shoup, D. (2003). Fare-free public transit at universities an evaluation. *Journal of Planning Education and Research, 23*(1), 69–82.
Chen, M., & Land, K. C. (1986). Testing the Health Belief Model: LISREL analysis of alternative models of causal relationships between health beliefs and preventive dental behavior. *Social Psychology Quarterly, 49*(1), 45–60.
Cohen, L., Manion, L., Morrison, K., & Wyse, D. (2010). *A guide to teaching practice*. New York: Routledge.
Higgins, R., & Collins, S. (2015). *Reflections on winning the smarter travel campus of the year award*. In: Proceedings of the ITRN2015, 27-28th August 2015, NUI Galway.
Conway, T. M., Dalton, C., Loo, J., & Benakoun, L. (2008). Developing ecological footprint scenarios on university campuses: A case study of the University of Toronto at Mississauga. *International Journal of Sustainability in Higher Education, 9*(1), 4–20.
Daggett, J., & Gutkowski, R. (2003). University transportation survey: Transportation in university communities. *Transportation Research Record: Journal of the Transportation Research Board, 1835*(1), 42–49.
Daniels, R., & Mulley, C. (2013). The paradox of public transport peak spreading: Universities and travel demand management. *International Journal of Sustainable Transportation, 7*(2), 143–165.
Elton, L. (2003). Dissemination of innovations in higher education. *Tertiary Education and Management, 9*, 199–214.
Finlay, J., & Massey, J. (2012). Eco-campus: Applying the ecocity model to develop green university and college campuses. *International Journal of Sustainability in Higher Education, 13*(2), 150–165.
Golledge, R., & Garling, T. (2007). *Cognitive maps and urban travel.* Retrieved January 21, 2014, from U.C. Transportation Center: http://www.uctc.net/papers/601.pdf
HDR Engineering. (2012, July). *Campus Mobility Plan-final report.* Retrieved December 2, 2015, from http://www2.acs.ncsu.edu/trans/about/documents/mobility-plan2012.pdf
Hess, S., Bierlaire, M., & Polak, J. (2004). *Estimation of value-of-time using Mixed Logit models* (No. TRANSP-OR-REPORT-2004-006).
Klockner, C. A., & Matthies, E. (2004). How habits interfere with norm-directed behaviour: A normative decision-making model for travel mode choice. *Journal of Environmental Psychology, 24*, 319–327.
Longo, G., Medeossi, G., & Padoano, E. (2015). Multi-criteria analysis to support mobility management at a University Campus. *Transportation Research Procedia, 5*, 175–185.
Louviere, J. J., Hensher, D. A., & Swait, J. (2000). *Stated choice methods*. Cambridge: Cambridge University Press.
MacDonald, R. A., & MacDonald, R. A. (2001). *Sustainable transportation and land use planning at Simon Fraser University: A case study of the Burnaby Mountain Campus* (No. MQ-51404 UMI). Simon Fraser University.
Maguire, M., Morgan, R., & Reiner, R. (2007). *The Oxford handbook of criminology* (4th ed.). Oxford: Oxford University Press.
MAX. (2007a). *Successful travel awareness campaigns and mobility management strategies*. Comprehensive State of the Art Report, Annex B1.1, Behaviour Change Models, European Commission, 6th Framework Programme, Integrated Transport Project 518368, Brussels.
MAX. (2007b). *Successful travel awareness campaigns and mobility management strategies*. Comprehensive State of the Art Report, Annex B1.2, Assessment Tools, European Commission, 6th Framework Programme, Integrated Transport Project 518368, Brussels.
MAX. (2009). *MaxSem: Max Self Regulation Model: Applying theory to the design and evaluation of Mobility Management Projects*. European Union: Max Partners.
McDonald, M., Hall, R., Schreffler, E., Gilliard, E., Sammer, G., Roider, O., & Klementschit, R. (2010, January). Cluster Report 5: Mobility Management, Deliverable D2.2, Civitas-Guard Project.

Nelson/Nygaard Consulting. (2012, January). *Campus Transportation Demand Management Plan.*
NIH. (2003). *NIH Guide. Maintenance of long term behavioral change.* Retrieved January 21, 2015, from http://grants.nih.gov/grants/guide/rfa-files/RFA-OB-03-003.html
Nutbeam, D. (2000). Health literacy as a public health goal: A challenge for contemporary health education and communication strategies into the 21st century. *Health Promotion International, 15*(3), 259–267.
Pitsiava-Latinopoulou, M., Basbas, S., & Gavanas, N. (2013). Implementation of alterative transport networks in university campuses: The case of Aristotle University of Thessaloniki, Greece. *International Journal in Higher Education, 14*(3), 310–323.
Politis, I., Papaioannou, P., & Basbas, S. (2012). Integrated choice and latent variable models for evaluating flexible transport mode choice. *Research in Transportation Business & Management, 3*, 24–38.
Prochaska, J. O., & DiClemente, C. C. (1983). Stages and processes of self-change of smoking: Toward an integrative model of change. *Journal of Consulting and Clinical Psychology, 51*, 390–395.
Prochaska, J. O., & Norcross, J. C. (2001). Stages of change. *Psychotherapy: Theory, Research, Practice, Training, 38*(4), 443.
PTA - Thessaloniki Public Transport Authority, Socioeconomic evaluation of the metro extention to the city of Kalamaria, Final Report, 2011.
Rose, G. (2008). Encouraging sustainable campus travel: Self-reported impacts of a university TravelSmart initiative. *Journal of Public Transportation, 11*(1), 85.
Rosenstock, I., Strecher, V., & Becker, M. (1994). The Health Belief Model and HIV risk behaviour change. In R. J. DiClemente & J. L. Peterson (Eds.), *Preventing AIDS: Theories and methods of behavioral interventions* (pp. 5–24). New York: Plenum Press.
Schlag, B., & Schade, J. (2010). Traffic and transportation psychology. In K. Button & P. Nijkamp (Eds.), *Transport dictionary*. Cheltenham: Edward Elgar Publishers.
Shaheen, S., Rodier, C. J., & Finson, R. S. (2003). University of California, Davis Long-Range Development Plan: A Davis Smart Mobility Model. California Partners for Advanced Transit and Highways (PATH).
Shannon, T., Giles-Corti, B., Pikora, T., Bulsara, M., Shilton, T., & Bull, F. (2006). Active commuting in a university setting: Assessing commuting habits and potential for modal change. *Transport Policy, 13*(3), 240–253.
Shiftan, Y. (2008). The use of activity-based modeling to analyze the effect of land-use policies on travel behavior. *Annals of Regional Science, 42*, 79–97.
Silva, P., & Ferreira, D. (2008, September). European best practice on sustainable mobility in university campus, T.a.T –Students Today, Citizens Tomorrow Project.
TAPESTRY. (2007). *TAPESTRY - travel awareness, publicity and education supporting a sustainable transport strategy in Europe*. European Commission, 5th Framework Programme, Brussels.
Toor, W., & Havlick, S. (2004). *Transportation and sustainable campus communities: Issues, examples, solutions*. Washington, DC: Island Press.
Train, K. (2003). *Discrete choice methods with simulation*. Cambridge: Cambridge University Press.
University of Essex. (2015). *Travel Plan-Estate Management Section.* Retrieved December 2, 2015, from http://www.essex.ac.uk/transport/documents/transport_sustainability.pdf
University of Liverpool John Moores. (2014, August). Green Travel Plans. Retrieved December 2, 2014, from http://www.ljmu.ac.uk/Sustainability/Sustainability_docs/LJMU_Green_Travel_Plan_August_2014.pdf
University of Warwick. (2007, June). *Main Campus Master Plan, Travel Plan and Car Management Strategy*. Retrieved December 2, 2015, from http://www2.warwick.ac.uk/about/campusdevelopment/application/6_uow_travel_plan_complete_document.pdf

Developing a Spatial Transferability Platform to Analyze National-Level Impacts of Connected Automated Vehicles

Ramin Shabanpour, Nima Golshani, Thomas S. Stephens, Joshua Auld, and Abolfazl Mohammadian

Abstract A recent application of the spatial transferability approach is to assess the potential impacts of the emerging connected automated mobility technology on people's travel behavior at the national level. While there are a few transportation simulation frameworks which can account for potential impacts of this technology in a simulated geographical context, there is yet to be any literature documenting disaggregated estimates of large-scale impacts of connected automated vehicles (CAVs) on travel behavior at the national level. Therefore, in order to provide a platform to assess national-level impacts of CAVs, this study develops a methodological framework based on transferability techniques, which uses data and models from a smaller geographical area—the POLARIS simulation results for the CAVs scenario in the Chicago metropolitan area—to generate disaggregate travel data at the national level. Comparison of the distributions of the transferred variables at the regional and the national contexts indicates that the platform is capable of transferring travel behavior indices to the national level with high level of accuracy.

1 Introduction

Spatial transferability of travel demand models refers to the application of models that are originally developed with data from one geographical region for forecasting purposes in another region. This approach has been frequently used in small-sized regions for the sake of cost and time savings, or in contexts where the required data for model development is not available. A recent application of the spatial transferability approach is to assess the potential impacts of the emerging connected

R. Shabanpour (✉) · N. Golshani · A. Mohammadian
Department of Civil and Materials Engineering, University of Illinois at Chicago, Chicago, IL, USA
e-mail: rshaba4@uic.edu; ngolsh2@uic.edu; kouros@uic.edu

T. S. Stephens · J. Auld
Argonne National Laboratory, Argonne, IL, USA
e-mail: tstephens@anl.gov; jauld@anl.gov

H. Briassoulis et al. (eds.), *The Practice of Spatial Analysis*,
https://doi.org/10.1007/978-3-319-89806-3_12

automated mobility technology on people's travel behavior at the national level. While connected autonomous vehicles (CAVs) are expected to have a profound impact on both people's travel behavior and transportation systems operation, their large-scale impacts at the national level are still unknown.

Research on the potential impacts of CAVs has been quickly maturing over the past few years; however, only very few studies have attempted to assess how their emergence would affect different aspects of travel behavior such as trip rate, travel time, and travel mode. Due to lack of experimental data on CAV use, these efforts are also limited to developing theoretical frameworks and speculation of behavioral changes. More specifically, while few researchers have attempted to analyze large-scale energy impacts of CAVs technology (see, for example, Brown et al. 2014; Stephens et al. 2016), there is yet to be any literature published on large-scale, disaggregate travel demand models that account for CAVs in the modeling procedure.

The current study aims to contribute to the literature by developing a methodological framework for assessing national-level impacts of CAV technology on people's travel behavior. The framework is based on spatial transferability techniques that use data and models from a smaller geographical area (i.e., the Chicago metropolitan area) and generate disaggregate data at the national level. In doing so, an advanced transportation systems simulation model, POLARIS (Auld et al. 2016, 2017), is employed to estimate the potential impacts of CAV technology both on travel demand and traffic flows in the Chicago metropolitan area. Currently, the main components of POLARIS are being modified to account for several potential CAVs impacts and for analyzing various related scenarios. The rich output of the CAV scenario analysis in the POLARIS framework includes information on individuals' detailed activity-travel patterns in the presence of CAV technology. Applying multiple statistical and data mining techniques, the simulation results are then transferred to the national-level context to evaluate the large-scale impacts of CAV technology. The general structure of the modeling framework is outlined in Fig. 1. In this framework, the baseline regional travel information is obtained from the POLARIS platform (Auld et al. 2016, 2017).

In the proposed framework, transferable variables—daily trip rate and travel time—are derived from POLARIS simulation results. A clustering algorithm is then applied to assign people into homogeneous groups through which various lifestyles are captured, followed by estimating joint models of the number of daily trips and total travel time within each cluster. Finally, using a Neural Network model, cluster membership rules are transferred to the national-level data (NHTS) and the estimated joint models are simulated within the corresponding clusters to develop a baseline national platform for analyzing CAV scenarios.

The performance of the proposed transferring approach is evaluated by comparing the distributions of target variables (total daily trip rates and travel time) in the regional and national context. The platform is found to be capable of transferring travel behavior indices to the national level with high level of accuracy. It is worth noting that the current study does not aim to evaluate the national-level implications of CAV technology, but rather, to develop a holistic framework which can accurately transfer the results of regional-level CAV scenarios to the national level. Indeed, the aim is to deliver a platform which can be utilized to test various policies

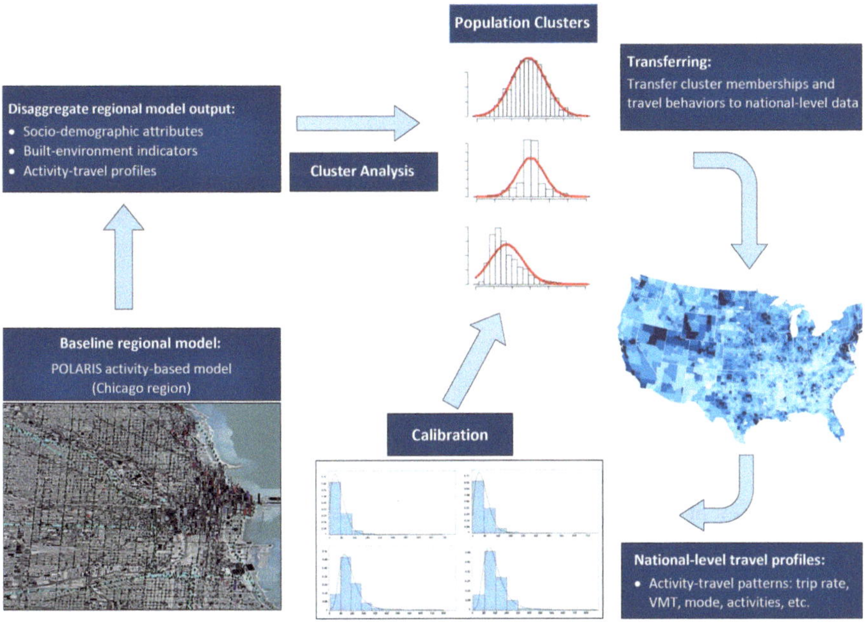

Fig. 1 Regional to national model transferability for CAV analysis (adapted from Shabanpour et al. 2017a)

and scenarios that may affect the use of CAV and their national impacts. Although the main motivation for developing such a transferability framework is to assess the national impacts of CAV scenarios, it can be used for many other types of national-level analyses.

2 Regional-Level Analysis

In this section, we elaborate on the baseline regional travel model and the simulation results. The baseline regional travel information is obtained from the POLARIS activity-based travel demand and network simulation framework (Auld et al. 2016, 2017), which is briefly introduced in the following.

2.1 *The POLARIS Framework*

The POLARIS modeling suite is an open-source agent-based modeling platform specifically designed for simulating large-scale transportation systems. The suite includes a discrete-event engine, memory management, visualization capabilities,

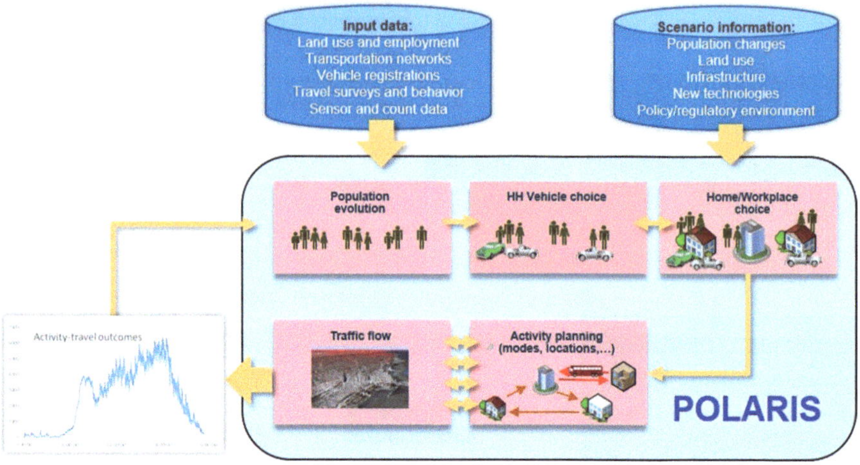

Fig. 2 Overview of POLARIS modeling framework (adapted from Shabanpour et al. 2017a)

and an extensible repository of common transportation objects which enables the rapid development and extension of integrated transportation simulations. The platform has been used to successfully simulate ITS interactions with regional demand, statewide long-distance passenger travel, and evacuations (Auld et al. 2016). POLARIS is designed as a continuously integrated activity-based model and network supply model, where individuals plan and schedule their activities dynamically, engage in simulated travel, and replan activities on the fly due to changing traffic conditions, new information or external control. The POLARIS model is able to generate disaggregate output, including individuals' record of daily activities and travels, travel mode and other choices, along with socio-demographics, land-use, and built-environment attributes. An overview of the POLARIS simulation framework is shown in Fig. 2.

The dynamic, integrated nature of POLARIS means that it is well suited for simulating vehicle connectivity and automation and their impact on individual travel behavior. Case studies for CAV impacts are currently being developed for the Chicago region. The current work, however, is focused on transferring baseline model results. For more information about the implementation of CAV scenarios in the POLARIS framework, the reader is referred to the authors' previous study (Auld et al. 2017). Disaggregate records from the POLARIS are analyzed and the household-level and individual-level attributes are derived through data mining techniques. Several land-use variables are also constructed to reflect built-environment settings. The extracted variables are as follows:

Demographic Information

Several demographic variables are derived from the POLARIS population synthesizer and are reformatted in accordance with the data types in the National Household Travel Survey (NHTS) database, which is the database that is used for the national-level analysis. These variables include age, gender, marital status, education level, job categories, and ethnicity. Several household-level variables are also generated using individual-level variables. These variables include the number of household members (of various age, race, education levels, and occupations). Other household-level variables such as household size, income, number of adults, vehicle ownership, and number of workers are directly derived from the population synthesizer.

Built-Environment Variables

Multiple land-use and built-environment indicators are also constructed to be used in the modeling process. These variables include housing density (number of housing units per square mile), employment density (number of employment opportunities per square mile), and population density (number of people per square mile), road density (total lengths of roadways per square mile), and block size (total length of roadways divided by the number of intersections in a tract).

Dependent Variables

In this study, two of the most important dimensions of individuals' travel behavior, total number of daily trips and daily travel time, are considered. Summary statistics of the variables derived from the population synthesizer and simulation results are presented in Table 1.

2.2 Population Segmentation

There has been strong evidence from the literature that travelers' lifestyle, attitudes, and perceptions significantly affect their travel behavior (Paulssen et al. 2014; Van Acker et al. 2016; Shabanpour et al. 2017b). Subsequently, travelers in a heterogeneous population may respond differently toward transport policies. Cluster analysis is one of the most widely used approaches to deal with the heterogeneity of travelers in terms of decision-making criteria toward various choices. In this study, a distance-based clustering method is used which classifies observations into relatively homogeneous collections by minimizing the variance across variables of interest within clusters and maximizing the variance between clusters (Li et al. 2013).

Table 1 Description of variables and summary statistics

Variable	Variable level	Mean	St. dev.
Age	–	45.13	24.15
HH size	–	2.91	1.44
HH vehicle	–	1.92	1.02
HH license	–	1.93	0.84
HH worker	–	1.45	0.94
Intersection density	–	161.10	88.94
Road density	–	502.89	544.74
Block size	–	2.43	1.86
Gender	Male	0.47	–
	Female	0.53	–
Education level	Less than high school	0.23	–
	High school	0.18	–
	Some college or Associate's degree	0.21	–
	University degree	0.38	–
Race	White	0.73	–
	African American	0.15	–
	Hispanic	0.08	–
	Asian	0.03	–
	American Indian, Alaskan Native	0.01	–
Job category	Sales/service	0.61	–
	Clerical/admin support	0.06	–
	Manufacturing/construction/maintenance	0.14	–
	Professional/managerial	0.19	–
Home type	House detached from any other house	0.75	–
	Duplex/townhouse	0.07	–
	Condominium/apartment	0.18	–
Home ownership	Own	0.86	–
	Rent	0.14	–
HH income	Less than $19,999	0.09	–
	$20,000–$34,999	0.10	–
	$35,000–$49,999	0.13	–
	$50,000–$59,999	0.08	–
	$60,000–$74,999	0.13	–
	$75,000–$99,999	0.18	–
	Greater than $100,000	0.29	–
Population density	Less than 99	0.02	–
	100–499	0.08	–
	500–999	0.04	–
	1000–1999	0.13	–
	2000–3999	0.19	–
	4000–9999	0.30	–
	10,000–24,999	0.18	–
	Greater than 25,000	0.06	–

(continued)

Table 1 (continued)

Variable	Variable level	Mean	St. dev.
Housing density	Less than 99	0.05	–
	100–499	0.12	–
	500–999	0.14	–
	1000–1999	0.23	–
	2000–3999	0.21	–
	4000–9999	0.16	–
	10,000–24,999	0.07	–
	Greater than 25,000	0.02	–

The clustering schema, however, is a critical issue. Theoretically, the further the population and the dependent variables are subdivided, the better homogeneous behavior can be observed within each cluster. However, the large number of clusters are difficult to work with. From the several types of available clustering methods that have been proposed in the literature, such as *K*-means, hierarchical, and two-step algorithms, this study applies the two-step clustering method (Chiu et al. 2001). It is argued that this algorithm converges faster than other techniques, specifically when dealing with large datasets, and it can simultaneously deal with both discrete and continuous variables (Chiu et al. 2001). Furthermore, one of the best features of this algorithm is that it can determine the optimal number of clusters on its own—a critical issue in other algorithms where the optimal number of clusters has to be determined separately using techniques such as the Gap Statistic.

The two-step clustering algorithm first divides the dataset into several sub-clusters based on the density of data points. In this step, the algorithm calculates the Bayesian information criterion for any data point and decides whether it should be merged with a previously formed cluster, or if it should remain as a new cluster. In the second step, the algorithm changes the criterion to a distance-based measure and combines the sub-clusters if the jump in the measure is greater than a specified threshold. The distance between any pair of clusters ($\mathrm{dist}_{\mathrm{ab}}$) can be written as (Chiu et al. 2001; Mohammadian and Zhang 2007):

$$\mathrm{dist}_{<\mathrm{a,b}>} = \eta_{\mathrm{a}} + \eta_{\mathrm{b}} - \eta_{<\mathrm{a,b}>} \tag{1}$$

where

$$\eta_{\upsilon} = -I\left[\sum_{c=1}^{C} \frac{1}{2}\log\left(\widehat{\sigma}_c^2 + \widehat{\sigma}_{\upsilon c}^2\right) + \sum_{d=1}^{D} \widehat{E}_{\upsilon d}\right] \tag{2}$$

$$\widehat{E}_{\upsilon d} = -\sum_{l=1}^{L_d} \frac{I_{\upsilon dl}}{I_{\upsilon}} \log \frac{I_{\upsilon dl}}{I_{\upsilon}} \tag{3}$$

where

I is the total number of observations in the dataset,
C and D are the total number of continuous and discrete variables, respectively,
$\widehat{\sigma}_c^2$ and $\widehat{\sigma}_{\upsilon c}^2$ are the estimated variances of the cth continuous variables in the dataset and in cluster υ, respectively,
L_d is the total number of categories for the dth discrete variable,
I_υ is the total number of observations in cluster υ, and
$I_{\upsilon dl}$ is the total number of observations in cluster υ that their dth categorical variable has the lth category.

Using the variables obtained through post-processing of the POLARIS simulation output, the two-step clustering method is applied with a 10% noise allowance to categorize a sample of 24,000 randomly selected individuals in the Chicago region into homogeneous clusters. The optimal number of clusters is determined using the Bayesian information criterion (BIC). At the end, the records are assigned to 11 clusters, each containing between 4.14 and 15.07% of the dataset. These results can be informative in distinguishing various lifestyles that are assumed to influence travelers' behavior. Table 2 presents brief summary statistics of the population clusters.

2.3 Joint Estimation of Trip Rate and Travel Time

As previously discussed, in each of the regional-level population clusters, a joint model of daily trip rates and travel time is estimated based on the POLARIS simulation results. Since the first component of the joint structure, the total number of daily trips, has an ordinal nature, the ordered probit model is applied to estimate this variable as:

$$\begin{aligned} U_i &= \beta X_i + \varepsilon_i \\ y_i &= j \quad \text{if } \mu_{j-1} < U_i < \mu_j \end{aligned} \tag{4}$$

where

U_i is an unobserved generalized utility function of the number of trips for observation i,
X_i is a vector of exploratory variables,
β is the vector of estimable parameters,
ε_i is the random error term that is assumed to be normally distributed with the cumulative density function of $\Phi(.)$,
j is the integer ordering that corresponds to y_i,
μ_j is the threshold that separates j and $j + 1$ categories, and
J is the total number of categories.

Table 2 Summary statistics[a] of population clusters at the regional-level analysis

Variables	Population clusters										
	Cluster 1	Cluster 2	Cluster 3	Cluster 4	Cluster 5	Cluster 6	Cluster 7	Cluster 8	Cluster 9	Cluster 10	Cluster 11
Gender	Female	Female	Female	Female	Female	Female	Female	Female	Female	Female	Female
Age	45.40	45.58	44.89	45.65	43.91	45.18	45.13	45.35	45.68	43.75	44.56
Education level	University degree	University degree	University degree	University degree	University degree	University degree	University degree	University degree	University degree	University degree	University degree
Job category	Professional[b]	Sale, service	Sale, service	Professional	Sale, service	Professional	Professional	Professional	Professional	Professional	Professional
Race	White	White	White	White	White	White	White	White	White	White	White
Home type	House[c]	House	House	House	House	Apartment, condominium	Apartment, condominium	House	Apartment, condominium	House	House
Home ownership	Own	Own	Own	Own	Own	Rent	Rent	Own	Own	Own	Own
HH income	>$100k	>$100k	>$100k	>$100k	$35k–$50k	<$20k	<$20k	>$100k	>$100k	>$100k	>$100k
HH size	2.85	4.02	3.13	2.96	1.63	1.72	1.93	3.21	2.84	4.24	3.66
HH vehicle	2.16	3.19	2.46	2.10	1.09	1.01	0.68	2.01	1.42	2.94	2.10
HH license	2.00	3.30	2.15	2.00	0.99	0.90	0.76	2.03	1.98	3.32	2.31
HH worker	1.38	2.47	1.52	1.44	0.60	0.64	0.73	1.50	1.65	2.67	1.84
Intersection density	82.61	113.21	19.53	137.85	109.82	200.60	255.16	195.52	273.33	196.40	244.21
Road density	75.00	137.20	12.99	196.06	147.82	542.24	1258.72	508.16	1290.83	494.93	1155.76
Block size	0.90	1.17	0.64	1.45	1.30	4.62	4.95	2.63	4.65	2.54	4.73
Population density	1000–1999	2000–3999	100–499	2000–3999	1000–1999	4000–9999	10,000–24,999	4000–9999	10,000–24,999	4000–9999	10,000–24,999
Housing density	500–999	1000–1999	100–499	1000–1999	2000–3999	2000–3999	4000–9999	2000–3999	10,000–24,999	2000–3999	4000–9999
Cluster size	2549	1727	2675	3118	1847	1711	2294	3709	1869	1019	2086
%	10.3	7.0	10.8	12.6	7.5	6.9	9.3	15.0	7.6	4.1	8.4

[a]Mode for categorical variables and average for continuous variables
[b]Professional, managerial, or technical
[c]One-family house, detached from any other houses

The probability of outcome j in observation i can be written as:

$$\begin{aligned} P(Y_i = 1) &= \Phi(\mu_i - \beta X_i) \\ P(Y_i = j) &= \Phi(\mu_i - \beta X_i) - \Phi(\mu_{j-1} - \beta X_i), \quad 1 < j < J \\ P(Y_i = J) &= 1 - \Phi(\mu_{J-1} - \beta X_i) \end{aligned} \tag{5}$$

As the second component of the analysis, travel time is treated as a continuous variable with a log-normal distribution which can be formulated as:

$$\ln(T_i) = aZ_i + v_i \tag{6}$$

where

T_i is the total travel time traveled by an individual,
Z_i is a vector of exploratory variables that affect total travel time,
α is a vector of estimable parameters, and
ν_i is a stochastic error term that is assumed to have a normal distribution.

In order to capture the shared unobserved effects on the two dependent variables, a bivariate normal distribution can be imposed on the error terms of Eqs. (4) and (6) as follows:

$$\begin{pmatrix} \varepsilon_i \\ v_i \end{pmatrix} N\left[\begin{pmatrix} 0 \\ 0 \end{pmatrix}, \begin{pmatrix} \sigma_\varepsilon^2 & \rho\sigma_\varepsilon\sigma_v \\ \rho\sigma_\varepsilon\sigma_v & \sigma_v^2 \end{pmatrix}\right] \tag{7}$$

where σ_ε^2 and σ_ν^2 are$$ the variances of the normal distributions corresponding to $\varepsilon$$$and ν with ρ as the correlation coefficient between the two random error terms. To avoid unnecessary complex analysis, Greene (2012) suggests to normalize one of the variances to one. Therefore, the bivariate normal distribution can be rewritten as:

$$\begin{pmatrix} \varepsilon_i \\ v_i \end{pmatrix} N\left[\begin{pmatrix} 0 \\ 0 \end{pmatrix}, \begin{pmatrix} 1 & \rho\sigma \\ \rho\sigma & \sigma^2 \end{pmatrix}\right] \tag{8}$$

Considering the correlation between the two dependent variables, the likelihood function of the joint ordered-continuous modes is formulated as follows (Farber et al. 2014)

$$
\begin{aligned}
L = \prod_{i=1}^{I} \Bigg\{ & m_{i1}\Phi(\beta X_i - \mu_1) \\
& + \sum_{j=2}^{J-1} m_{ij} \left\{ \left[\frac{1}{\sigma T_i}\phi\left(\frac{\ln(T_i) - \alpha Z_i}{\sigma}\right)\Phi\left(\frac{\mu_j - \beta X_i - \rho\left(\frac{\ln(T_i) - \alpha Z_i}{\sigma}\right)}{\sqrt{1-\rho^2}}\right)\right] \right. \\
& \left. - \left[\frac{1}{\sigma T_i}\phi\left(\frac{\ln(T_i) - \alpha Z_i}{\sigma}\right)\Phi\left(\frac{\mu_{j+1} - \beta X_i - \rho\left(\frac{\ln(T_i) - \alpha Z_i}{\sigma}\right)}{\sqrt{1-\rho^2}}\right)\right] \right\} \\
& + m_{iJ} \left[\frac{1}{\sigma T_i}\phi\left(\frac{\ln(T_i) - \alpha Z_i}{\sigma}\right)\Phi\left(\frac{\mu_{J-1} - \beta X_i - \rho\left(\frac{\ln(T_i) - \alpha Z_i}{\sigma}\right)}{\sqrt{1-\rho^2}}\right)\right] \Bigg\}
\end{aligned} \tag{9}
$$

2.4 Estimation Results

As mentioned earlier, cluster-specific joint ordered-continuous models are developed to estimate daily trip rates and travel time for members of each cluster (11 models in total). From all the relevant variables and variable interactions that have been tested, the significant coefficients are presented in Table 3. Due to space limitations, it is not possible to present the estimation results of all 11 cluster-specific models; thus, the estimation results of four clusters are presented only as indicative examples. In the estimation results, positive signs of the coefficients in the travel time model indicate that increased in the corresponding variables directly increase the daily travel time. With respect to the ordered model, however, a positive sign indicates that increases in the corresponding variable raise the probability of the last outcome (i.e., 11+ trips) and, consequently, lower the probability of the first outcome (i.e., 0 trips).

Table 3 reveals that a wide range of determinants including personal demographic information (e.g., age, employment status, education level), household-related variables (e.g., number of vehicles, household size, number of household workers), and built-environment characteristics (e.g., population density, road density, housing density) significantly affect individuals' daily trip rates and total travel time. The results also indicate that the sign, magnitude, and significance level of the estimated parameters vary across clusters. For instance, household size is positively associated with individuals' total daily travel time in cluster 5, while it reduces the total travel time in cluster 4. This heterogeneous behavior can be the result of differences in

Table 3 Estimation results of joint model of daily trip rate and travel time

	Cluster 1		Cluster 4		Cluster 5		Cluster 11	
Variable	Parameter	*t*-stat	Parameter	*t*-stat	Parameter	*t*-stat	Parameter	*t*-stat
Travel time								
Constant	5.48***	10.81	1.40***	25.63	5.03***	7.83	2.69***	12.46
Employment status: unemployed	−1.20***	−11.39	−0.98***	−8.76	−0.71***	−6.31	−0.84***	−5.09
Job category: professional	–	–	–	–	0.16*	1.78	–	–
Education: less than high school	–	–	–	–	0.15**	2.07	0.24***	2.81
Race: white	−0.14*	−1.95	–	–	0.23**	2.26	–	–
Income: more than $75k	–	–	0.48***	3.54	–	–	0.33***	4.07
HH vehicle: none	0.32***	3.88	0.35***	2.89	–	–	–	–
HH size	–	–	−0.20***	−4.00	0.10***	2.98	0.13**	2.16
HH size: greater than 3	–	–	–	–	−0.28**	−2.19	–	–
HH worker: greater than 2	–	–	–	–	0.41**	2.14	0.21*	1.93
Home-work distance: more than 10 mi	0.64***	5.08	0.95***	8.12	0.55***	4.21	0.46***	7.03
Housing density: greater than 4000	−1.25***	−2.64	–	–	–	–	–	–
Block size	−0.06*	−1.94	−0.19**	−2.51	–	–	–	–
Road density: more than 300	–	–	0.18***	2.63	–	–	0.55**	2.23
Intersection density	–	–	–	–	−0.19*	−1.69	–	–
Trip rate								
Constant	1.47***	12.56	1.49***	14.17	0.78***	3.25	0.91***	8.66
Age: less than 18	−0.07**	−1.96	–	–	–	–	–	–
Employment status: unemployed	−0.55***	−8.01	−0.55***	−8.21	−0.47***	−5.91	−0.28***	−5.07
Job category: sale/service	0.09**	2.14	–	–	–	–	0.12*	1.86
Job category: clerical/admin support	–	–	–	–	−0.22***	−2.64	–	–
Education: university degree	−0.06**	−2.23	–	–	–	–	–	–
Race: African American	–	–	–	–	−0.14*	−1.65	−0.23***	3.05
Income: more than $75k	–	–	0.21***	2.99	0.20***	4.39	–	–

Income: more than $100k	0.16*	1.96	–	–	–	–	−0.08***	−3.29
HH vehicle	0.08***	2.91	0.05**	2.36	0.09***	3.14	0.11*	1.66
HH size	–	–	−0.09***	−3.14	−0.08***	−3.29	–	–
HH size: greater than 3	0.13**	2.31	–	–	–	–	–	–
HH worker	−0.06**	−2.15	−0.08***	−2.69	–	–	−0.17***	−3.05
Home type: house	0.08**	2.32	0.06**	2.15	–	–	–	–
Home-work distance: less than 1 mi	0.14***	3.77	0.12***	2.86	0.32***	6.61	–	–
Block size	−0.07***	−4.38	−0.10***	−2.66	–	–	−0.21***	−2.89
Road density: more than 300	–	–	–	–	0.24*	1.67	0.17***	3.33
Intersection density: more than 170	–	–	–	–	0.45***	2.66	–	–
Thresholds parameters								
μ_2	0.30***	6.46	0.20***	5.05	0.18***	2.60	0.29***	7.71
μ_3	1.16***	22.45	1.02***	21.34	1.40***	15.79	1.51***	14.37
μ_4	1.38***	25.97	1.23***	25.35	1.66***	18.56	1.79***	24.09
μ_5	1.66***	29.57	1.51***	29.75	2.00***	21.72	1.16***	33.94
μ_6	1.83***	31.26	1.68***	31.98	2.18***	23.21	2.08***	16.96
μ_7	2.01***	32.54	1.88***	34.09	2.41***	24.76	2.92***	18.24
μ_8	2.16***	33.24	2.01***	35.07	2.61***	25.79	1.88***	31.57
μ_9	2.33***	33.38	2.17***	35.79	2.77***	26.36	2.50***	30.37
μ_{10}	2.44***	33.04	2.29***	35.92	2.97***	26.58	3.16***	34.00
Model specification								
σ: standard deviation of the error term	2.03***	47.34	1.99***	49.17	1.51***	39.47	1.84***	37.54
ρ: correlation coefficient	0.88***	111.38	0.89***	137.40	0.82***	56.62	0.83***	40.99
Log-likelihood at convergence	−5421		−5920		−3246		−4864	

Note: * significant at 90%, ** significant at 95%, *** significant at 99%

common attributes of individuals in the two clusters. From Table 2, it is evident that cluster 4 comprises households with more licensed drivers and vehicles compared to cluster 5; this can be interpreted as a higher chance of household members sharing out-of-home responsibilities, which may reduce the total daily travel time of individuals (Shabanpour et al. 2018).

Table 3 also indicates that higher distances between home and work locations increase the total daily travel time in all clusters. Also, we found that unemployed individuals are associated with lower trip rates and travel times. Furthermore, the correlation coefficient of the error terms in the joint models is statistically significant which supports the adoption of the joint modeling approach. This is a significant finding which implies the existence of unobserved shared factors that influence both daily trip rates and total travel time.

3 National-Level Analysis

Thus far, the target variables for each population cluster in the regional context have been estimated. This section is devoted to describing how the regional results can be transferred to the national level. The last sub-section will compare the results of origin and destination contexts, i.e., the regional and the national levels.

3.1 National Data Source

The National Household Travel Survey (NHTS) 2009 (FHWA 2009) was used to obtain national-level population information. Sponsored by the Bureau of Transportation Statistics (BTS) and the Federal Highway Administration (FHWA), NHTS is the nation's inventory of daily and long-distance travel. The dataset includes socio-demographic characteristics of households and persons, and detailed information on daily and longer-distance travel for all trip purposes and by all modes of travel. The dataset includes information of 308,901 individuals and almost 382,000 trips. Out of the 150,147 valid households in the final 2009 NHTS dataset, 25,510 households are in the national sample, while the remaining 124,637 households are from the 20 add-on areas. The 2009 NHTS data reflect average travel behavior in the nation and have been designed to provide statistically valid estimates of trip rates and travel statistics at the national level, at the census tract level, and at the seven area-type levels (divided by population size and the availability of transit services). The dataset includes:

- Household information on the relationships among household members, home ownership and type, income, location characteristics, and other demographic information.

- Individual-level demographic characteristics in sampled households including age, education level, worker status, driver status, annual miles driven, and disabilities impacting travel.
- Information on household vehicle, including year, make, model, odometer reading (mileage accrual) and estimates of annual miles, length of vehicle ownership, and fuel costs.
- Data about drivers, including information on travel as part of work.
- Data about one-way trips taken during a designated 24-h period (the household's designated travel day), including the time the trip began and ended, length of trip, composition of the travel party, mode of transportation, purpose of the trip, and the specific vehicle used (if a household vehicle).
- Information to describe characteristics of the geographic area in which the sample household and workplace of sample persons are located.
- The number of transit, walk, bike, and motorcycle trips made over the previous week or month.

3.2 *Cluster Membership Transferability*

To transfer cluster membership of individuals from the regional-level dataset to the national-level dataset, a Neural Network (NN) model was employed. Neural Network models are among the most widely used machine learning algorithms, which consist of an input layer, an output layer, and a potentially unlimited number of intermediate hidden layers to improve the learning ability of the model (Golshani et al. 2018). As depicted in Fig. 3, the input layer contains a set of nodes representing the input variables (i.e., demographics and land-use information) and the output layer represents the classification results (i.e., cluster memberships). To avoid overtraining, the regional-level dataset of 24,000 individual records was partitioned to provide a randomly chosen test set of 7200 (30%). The remaining records were further partitioned into a 14,400 (60%) record training set and a 2400 (10%) record validation set to be used to test for overfitting in the training process. To assess the predictive potential of the trained NN, the test dataset was introduced to the trained network as input and compared with the actual desired response. Extensive testing efforts were undertaken to obtain the best networks with regard to prediction potential. The final network that generated the best model fit offered 95.8% accuracy on the test set (as shown in Table 4). The NN model calibrated on the regional dataset was then used to simulate cluster membership for each individual in the national-level dataset.

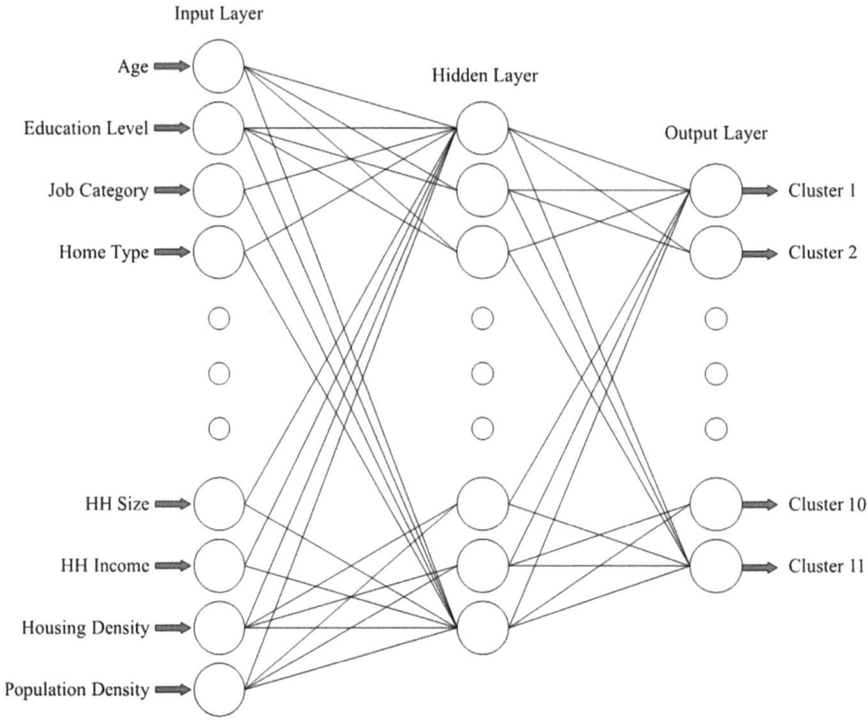

Fig. 3 Schematic design of the developed Neural Network model

3.3 Transferring Travel Statistics

By transferring the cluster membership rules to the national dataset, the national observations were assigned to 11 clusters—the same as those that are formed based on the regional dataset. The underlying assumption is that the members of each cluster have the same travel patterns in the regional and the national context. Thus, the estimated cluster-specific joint models are simulated in the national dataset to predict the total daily travel time and daily trip rates of each member in the corresponding clusters.

Following that, travel statistics of regional observations (observed data) and national observations (predicted data) are compared to validate the transferability model and evaluate its performance. Figure 4 presents total daily travel time for members of the four clusters that were considered in the last section. In each dataset, the best-fitted statistical distribution representing the trend of the target variable is also determined. Comparison of the fitted distributions in each cluster indicates that the proposed approach is able to accurately transfer individuals' total daily travel time to the national level. Furthermore, Fig. 5 presents the distributions of the total daily trip rates for members of the three clusters. Similar to the daily travel time, promising results for transferring daily trip rates were found.

Table 4 Predicted versus observed cluster memberships in the final Neural Network model[a]

Sample	Observed	Predicted											
		1	2	3	4	5	6	7	8	9	10	11	Percent correct
Training	1	**1750**	29	1	1	12	0	0	0	0	0	0	**97.6**
	2	0	**1191**	5	4	5	0	0	0	0	0	0	**98.8**
	3	0	12	**1863**	0	7	0	0	0	0	0	0	**99.0**
	4	12	6	0	**2166**	9	0	0	0	0	0	0	**98.8**
	5	14	1	4	21	**1233**	14	0	0	0	0	0	**95.8**
	6	0	2	0	0	36	**1129**	19	25	0	1	0	**93.2**
	7	0	0	0	0	3	9	**1527**	0	20	0	10	**97.3**
	8	0	0	0	9	0	13	0	**2526**	0	78	3	**96.1**
	9	0	0	0	0	0	0	46	0	**1238**	0	33	**94.0**
	10	0	6	0	0	0	1	0	8	0	**709**	0	**97.9**
	11	0	0	0	0	0	0	40	7	32	1	**1419**	**94.7**
	Overall percent	10.2	7.2	10.9	12.7	7.5	6.7	9.5	14.8	7.4	4.6	8.5	**96.4**
Testing	1	**736**	15	1	1	3	0	0	0	0	0	0	**97.4**
	2	2	**513**	1	3	3	0	0	0	0	0	0	**98.3**
	3	1	7	**782**	0	3	0	0	0	0	0	0	**98.6**
	4	8	4	0	**909**	2	0	0	2	0	0	0	**98.3**
	5	14	4	2	6	**522**	12	0	0	0	0	0	**93.2**
	6	0	1	0	0	13	**450**	23	11	0	1	0	**90.2**
	7	0	0	0	0	2	3	**698**	0	19	0	3	**96.3**
	8	0	0	0	3	0	2	0	**1057**	0	15	3	**97.9**
	9	0	0	0	0	0	1	24	0	**510**	0	17	**92.4**
	10	0	5	0	0	0	0	0	3	0	**287**	0	**97.3**
	11	0	0	0	0	0	0	19	3	16	0	**549**	**93.5**
	Overall percent	10.4	7.5	10.8	12.6	7.5	6.4	10.5	14.7	7.4	4.2	7.9	**95.8**

[a]Dependent variable: cluster membership
The bold values indicate the correctly predicted cluster memberships

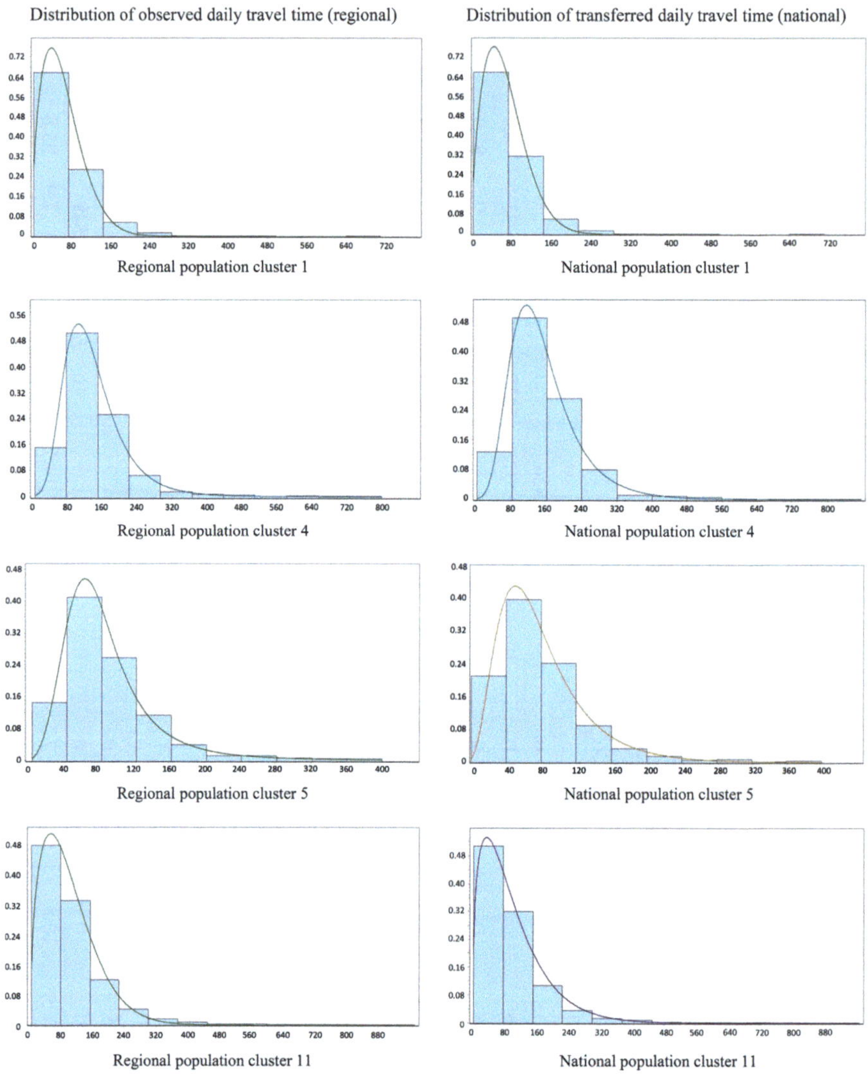

Fig. 4 Comparison of observed and transferred daily travel time

4 Conclusion

This study developed a methodological framework based on the spatial transferability approach to assess the national-level impacts of CAV technology. It has not been applied to evaluate the national-level implications of CAV technology yet, but rather, it aims to develop a holistic framework which can accurately transfer the

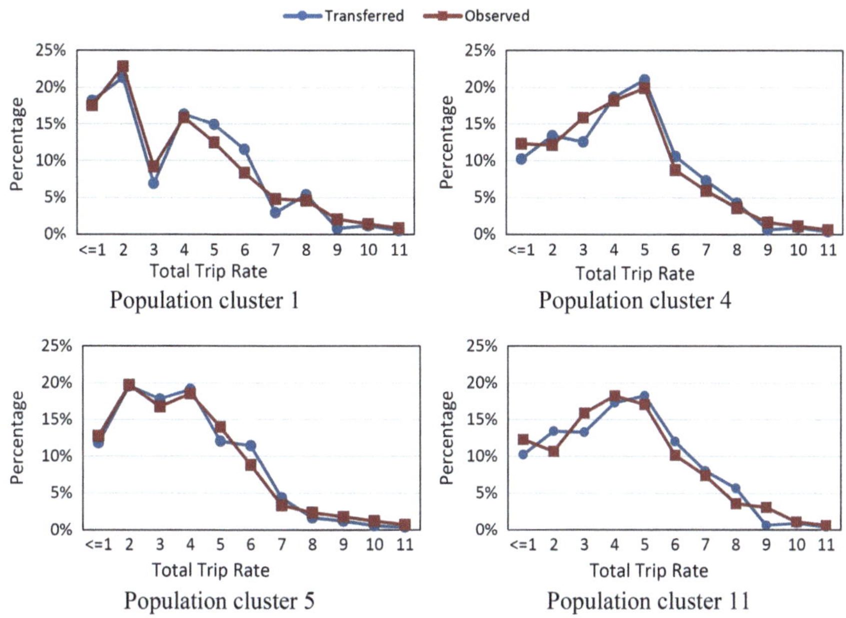

Fig. 5 Comparisons of observed and transferred daily trip rates

results of regional-level CAV scenarios to the national level. In this framework, the POLARIS transportation systems simulation model was used to estimate the potential impacts of CAV technology both on travel demand and traffic flows in the Chicago metropolitan area. Applying multiple data mining and statistical models, the simulation results on total daily trip rates and travel time were transferred to the national-level context to evaluate the large-scale impacts of CAV technology. Overall, the transferability results seem quite promising and the proposed platform proves to be capable of satisfactorily transferring the two travel variables from the regional to the national level. This baseline platform can be utilized to evaluate the national impacts of CAV scenarios. Although the main motivation for developing such a transferability framework was to assess the national impacts of CAV scenarios, it is suggested that it can be used for many other types of national-level analyses.

Acknowledgments The authors gratefully acknowledge the sponsorship of the Systems and Modeling for Accelerated Research in Transportation (SMART) Mobility Laboratory Consortium, an initiative of the Energy Efficient Mobility Systems (EEMS) Program, managed by David Anderson of the Vehicle Technologies Office of the U.S. Department of Energy. This study was conducted under Contract No. DE-AC02-06CH11357 to Argonne National Laboratory, a U.S. Department of Energy laboratory managed by UChicago Argonne, LLC. The authors are solely responsible for the findings of this research which do not necessarily represent the views of the U.S. Department of Energy or the United States Government.

References

Auld, J., Hope, M., Ley, H., Sokolov, V., Xu, B., & Zhang, K. (2016). POLARIS: Agent-based modeling framework development and implementation for integrated travel demand and network and operations simulations. *Transportation Research Part C: Emerging Technologies, 64*, 101–116.

Auld, J., Sokolov, V., & Stephens, T. S. (2017). Analysis of the effects of connected–automated vehicle technologies on travel demand. *Transportation Research Record, 2625*, 1–8. https://doi.org/10.3141/2625-01.

Brown, A., Gonder, J., & Repac, B. (2014). *An analysis of possible energy impacts of automated vehicle* (pp. 137–153). Cham: Springer. https://doi.org/10.1007/978-3-319-05990-7_13.

Chiu, T., Fang, D., Chen, J., Wang, Y., & Jeris, C. (2001). A robust and scalable clustering algorithm for mixed type attributes in large database environment. In: *Proceedings of the 7th ACM SIGKDD International Conference on Knowledge Discovery and Data Mining* (pp. 263–268). http://doi.acm.org/10.1145/502512.502549.

Farber, S., Bartholomew, K., Li, X., Paez, A., & Nurul Habib, K. M. (2014). Assessing social equity in distance based transit fares using a model of travel behavior. *Transportation Research Part A: Policy and Practice, 67*, 291–303. https://doi.org/10.1016/j.tra.2014.07.013.

FHWA. (2009). *U.S. Department of Transportation, Federal Highway Administration, 2009 National Household Travel Survey*. [WWW Document]. Retrieved 3.1.17, from http://nhts.ornl.gov.

Golshani, N., Shabanpour, R., Mahmoudifard, S. M., Derrible, S., & Mohammadian, A. (2018). Modeling travel mode and timing decisions: Comparison of artificial neural networks and copula-based joint model. *Travel Behaviour and Society, 10*, 21–32. https://doi.org/10.1016/J.TBS.2017.09.003.

Greene, W. H. (2012). *Econometric analysis* (7th ed.). Boston: Pearson Education.

Li, Z., Wang, W., Yang, C., & Ragland, D. R. (2013). Bicycle commuting market analysis using attitudinal market segmentation approach. *Transportation Research Part A: Policy and Practice, 47*, 56–68. https://doi.org/10.1016/j.tra.2012.10.017.

Mohammadian, A., & Zhang, Y. (2007). Investigating transferability of National Household Travel Survey Data. *Transportation Research Record, 1993*, 67–79. https://doi.org/10.3141/1993-10.

Paulssen, M., Temme, D., Vij, A., & Walker, J. L. (2014). Values, attitudes and travel behavior: A hierarchical latent variable mixed logit model of travel mode choice. *Transportation (Amst)., 41*, 873–888. https://doi.org/10.1007/s11116-013-9504-3.

Shabanpour, R., Auld, J., Mohammadian, A. K., & Stephens, T. (2017a). Developing a platform to analyze behavioral impacts of connected automated vehicles at the national level. In: *Proceedings of the 96th Annual Meeting of the Transportation Research Board (TRB)*, Washington, DC.

Shabanpour, R., Golshani, N., Derrible, S., Mohammadian, A., & Miralinaghi. (2017b). Joint discrete-continuous model of travel mode and departure time choices. *Transportation Research Record: Journal of the Transportation Research Board, 2669*, 41–51. https://doi.org/10.3141/2669-05.

Shabanpour, R., Golshani, N., Tayarani, M., Auld, J. & Mohammadian, A.K. (2018). Analysis of telecommuting behavior and impacts on travel demand and the environment. *Transportation Research Part D: Transport and Environment, 62*, 563–576. https://doi.org/10.1016/j.trd.2018.04.003.

Stephens, T. S., Gonder, J., Chen, Y., Lin, Z., Liu, C., & Gohlke, D. (2016). *Estimated bounds and important factors for fuel use and consumer costs of connected and automated vehicles*. National Renewable Energy Laboratory Technical Report NREL/TP-5400-67216. Retrieved from https://www.nrel.gov/docs/fy17osti/67216.pdf.

Van Acker, V., Goodwin, P., & Witlox, F. (2016). Key research themes on travel behavior, lifestyle, and sustainable urban mobility. *International Journal of Sustainable Transportation, 10*, 25–32. https://doi.org/10.1080/15568318.2013.821003.

Part III
Integrated Land Use/Transport Analysis

The Impact of Public Transport Infrastructure on Residential Land Value: Using Spatial Analysis to Uncover Policy-Relevant Processes

Dimitris Potoglou, Hanna Maoh, Yiming Wang, and Scott Orford

Abstract This chapter reports the findings of a recent study, which aims to quantify the land value benefits of public transport infrastructure. The empirical investigation focuses on central Cardiff, Wales, where data comprising approximately 9000 property sale records were collected in the period 2000–2009. Through an exploratory and explanatory spatial data analysis, it was found that: (a) there is an underlying bus stop-oriented spatial dependence of the values of residential land, which varies based on the types of housing property (e.g. detached house, semi-detached house, flat) across different areas in Cardiff; (b) after controlling for spatial dependence, there is still a significant positive correlation between bus stops and change in land value. Most importantly, different types of property seem to benefit differently from land value uplift due to bus stop locations. The highest benefits are identified for flats and semi-detached homes followed by terraced and detached homes. The corresponding price gains range from 1.4% for flats to 0.7% for detached homes.

1 Introduction

Contemporary Anglo-American human geography has been lamented as lacking policy relevance or even uninterested in dealing with practical policy issues (Martin 2001). However, Professor Pavlos Kanaroglou demonstrated through his academic career the many significant real-world changes a rigorously trained and

D. Potoglou (✉) · S. Orford
School of Geography and Planning, Cardiff University, Cardiff, Wales, UK
e-mail: potogloud@cardiff.ac.uk; orfords@cardiff.ac.uk

H. Maoh
Department of Civil and Environmental Engineering, University of Windsor, Windsor, ON, Canada
e-mail: maohhf@uwindsor.ca

Y. Wang
School for Policy Studies, University of Bristol, Bristol, UK
e-mail: yiming.wang@bristol.ac.uk

H. Briassoulis et al. (eds.), *The Practice of Spatial Analysis*,
https://doi.org/10.1007/978-3-319-89806-3_13

practice-minded human geographer could bring to the broader society as well as the local community. One of his intellectual legacies is the policy orientation of his research, which often applies practical human geography skills and knowledge to address the most realistic societal problems.

As per the same spirit, this chapter contributes a recent study regarding the local public bus system in Cardiff, the capital city of Wales in the UK. Like many other major cities in the UK, Cardiff has witnessed substantial funding cuts for local public services due to various austerity measures by the central government, alongside rising living cost for ordinary citizens as a result of fast increasing house prices in recent years (Wang et al. 2015). How to mobilise the local property market dynamics to help fund the local civil services, including but not limited to public transport infrastructure, is thus a top issue of the local policy agenda.

Land value capture (LVC) has been one of the policy initiatives under discussion. Theoretically, the basic land economic principles underpinning LVC can be traced back in time to George (1879) and Riccardo (1891), for instance. Empirically, LVC is also a well supported idea by a large volume of evidence collated worldwide, most of which confirms the significant land value contribution by physical infrastructure, in general, and public transportation facilities, in particular, towards privately owned real estate properties (Bollinger and Ihlanfeldt 1997; Debrezion et al. 2007; Doherty 2004; Hess and Almeida 2007; Ryan 1999; Rybeck 2004; Smith and Gihring 2006; Wang et al. 2016). Flourishing academic scholarship in this field has triggered policy interest in financing public transport infrastructures by capturing the lifted land value through a variety of fund-raising channels, ranging from land value taxation to various types of private finance initiatives (Wang and Baddeley 2016).

Nevertheless, how to accurately appraise the residential land value betterment contributed by public transit facilities remains as technical challenge due to the innate geographic as well as price segmentation remains as the property market, let alone the complex spatio-temporal heterogeneity and autocorrelation associated with most house price data, which tend to vary significantly both with time and location. Adding to the challenge is the question of how to communicate the findings of advanced spatial statistical analysis to the general public and, ultimately, to let the numbers make policy sense.

In this chapter, we aim to demonstrate a quantitative human geography approach to address the technical and communication challenges mentioned above. Primarily employing a spatial error model to assess the extent of residential land value uplift by a sample of bus stops located within the City of Cardiff and accounting for spatial autocorrelation of property prices, it was found that different types of property benefit differently from adjacent public transport infrastructure. The model results were visualised through digital mapping using geographic information systems (GIS). While this study is mainly intended to inform local public policy with regard to value capture finance, a paralleled intellectual objective is to prompt an urgently needed restoration of policy orientation for contemporary British human geography so that the discipline can continue to be taken as a serious and respected endeavour.

The remainder of this chapter is structured as follows. The next section presents the setup of the statistical model, which is followed by data analysis and empirical results. The numerical findings are then visualised and discussed, unravelling their policy as well as academic implications. The last section concludes the chapter and provides some ideas for future research.

2 Research Design

2.1 Research Questions and Hypotheses

This study focuses on the empirical and technical aspects related to levying land value tax. It aims to explore how local public infrastructure may contribute to the adjacent property values and address the according policy implications. Infrastructure may include a number of physical facilities; however, the primary focus in this study is on the relationship between land values and public transport infrastructure.

The core research question is to what extent the presence of public transport infrastructure is associated with adjacent land values and how to quantify this relationship in monetary terms. Due to a lack of systematically collected and up-to-date data on land values in the UK, this research uses property prices as a proxy. Existing approaches are characterised by two shortcomings. Firstly, conventional econometric (e.g. location intensive or dwelling specific) models applied to estimate land values tend to overlook the role of transport infrastructure. Secondly, existing models rarely account for spatial relationships (i.e. spatial autocorrelation) in the observed land values, thus resulting in inefficient and biased estimates.

Given the research question, two hypotheses will be tested in this study:

(a) There is an underlying spatial dependence on land values, which varies depending on the type of the property (e.g. detached house, semi-detached house, flat);
(b) After controlling for spatial dependence, there is still a significant positive relationship between public transport (bus, train) and land values.

2.2 Exploratory Tools for Spatial Dependence (Moran's I, Kernel Estimation)

Given the point pattern nature of the property data, the Moran's I statistic was applied to examine the presence of spatial autocorrelation among the price of all properties n. Spatial autocorrelation arises when the price of a given property is affected by the price of neighbouring properties. The Moran's I statistic is calculated as follows:

$$I = \frac{n\sum_{i=1}^{n}\sum_{j=1}^{n} W_{ij}(y_i - \bar{y})(y_j - \bar{y})}{\sum_{i=1}^{n}\sum_{j=1}^{n} W_{ij}\sum_{i=1}^{n}(y_i - \bar{y})^2} \tag{1}$$

where.

W_{ij} is a weight that is set to unity if property i is a neighbour of property j;

y_i and y_j are the prices of properties i and j, respectively;

$\bar{y}$ is the mean price of all properties.

When running the Moran's I test, the null hypothesis is that there is no spatial autocorrelation among neighbouring properties. A statistically significant and positive Moran's I statistic will lead to the rejection of the null hypothesis of no spatial autocorrelation, suggesting that the price values associated with the point pattern are spatially autocorrelated (Bailey and Gatrell 1995). In other words, the price values of properties exhibit clustering over space.

Another test to explore the presence of spatial clustering is to estimate a kernel surface based on the values of the properties forming the point pattern. In kernel estimation, space is divided into a finite number of grid cells s for which intensity $\mu(s)$ is calculated. The latter is based on estimating a bivariate probability density from a sample of properties at locations (s_1, s_2, s_3, ..., s_m) that fall within radius μ from the centre of grid cell s. As shown in Maoh et al. (2010), the kernel density can be calculated as follows:

$$\mu_\tau(s) = \frac{\sum_{i=1}^{m} k\left(\frac{s-s_i}{\tau}\right)y_i}{\left(\frac{s-s_i}{\tau}\right)} \tag{2}$$

where.

$k()$ is the kernel function formulated as a bivariate probability density;

y_i is the value of property i at location s_i; and

τ is a buffer distance defining the maximum radius between location s and the m neighbouring properties at locations (s_1, s_2, s_3, ..., s_m). Visual inspection of the estimated kernel surface will allow the analyst to identify clusters of high property prices.

2.3 Spatial Regression Models

Building on the work of Wang et al. (2015), a hedonic regression model is introduced to provide further insights about the relationship between bus stop location and property values. In line with existing research, the following linear semi-log function form is assumed:

$$\log(Y) = XB + \varepsilon \tag{3}$$

where.

Y is a $(n \times 1)$ vector representing the value of the n modelled properties;

X is a $((k + 1) \times n)$ matrix with the first column set to unity while the remaining k columns represent the covariates used in the specification of the model;

B is a vector of $k + 1$ parameters to be estimated; and

ε is a $(n \times 1)$ vector of error terms associated with the estimated price of the n modelled properties.

The natural logarithm of the price $\log(Y)$ can be set up as a function of a number of hedonic control variables, temporal variables, jurisdictional variables, and transportation infrastructure variables. Hedonic control variables represent certain property attributes such as the floor area of the property, whether the property is freehold, newly built and the location of the property with respect to the central business district. Wang et al. (2015) also considered the type of property (i.e., detached, semi-detached, terraced, or flat). However, this is not needed in this study as each property type will be modelled separately.

In terms of temporal variables, the year in which the property was sold can be used in the right-hand side of Eq. 3 by setting it up as a categorical variable that takes on the value 1 if the property was sold in year t and zero (0) otherwise. Typically, sales that occurred earlier in time will tend to have a reduced effect on the value of the property, other things being equal. Jurisdictional variables are set up as dummy-coded variables to identify the municipality within which the property is located. These variables were used by Wang et al. (2015) as proxy to capture the average local environment effects (e.g. residential amenities or presence of schools). Finally, transportation infrastructure refers to spatial accessibility to public transportation infrastructure. Table 1 provides a list of the covariates used in the specification of the models.

While Eq. 3 could be estimated via the ordinary least square (OLS) method, the spatial nature of the data might reduce the ability of the OLS to produce unbiased estimates. More specifically, OLS assumes that values of the dependent variable and the error terms are independent. However, the presence of spatial autocorrelation in

Table 1 List of explanatory variables

Covariate	Description	Expected sign
FREEHOLD	1 if property is on freehold tenure, 0 otherwise	(+)
NEW	1 if property is newly built when sold, 0 otherwise	(+)
FLOOR AREA	Floor area of a property in square metres (m^2)	(+)
LOGDCBD	Natural logarithm of distance to the central business district in metres (m)	(+)
YEARt	1 if property was sold in year t, 0 otherwise ($t = 2000, 2001, 2002, \ldots, 2009$); year $t = 2009$ is set as reference	(−)
BS(k)	Number of bus stops within k metres (m) of the property ($k = 300, 400, 500, 750, 1000, 1500$)	(+)

$\log(Y)$ or the error terms ε could lead to severe bias in the estimated parameters. To rectify this problem, spatial regression models have been proposed in the past (e.g. Efthymiou and Antoniou 2013; LeSage and Pace 2009; Willey et al. 2011; Spinney et al. 2014; Maoh and Kanaroglou 2007). Two types of spatial models have been widely used: (1) Spatial lagged (SAR) model and (2) Spatial error (SEM) model. The spatial lag model is utilised if the dependent variable exhibits spatial autocorrelation. In such case, the conventional regression model shown in Eq. 3 is modified to the following:

$$\log(Y) = XB + \rho W \log(Y) + \varepsilon \tag{4}$$

where.

W is a weight matrix, as in the case of the Moran's I equation; and.

ρ is a spatial lag parameter to be estimated. A positive and significant value of ρsuggests the presence of spatial autocorrelation in the dependent variable $\log(Y)$.

In Eq. 3, if the error terms ε are spatial autocorrelated, then the spatial error model could be used. Here, Eq. 3 is modified to the following:

$$\log(Y) = XB + \lambda W u + \varepsilon \tag{5}$$

where.

u is a $n \times 1$ vector of spatially autocorrelated error terms,

W is a weight matrix as defined earlier; and.

λ is an autoregressive parameter to be estimated. A positive and highly significant value of λ suggests the presence of autocorrelation in the error terms of Eq. 3.

3 Study Area and Data

The geographical focus of this study was the central area in the City of Cardiff, the capital of Wales, UK. The data in this research came from two sources. Firstly, the residential property sales were obtained from HM Land Registry for England and Wales and included: address of the property, price, date the property was sold, property type (i.e. detached, semi-detached, terraced, or flat), tenure (freehold or leasehold), and whether it was newly built (Land Registry 2017). Secondly, public transport infrastructure corresponded to the Cardiff Bus, the major public transport operator in the City of Cardiff. Bus stop locations of Cardiff Bus were retrieved from the 2007 National Public Transport Access Node (NaPTAN-v2.2) dataset (Department for Transport 2017) and are shown in Fig. 1. As detailed in Wang et al. (2015), the number of bus stops within specified radii of each property (i.e. 300 m, 400 m, 500 m, 750 m, 1000 m, and 1500 m) was used as the proxy accessibility measure to the public transport service in Cardiff.

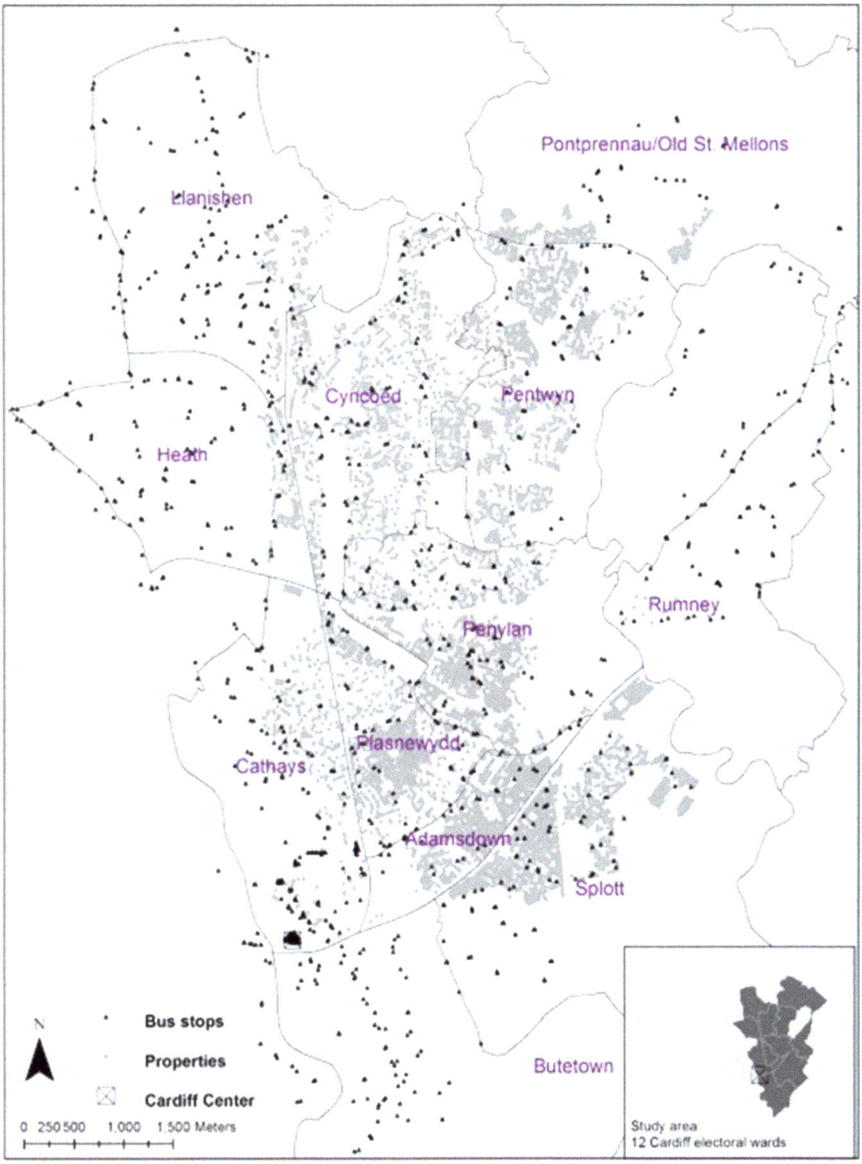

Fig. 1 Location of properties and bus stops in the study area

The property sales available in the dataset covered an area of 6 km by 4 km across Cardiff's 12 electoral wards (see also, Fig. 1). The available property sales data were representative of the housing stock in the City and included typical Victorian and Edwardian terraced houses, flats in converted buildings and suburban-style semi-detached and detached houses, alongside a number of bungalows and purpose-built flats. The working dataset included over 9600 residential properties within the area that were sold between 2000 and 2009. Following tests on leverage points and outliers (Cook's Distance), observations violating the assumptions of a typical hedonic model were already screened out from further analysis (Wang et al. 2015). The floor area of each property was computed by Orford (2010). The variable floor area is the most important structural feature in determining the price of the property and is highly correlated with other attributes such as the number of rooms or size of the garden, etc. (Sirmans et al. 2006; Xiao et al. 2016). Including floor area as explanatory variable in a residential property price model helped mitigate, at some level, any omitted structural attributes of the property.

Table 2 presents summary statistics of the sales price and the explanatory variables by property type in the sample. As expected, detached houses are notably the

Table 2 Descriptive statistics of sales price and key explanatory variables

	Mean (standard deviation)				
Dependent variable: Definition	Detached	Flat	Semi-detached	Terraced	All properties
Property sale price (£)	256,836 (132249)	88,606 (32481)	159,932 (77502)	118,303 (57647)	143,937 (87267)
Natural logarithm of the sale price—Log(price)	12.3 (0.5)	11.3 (0.4)	11.9 (0.5)	11.6 (0.5)	11.72 (0.5)
Covariates: Definition					
No. of bus stops within 300 m of the property—BS(300)	6.6 (4.0)	8.1 (3.2)	6.8 (3.8)	7.9 (3.6)	7.5 (3.7)
No. of bus stops within 400 m of the property—BS (400)	11.6 (6.0)	15.4 (6.6)	11.9 (5.3)	14.2 (4.9)	13.3 (5.3)
No. of bus stops within 500 m of the property—BS (500)	17.8 (8.4)	22.0 (8.6)	18.3 (7.5)	22.2 (6.6)	20.7 (7.4)
No. of bus stops within 750 m of the property—BS (750)	39.5 (15.0)	50 (13.8)	39.1 (14.8)	48.8 (13.6)	45.3 (14.8)
No. of bus stops within 1000 m of the property—BS (1000)	69.0 (23.4)	85.5 (22.8)	66.4 (24.5)	83.9 (24.1)	77.8 (25.4)
No. of bus stops within 1500 m of the property—BS (1500)	149.4 (44.4)	188.1 (60)	144.4 (50.7)	180.1 (58.2)	167.7 (57.5)
Floor area of a property in square metres (m^2) (FLOOR AREA)	134.5 (52.2)	103.9 (33.5)	100.9 (38.1)	96.9 (34.4)	102.6 (39.6)
Distance to the central business district in metres (m) (LOGDCBD)	4035.9 (1192.5)	3099.2 (1399.1)	3885 (1280.5)	2787 (1317.3)	3225.9 (1408.5)
NEW build properties (NEW)	9.6%	4.1%	8.3%	0.3%	3.6%
FREEHOLD properties (FREEHOLD)	97.9%	5.7%	89.9%	94.8%	90.5%
Sample size (%)	1120 (12%)	366 (4%)	2478 (26%)	5689 (59%)	9653

most expensive of all properties, on average, and they are also the largest. Flats are the cheapest types of property despite being the same average size as terraces and semi-detached houses. The average distances to the CBD of each property reflect the structure of a typical British city, with Victorian terraces closest to the city centre and suburban detached houses and semi-detached properties the furthest away. The largest stock of newly built properties includes detached and semi-detached houses whereas the smallest proportion of newly built properties corresponds to terraced houses. Detached and terraced houses represent the highest proportions of freehold properties; that is, the owner of the property owns both the land and building in perpetuity.

4 Results

4.1 Spatial Dependence on Property Land Values

Table 3 shows that positive spatial autocorrelation is present in the Log(price) variable as expected and is also present when the sample is split by property type. The magnitude of the Moran's I statistics suggests that the degree of spatial autocorrelation is similar in all property types with the exception of flats where it is much weaker, reflecting the more dispersed nature of flats within the housing stock.

This effect is also demonstrated in the heat maps (Kernel) in Fig. 2, which reveals the clusters of high-value properties of each type are concentrated into 5 out of the 12 wards in the sample. The high-price values of detached and semi-detached clusters are found across suburban wards—although there are notable high-value clusters of semi-detached housing in wards closer to the CBD—and high-value terraced clusters within the inner-city wards. The significant spatial autocorrelation in the dependent variable suggests that an OLS regression model would be inappropriate in this instance, and a spatial regression model would be a better specification.

Table 3 Testing for spatial autocorrelation in sales Log(price) by property type

Spatial autocorrelation	Detached	Flat	Semi-detached	Terrace	All properties
Moran's I	0.512	0.240	0.513	0.405	0.534
$E[I]$	−0.001	−0.003	0.000	0.000	−0.0001
Mean	−0.001	−0.002	−0.001	0.000	−0.0001
S.D.	0.016	0.024	0.011	0.007	0.004
z-value	33.028	10.217	46.548	59.656	123.17
p-value	0.001	0.001	0.001	0.001	0.001

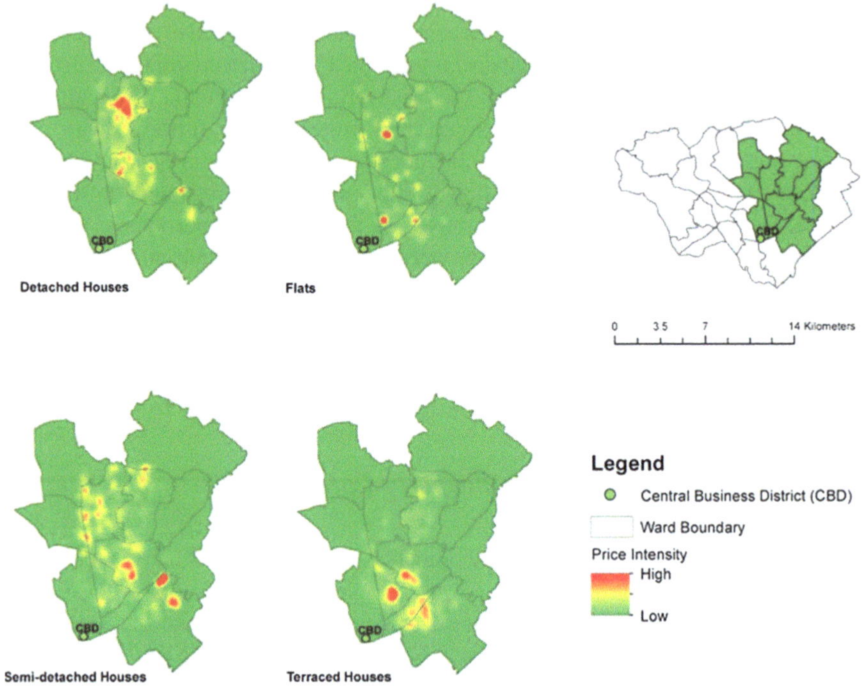

Fig. 2 Kernel estimation of property sale price by type

4.2 *The Role of Public Transport Infrastructure on Land Values (by Property Type)*

Tables 5, 6, 7, and 8 in the Appendix report model coefficients for the SEMs estimated for the six bus stop radii for each of the four property types. The OLS and SAR model coefficients are not reported here for reasons of brevity but their AIC statistics are presented in Table 4. These show that the SEMs have overall smaller AIC values than the OLS and SAR model equivalents. The SEM models are therefore considered a better fit and hence are the focus of reported findings.

For detached and semi-detached properties, the best model fit was for the radius of 1500 m, reflecting the lower density suburban areas where this property type predominates. This compares to the 500 m radius for flats and the 1000 m radius for terraces reflecting the higher density inner-city areas and the propensity of flats to be located along main roads. The BS(k) variables are significant in all models with the exception of the 300 m radii in the detached model and the 400 m radii in the semi-detached model.

The largest estimated BS(k) values occur at the 300 m radii in the semi-detached model ($B = 0.012$), at the 400 m radii in the detached model ($B = 0.07$), and at the

Table 4 OLS, SAR, and SEM model comparison based on the AIC statistic

Property type	Model	$k = 300$ M	$k = 400$ M	$k = 500$ M	$k = 750$ M	$k = 1000$ M	$k = 1500$ M
Detached	OLS	390.1	352.1	317.8	262.9	229.0	130.8
	SAR	−204.9	−216.3	−221.1	−225.0	−230.3	−239.9
	SEM	−305.4	−309.5	−310.7	−310.4	−318.7	−336.9
Semi-detached	OLS	921.9	877.1	831.6	791.8	637.0	456.1
	SAR	−656.0	−656.2	−666.4	−665.7	−709.8	−749.3
	SEM	−1068.1	−1051.5	−1056.0	−1058.9	−1091.6	−1110.9
Terrace	OLS	1684.3	1594.9	1354.4	1276.6	1044.6	1244.0
	SAR	−1486.7	−1502.6	−1558.6	−1610.7	−1675.8	−1600.6
	SEM	−2291.7	−2293.3	−2320.7	−2328.4	−2350.3	−2331.3
Flat	OLS	35.7	33.4	23.6	−15.4	−0.1	−16.2
	SAR	−67.5	−71.9	−77.1	−93.4	−82.8	−98.0
	SEM	−169.9	−177.1	−186.4	−168.2	−169.3	−179.0

500 m radii for both the flat model ($B = 0.014$) and the terrace model ($B = 0.010$). These estimates suggest that semi-detached and detached properties have to be closer to bus stops than terrace properties and flats for the maximum benefit to be capitalised into property values. While the estimated B-values are similar for flats, terrace and semi-detached properties, suggesting an equivalent impact on property prices, it is notably larger for detached houses (the estimated B-value is five times larger than the B-value for flats) suggesting a much larger impact on price.

It is possible to backtransform the estimated coefficients of the number of bus stops variable, BS(k) , so as to calculate the marginal increase in land value as a result of adding an extra bus stop within a certain radius around a property. For example, it is found that the highest gain in land value price of a detached home occurs when an extra bus stop is placed within 400 m of the property and is equal to $(e^{0.007}-1) \times 100 = 0.7\%$ of its sale price. Flats and semi-detached homes obtain the highest gains when an additional bus stop is placed within 500 and 300 m of each property type and correspond to 1.4% and 1.2% of their sale price, respectively. Finally, terrace homes benefit around 1.05% of their sale price when an extra bus stop is placed within 500 m of this type of property. It is worth noting here that although all models suggest that the availability of public transport infrastructure significantly promotes land value and, thereby, raise property values, the corresponding parameter estimates of the bus stop variables and their backtransformations, subsequently, suggest that properties benefit differently according to their type and within different catchment areas (distance buffers). A summary of these effects is shown in Fig. 3.

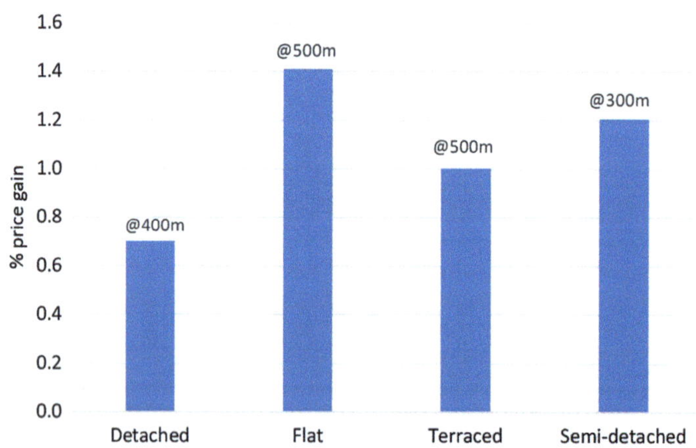

Fig. 3 Maximum % price gains from an additional bus stop by property type

5 Conclusion

This study explored the association between public transport infrastructure and land values for different types of property. The motivation for this study arises from the potential viability of levying a land value tax within a local UK context. This topic was approached from an empirical and technical aspect by examining the contribution of local public transport infrastructure, manifested as the number of bus stops within different radii, onto land values proxied by property price for different types of property. As expected, land values of different property types exhibit different levels of clustering—high values, in particular, tend to cluster together. In OLS hedonic regression type of models, the effect of clustering tends of be overlooked and as a result estimates may be biased or inconsistent. In this study, following an exploratory analysis, it was possible to detect the degree of spatial clustering (autocorrelation) and account for this effect while quantifying the gains in property price from adjacent public transport infrastructure by property type.

The key conclusion is that different types of property benefit differently in terms of price gains and within different catchment areas. These findings help refine our understanding of the relationship between land use and public transport infrastructure and provide robust evidence on the monetary value-added onto land because of public transport infrastructure. The latter becomes increasingly important for policy decision-making in recent years as governments are in need of fair taxation policies. The latter has revived the discussion around land value tax, a levy based on the exact premise that properties benefit from adjacent public infrastructure funded by tax payers.

Appendix: Spatial Error Model Estimates by Property Type

Table 5 Detached

	$k = 300$ M		$k = 400$ M		$k = 500$ M		$k = 750$ M		$k = 1000$ M		$k = 1500$ M	
	B	*t*-stats	*B*	*t*-stats	*B*	*t*-stats	*B*	*t*-stats	*B*	*t*-stats	*B*	*t*-stats
LAMBDA	0.820	39.02	0.814	37.86	0.809	37.09	0.800	35.60	0.783	33.19	0.760	30.25
CONSTANT	10.475	12.92	9.938	12.14	9.695	11.70	9.265	10.56	8.442	10.09	7.104	8.82
FREEHOLD	0.267	6.81	0.266	6.80	0.266	6.79	0.269	6.83	0.266	6.74	0.270	6.85
NEW	0.066	2.53	0.068	2.61	0.070	2.67	0.069	2.63	0.074	2.83	0.078	2.99
FLOOR AREA	0.004	25.93	0.004	25.81	0.004	26.10	0.004	25.98	0.004	26.05	0.004	26.30
LOGDCBD	0.177	1.84	0.233	2.42	0.259	2.67	0.303	3.01	0.384	4.05	0.515	5.74
YEAR1	−0.824	−7.72	−0.813	−7.60	−0.820	−7.67	−0.819	−7.64	−0.822	−7.65	−0.829	−7.71
YEAR2	−0.673	−6.27	−0.662	−6.15	−0.668	−6.21	−0.670	−6.21	−0.670	−6.20	−0.679	−6.28
YEAR3	−0.536	−5.01	−0.526	−4.91	−0.532	−4.96	−0.531	−4.94	−0.532	−4.95	−0.542	−5.03
YEAR4	−0.339	−3.17	−0.327	−3.05	−0.335	−3.12	−0.334	−3.11	−0.337	−3.12	−0.342	−3.18
YEAR5	−0.219	−2.05	−0.207	−1.93	−0.214	−2.00	−0.215	−2.00	−0.216	−2.01	−0.225	−2.09
YEAR6	−0.166	−1.55	−0.152	−1.42	−0.159	−1.48	−0.162	−1.50	−0.162	−1.50	−0.169	−1.57
YEAR7	−0.116	−1.09	−0.104	−0.97	−0.110	−1.03	−0.112	−1.04	−0.111	−1.03	−0.122	−1.13
YEAR8	−0.072	−0.67	−0.061	−0.57	−0.068	−0.64	−0.067	−0.63	−0.069	−0.64	−0.074	−0.69
BS(k)	0.002	0.53	0.007	2.13	0.006	2.44	0.004	2.55	0.005	4.17	0.004	6.47
$N = 1120$												
R-squared	0.845		0.845		0.844		0.844		0.844		0.845	
AIC	−305.426		−309.508		−310.657		−310.423		−318.696		−336.904	

Table 6 Flats

	$k = 300$ M		$k = 400$ M		$k = 500$ M		$k = 750$ M		$k = 1000$ M		$k = 1500$ M	
	B	*t*-stats	*B*	*t*-stats	*B*	*t*-stats	*B*	*t*-stats	*B*	*t*-stats	*B*	*t*-stats
LAMBDA	0.831	22.56	0.840	23.74	0.841	23.80	0.811	20.40	0.813	20.59	0.802	19.54
CONSTANT	10.647	13.40	10.020	11.74	9.377	10.78	9.608	10.61	8.950	8.44	6.902	5.85
FREEHOLD	0.388	9.41	0.402	9.83	0.385	9.58	0.388	9.33	0.395	9.49	0.390	9.51
NEW	0.049	0.85	0.043	0.77	0.068	1.23	0.081	1.40	0.080	1.37	0.077	1.36
FLOOR AREA	0.002	5.05	0.002	5.68	0.002	5.05	0.002	5.01	0.002	5.16	0.002	4.91
LOGDCBD	0.073	0.74	0.137	1.30	0.208	1.95	0.185	1.74	0.250	2.09	0.468	3.56
YEAR1	−0.688	−7.65	−0.696	−7.85	−0.716	−8.16	−0.696	−7.68	−0.682	−7.55	−0.669	−7.47
YEAR2	−0.685	−7.69	−0.687	−7.81	−0.705	−8.11	−0.694	−7.73	−0.685	−7.64	−0.661	−7.43
YEAR3	−0.439	−4.89	−0.455	−5.14	−0.466	−5.33	−0.453	−5.01	−0.435	−4.82	−0.423	−4.73
YEAR4	−0.191	−2.12	−0.201	−2.27	−0.217	−2.47	−0.201	−2.22	−0.187	−2.07	−0.173	−1.93
YEAR5	−0.032	−0.35	−0.039	−0.44	−0.046	−0.52	−0.033	−0.36	−0.016	−0.17	−0.006	−0.07
YEAR6	−0.008	−0.08	−0.013	−0.14	−0.017	−0.19	−0.008	−0.09	−0.002	−0.02	0.002	0.02
YEAR7	0.050	0.56	0.044	0.50	0.028	0.32	0.046	0.52	0.062	0.69	0.081	0.91
YEAR8	0.087	0.98	0.081	0.93	0.064	0.75	0.071	0.80	0.088	1.00	0.096	1.09
BS(k)	0.010	2.49	0.012	3.68	0.014	4.83	0.005	2.25	0.004	2.45	0.004	4.07
$N = 366$												
R-squared	0.811		0.815		0.820		0.808		0.809		0.813	
AIC	−169.889		−177.108		−186.435		−168.242		−169.262		−179.001	

Table 7 Terraced

	$k = 300$ M		$k = 400$ M		$k = 500$ M		$k = 750$ M		$k = 1000$ M		$k = 1500$ M	
	B	*t*-stats	*B*	*t*-stats	*B*	*t*-stats	*B*	*t*-stats	*B*	*t*-stats	*B*	*t*-stats
LAMBDA	0.831	86.08	0.829	85.09	0.820	81.42	0.815	79.86	0.806	76.66	0.817	80.54
CONSTANT	12.314	44.41	12.167	42.59	11.491	38.59	10.252	25.59	8.965	19.45	9.489	19.58
FREEHOLD	0.154	12.45	0.155	12.51	0.155	12.50	0.157	12.63	0.154	12.46	0.156	12.63
NEW	0.167	3.60	0.170	3.66	0.169	3.64	0.169	3.63	0.174	3.74	0.172	3.71
FLOOR AREA	0.002	17.89	0.002	17.97	0.002	18.08	0.002	17.78	0.002	17.90	0.002	18.25
LOGDCBD	−0.097	−2.81	−0.083	−2.36	−0.015	−0.42	0.121	2.66	0.258	5.03	0.206	3.80
YEAR1	−0.900	−37.84	−0.901	−37.86	−0.901	−37.83	−0.903	−37.89	−0.901	−37.80	−0.901	−37.84
YEAR2	−0.782	−33.16	−0.783	−33.18	−0.782	−33.11	−0.784	−33.20	−0.783	−33.09	−0.783	−33.16
YEAR3	−0.581	−24.69	−0.582	−24.71	−0.581	−24.66	−0.584	−24.76	−0.583	−24.71	−0.583	−24.74
YEAR4	−0.350	−14.84	−0.351	−14.86	−0.350	−14.81	−0.353	−14.93	−0.351	−14.83	−0.351	−14.85
YEAR5	−0.168	−7.03	−0.169	−7.07	−0.167	−7.00	−0.171	−7.18	−0.169	−7.05	−0.169	−7.07
YEAR6	−0.083	−3.48	−0.084	−3.52	−0.084	−3.52	−0.086	−3.57	−0.085	−3.54	−0.085	−3.53
YEAR7	−0.063	−2.67	−0.063	−2.69	−0.063	−2.65	−0.065	−2.76	−0.063	−2.68	−0.063	−2.69
YEAR8	−0.015	−0.64	−0.016	−0.69	−0.015	−0.64	−0.018	−0.76	−0.015	−0.65	−0.016	−0.69
BS(k)	0.006	2.72	0.006	3.07	0.010	6.32	0.008	7.07	0.007	8.87	0.003	7.20
$N = 5689$												
R-squared	0.850		0.850		0.850		0.850		0.850		0.850	
AIC	−2291.680		−2293.330		−2320.670		−2328.380		−2350.320		−2331.300	

Table 8 Semi-detached

	k = *300 M*		*k* = *400 M*		*k* = *500 M*		*k* = *750 M*		*k* = *1000 M*		*k* = *1500 M*	
	B	*t-stats*	*B*	*t-stats*	*B*	*t-stats*	*B*	*t-stats*	*B*	*t-stats*	*B*	*t-stats*
LAMBDA	0.850	65.06	0.857	67.67	0.849	64.64	0.844	63.01	0.828	58.17	0.812	53.86
CONSTANT	11.380	20.98	11.751	20.46	11.359	20.29	10.836	18.01	9.403	16.17	8.249	14.03
FREEHOLD	0.139	10.12	0.139	10.14	0.140	10.18	0.141	10.24	0.141	10.28	0.144	10.47
NEW	0.085	4.31	0.082	4.16	0.085	4.32	0.087	4.42	0.095	4.85	0.099	5.04
FLOOR AREA	0.003	22.34	0.003	22.28	0.003	22.32	0.003	22.14	0.003	22.12	0.003	22.23
LOGDCBD	0.032	0.49	−0.003	−0.05	0.034	0.51	0.089	1.27	0.235	3.51	0.355	5.34
YEAR1	−0.800	−22.44	−0.799	−22.40	−0.798	−22.32	−0.799	−22.32	−0.802	−22.43	−0.807	−22.51
YEAR2	−0.668	−18.93	−0.669	−18.94	−0.669	−18.89	−0.669	−18.89	−0.674	−19.03	−0.677	−19.08
YEAR3	−0.501	−14.30	−0.500	−14.27	−0.499	−14.22	−0.500	−14.22	−0.503	−14.31	−0.507	−14.40
YEAR4	−0.272	−7.77	−0.272	−7.77	−0.272	−7.76	−0.272	−7.75	−0.275	−7.84	−0.279	−7.94
YEAR5	−0.129	−3.72	−0.130	−3.75	−0.129	−3.72	−0.129	−3.73	−0.133	−3.83	−0.137	−3.95
YEAR6	−0.062	−1.76	−0.062	−1.76	−0.061	−1.73	−0.062	−1.74	−0.065	−1.83	−0.068	−1.92
YEAR7	−0.040	−1.15	−0.041	−1.17	−0.040	−1.16	−0.042	−1.20	−0.043	−1.24	−0.048	−1.38
YEAR8	0.003	0.09	0.003	0.08	0.003	0.08	0.003	0.09	0.001	0.03	−0.004	−0.12
BS(k)	0.012	4.13	0.000	−0.04	0.004	2.26	0.004	2.98	0.006	6.90	0.004	8.80
N = 2478												
R-squared	0.859		0.859		0.858		0.858		0.859		0.859	
AIC	−1068.090		−1051.520		−1055.990		−1058.850		−1091.590		−1110.860	

References

Bailey, T. C., & Gatrell, A. C. (1995). *Interactive spatial data analysis*. New York, NY: Longman Scientific & Technical; J. Wiley, Essex, England.

Bollinger, C. R., & Ihlanfeldt, K. R. (1997). The impact of rapid rail transit on economic development: The case of Atlanta's MARTA. *Journal of Urban Economics, 42*(2), 179–204.

Debrezion, G., Pels, E., & Rietveld, P. (2007). The impact of railway stations on residential and commercial property value: A meta-analysis. *The Journal of Real Estate Finance and Economics, 35*(2), 161–180.

Department for Transport. (2017). *NaPTAN*. Retrieved 10 April, 2017, from http://naptan.dft.gov.uk/naptan/schema/schemas.htm

Doherty, M. (2004). *Funding public transport development through land value capture programs*. Ultimo: Institute for Sustainable Futures.

Efthymiou, D., & Antoniou, C. (2013). How do transport infrastructure and policies affect house prices and rents? Evidence from Athens, Greece. *Transportation Research Part A: Policy and Practice, 52*(Supplement C), 1–22.

George, H. (1879). *Progress and poverty*. New York: Schalkenberg.

Hess, D. B., & Almeida, T. M. (2007). Impact of proximity to light rail rapid transit on station-area property values in Buffalo, New York. *Urban Studies, 44*(5–6), 1041–1068.

Land Registry. (2017). Retrieved 10 April, 2017, from http://landregistry.data.gov.uk/app/ppd

LeSage, J. P., & Pace, R. K. (2009). *Introduction to spatial econometrics*. London: CRC Press.

Maoh, H., & Kanaroglou, P. (2007). Geographic clustering of firms and urban form: A multivariate analysis. *Journal of Geographical Systems, 9*, 29–52.

Maoh, H., Koronios, M., & Kanaroglou, P. (2010) *Modeling type choice outcomes of housing development projects in the City of Hamilton, Canada*. TRB 89th Annual Meeting Compendium of Papers DVD, 2010, Washington, DC.

Martin, R. (2001). Geography and public policy: The case of the missing agenda. *Progress in Human Geography, 25*(2), 189–210.

Orford, S. (2010). Towards a data-rich infrastructure for housing-market research: Deriving floor-area estimates for individual properties from secondary data sources. *Environment and Planning. B, Planning & Design, 37*(2), 248–264.

Riccardo, D. (1891). *Principles of political economy and taxation*. London: G Bell and Sons.

Ryan, S. (1999). Property values and transportation facilities: Finding the transportation-land use connection. *Journal of Planning Literature, 13*(4), 412–427.

Rybeck, R. (2004). Using value capture to finance infrastructure and encourage compact development. *Public Works Management & Policy, 8*(4), 249.

Sirmans, G. S., MacDonald, L., Macpherson, D. A., & Zietz, E. N. (2006). The value of housing characteristics: A meta analysis. *The Journal of Real Estate Finance and Economics, 33*(3), 215–240.

Smith, J. J., & Gihring, T. A. (2006). Financing transit systems through value capture. *American Journal of Economics and Sociology, 65*(3), 751–786.

Spinney, J. E. L., Maoh, H., & Kanaroglou, P. (2014). Location and land values: Comparing the accuracy and fairness of mass appraisal models. *Canadian Journal of Regional Science, 37*(1/3), 19–26.

Wang, Y., & Baddeley, M. (2016). The problem of land value betterment: A simplified agent-based test. *The Annals of Regional Science, 57*(2–3), 413–436.

Wang, Y., Potoglou, D., Orford, S., & Gong, Y. (2015). Bus stop, property price and land value tax: A multilevel hedonic analysis with quantile calibration. *Land Use Policy, 42*, 381–391.

Wang, Y., Feng, S., Deng, Z., & Cheng, S. (2016). Transit premium and rent segmentation: A spatial quantile hedonic analysis of Shanghai Metro. *Transport Policy, 51*, 61–69.
Willey, K., Maoh, H., & Kanaroglou, P. (2011). Exploring and modelling the level of service of urban public transit: The case of the greater Toronto and Hamilton area, Canada. *Transportation Letters, 3*, 77–89.
Xiao, Y., Orford, S., & Webster, C. J. (2016). Urban configuration, accessibility, and property prices: A case study of Cardiff, Wales. *Environment and Planning. B, Planning & Design, 43*, 108–129.

California Business Establishment Evolution and Transportation Provision

Adam W. Davis, Jae Hyun Lee, Elizabeth C. McBride, Srinath Ravulaparthy, and Konstadinos G. Goulias

Abstract This research filled a gap in empirically supported knowledge linking the survival and economic success of business establishments to locational characteristics including access to transportation facilities. This relationship was studied for the entire State of California while controlling in a statistically robust way for a variety of factors influencing business life cycle events, such as closures, formation/birth, and relocation. Using panel data analysis methods and data spanning two decades, the external (local context, transportation network) and internal (business type, number of employees, age of the establishment) factors that influence the success and failure of business establishments were examined. Using panel analysis, manufacturing and retail business establishments were studied in particular illustrating differences and commonalities in event occurrence and the factors that impact these events.

1 Introduction and Literature

This chapter seeks to fill a gap in empirically supported knowledge linking the survival and economic success of business establishments to local land use and access to the transportation system that serves these establishments. This relationship is investigated for the entire State of California over the last two decades while controlling in a statistically robust way for a variety of factors influencing business life cycle events, such as closures, formation/birth, and relocation. This is accomplished by combining longitudinal business establishment population event data,

A. W. Davis · J. H. Lee · E. C. McBride · S. Ravulaparthy
University of California, Santa Barbara, Santa Barbara, CA, USA
e-mail: awdavis@geog.ucsb.edu; lee@geog.ucsb.edu; emcbride@umail.ucsb.edu; srinath@geog.ucsb.edu

K. G. Goulias (✉)
Department of Geography and GeoTrans Lab, University of California, Santa Barbara, Santa Barbara, CA, USA
e-mail: goulias@geog.ucsb.edu

H. Briassoulis et al. (eds.), *The Practice of Spatial Analysis*,
https://doi.org/10.1007/978-3-319-89806-3_14

various transportation access and level of service indicators, the presence and density of neighboring businesses, and geographical market size from available US Census data.

The spatial distribution of economic activities has a profound impact on urban organization and development. Business establishments provide services and employment opportunities that influence the locational patterns of households, the behavior of individuals seeking to purchase goods and services, and the revenues of local jurisdictions. The spatial distribution of existing business establishments also affects location decisions of other business establishments and impacts the regional transportation network in the form of accessibility, traffic circulation, and possible congestion. In this context, integrated models of land-use and transportation are used to further analyze the impacts of these changes in regional planning and policy. Increasingly among researchers and practitioners, there is a heavy push towards more disaggregate modeling of these integrated systems. This type of modeling is based on the behaviors of relevant market agents such as households, persons, business establishments, and land developers that make decisions regarding their locations as well as personal travel and the movement of goods and services (Waddell et al. 2007; Strauch et al. 2005; Miller et al. 2010; Hunt and Abraham 2005; Ravulaparthy et al. 2017).

One of the most critical dimensions of these systems is the spatial distribution of economic activities, which is significantly determined by geographical movements of business establishments, together with business establishment formation and expansion, decline and closure. An approach to study these processes is to track the life cycle events of business establishments in a region and then develop simulation software that replicates the evolution of these business establishments while associating this evolution with urban planning policies (van Wissen 2000; Hunt et al. 2003; Elegar and Miller 2006; Kumar and Kockelman 2008; Moeckel 2009; de Bok and van Oort 2011; Maoh and Kanaroglou 2013). A fundamental aim of these model systems is to accurately describe the triggers underlying the dynamics in the spatial distribution of economic activities in a region. Therefore, to better represent business establishment dynamics, the spatial environment, and the location of these business establishments and their evolution process must be explicitly accounted for and depicted by these modeling efforts.

A few studies in this context have recognized the importance of the spatial dimension of business establishments represented as demographic processes (Bodenmann and Axhausen 2010; de Bok and van Oort 2011; Manzato et al. 2010a, b; Maoh and Kanaroglou 2007, 2009, 2013; Ravulaparthy et al. 2017) and analyzed the spatial characteristics in the form of accessibility to infrastructure, agglomeration economies, and regional effects and their impacts on business establishment decision-making behavior (Chen et al. 2011; Lei et al. 2012; Ravulaparthy and Goulias 2011). However, these studies address a limited selection of business establishment life cycle events. Manzato et al. (2010a, b) include space in modeling only business establishment survival rates, while Maoh and Kanaroglou (2007, 2009, 2013) model the spatial dimension of both business establishment migration and business establishment dissolution. In addition, these studies simplify the spatial representation framework in their modeling efforts, thereby ignoring important aspects of spatial effects (e.g., relative location in the region) on business establishment life cycle events.

Ravulaparthy (2013) rectified some of these issues and thoroughly investigated locational impacts on business establishment life cycle events by explicitly representing the relative importance of a location in the network across multiple spatial scales through the notion of roadway network centrality indicators. For this purpose, he examined the quality and locational advantage of a business establishment in terms of its closeness to other businesses, intermediacy in the paths of travelers, straightness of the paths along which each business is located, and the reach of other businesses by also accounting for the configuration and connectivity of the regional transportation network. These indicators of centrality were proven to be significant factors in business establishment financial success, the probabilities of relocating, dissolving, and in the creation of new business establishments. This analysis showed centrality to be important even when controlling for the fundamental linkages of business establishment internal and external factors on business establishment relocation, dissolution, and formation. Moreover, the influence of external factors to each business establishment depends heavily on the type of business and its locational needs (e.g., proximity to a freeway for a manufacturing firm versus a restaurant). To keep the analysis feasible, Ravulaparthy and Goulias (2014) performed this research exclusively with data from Santa Barbara County.

The basic data ingredients for the methods presented in this chapter are: (a) the longitudinal record of all business establishments in California in the NETS database; (b) highway and other transportation infrastructure locations for the entire State of California; (c) US Census data population counts at the block level. In Davis et al. (2016), a case study using fine-grained accessibility data for the Los Angeles area are also included. Merging and fusing data from different sources presents some challenges, which are solved by using business establishment locations as the fixed points of analysis.

The basic methodological ingredients for the study in this chapter are adapted from the work of Ravulaparthy and Goulias (2014) in a Santa Barbara case study estimating regression models for the probability of birth, dissolution, and relocation. New methods are presented here to extract land use and competition metrics from the business establishment data and handle the massive amount of data for the entire State of California. The key contribution of this research study is in operationalizing and unifying the internal, regional, and locational factors that affect business establishment survival to more conclusively identify the role of the transportation system and land use in economic development and business establishment success. The questions addressed are the following:

1. In what ways does access to transportation infrastructure affect the success, failure, and relocation of businesses?
2. In what ways does local land use affect the success, failure, and relocation of businesses? Which types of mixed-use environments are beneficial to business establishment success?

3. How do the effects of land use and transportation accessibility vary across different business types?
4. How do the factors that predict the formation of new business establishments relate to the factors that predict the dissolution of businesses?

2 Data Overview and Processing

This study investigates the success and failure of business establishments as a function of attributes of their local environment. To do this, a range of data sources, including a comprehensive record of business establishment life history events, major infrastructure locations, and census data were brought together. Figure 1 shows an overview of the data processing steps.

3 NETS Data Structure

The main data source for this project is the 2013 NETS database. This database contains geo-coded firm-level records for 6.7 million business establishments in California with longitudinal information about their industry type, location, headquarters, and performance over the period of 1990–2013. The NETS database is constructed by taking a series of "snapshots" based on the Dun and Bradstreet archival national establishment data (Walls 2007). The unit of observation in the NETS database is a business establishment that produces goods or services at a single physical location—for example, a single store. This database tracks every establishment from its formation to its dissolution, through any physical moves it makes, capturing any changes in ownership and business type. NETS records information on location of the establishment, employment, sales, and industry type for each year, a 6-digit North American Industrial Classification System (NAICS) code for each year the business existed, and a record of relocation events. Table 1 lists all business classes used in this study along with the number of unique businesses and total observations for each class.

In an extensive task of data processing, a full annual record of the internal and external factors hypothesized to affect business establishments in each category in each year was created. A number of other variables were calculated for use in the final model specification (Davis et al. 2016), namely a variable indicating whether the US economy was in recession for at least one quarter in the given year (1990, 1991, 2001, 2008, and 2009), firm age, age squared, and age cubed (this enables models to incorporate a polynomial approximation of the effects of business age on firm success). An approximate measure of business efficiency is calculated by dividing a business's sales by its employees (this variable is divided by 10,000 to bring coefficients more in line with those of other variables). Size categories were calculated from employee counts to address the nonlinear impact of business size on

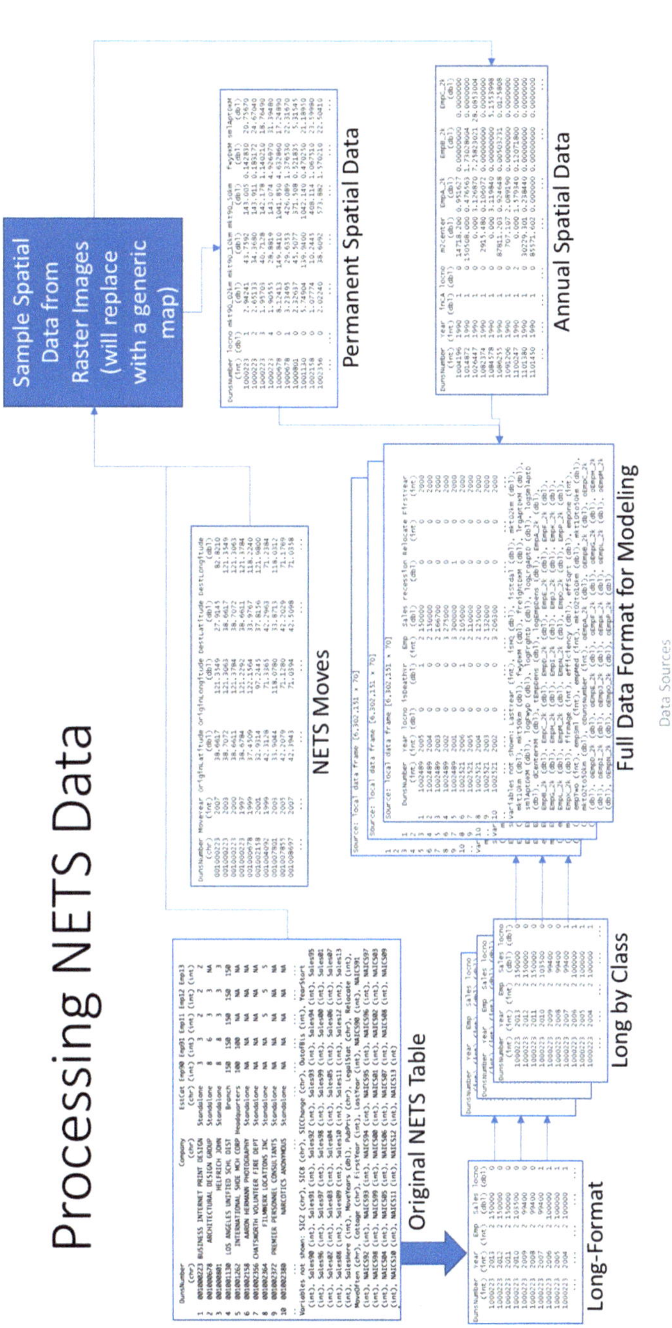

Fig. 1 NETS data processing overview

Table 1 Business establishment categories

Category	2-digit NAICS	Establishments (Unique DUNS)	Observations (DUNS:Years)
Agriculture, forestry, fishing, and hunting	11	78,990	728,397
Mining	21	6676	55,160
Utilities	22	7061	66,747
Construction	23	550,750	3,862,499
Manufacturing	31, 32, 33	343,107	2,916,915
Wholesale trade	42	431,344	3,005,409
Retail trade	44, 45	928,089	6,379,413
Transportation and warehousing	48, 49	169,354	1,111,009
Information	51	212,048	1,337,209
Professional, scientific, management, administrative, and waste services	54, 55, 56	1,975,382	12,129,165
Health care	62	466,341	3,700,886
Arts, entertainment, recreation, accommodation and food services	71, 72	346,838	2,871,635
Other services	81	699,929	5,259,591
Finance, insurance, real estate, and rental and leasing [FIRE]	52, 53	691,557	4,921,333
Public Administration and Armed Forces	92	29,665	247,909
Educational services	61	84,159	787,201
Undefined	99	9344	45,816

business success. Specific cutoffs were chosen by consulting the histograms of business size: one employee, two employees, three to eight employees, nine to twenty employees, and more than twenty employees; in each model, the reference case is large establishments (>20 employees). Event occurrence (business establishment formation, dissolution, and relocation) were also confirmed by checking the First Year and Last Year columns in the original dataset and looking for zero values in employee counts. It should also be noted that for all spatial variable processing steps, all raster datasets were snapped to the same grid. This makes the extraction process more efficient and makes it possible to check data quality in a consistent way and produce combined raster surfaces from model outputs.

4 Activity Density/Land Use Surface Estimation

Local land use is likely to play a major role in the success or failure of business establishments. Business establishment data can be used to compute multiple possible measures of land use, and it is likely impossible to represent all aspects of land use with a single variable. Diverse, high density environments foster agglomeration economies that help some businesses thrive, but they also likely drive up real estate rents, pushing out less successful businesses. The presence of related businesses may result in collaboration or intense competition. To measure land use, employee density in multiple business categories was used, which enables the investigation of both the direct effects of density and the effects of collaboration and competition. A summary view of the US Bureau of Economic Analysis Input-Output tables (U.S. Bureau of Economic Analysis 2014) shows that the direct financial relationships between groups of businesses were significant and diverse (see Fig. 2); some of these interactions likely require spatial proximity. The variation of land use preferences among different business classes was of particular interest in this study. Densities of other business establishments surrounding each business were computed using a kernel smoothing process with 2 km bandwidth (for more detailed reasoning of these spatial analysis steps, see Davis et al. 2016). Figure 3 shows the resulting density surface when aggregated to a block-group level for display purposes. This density generation process was repeated for each year for each of the 16 business categories.

5 Other Spatial Variables

Access to customers and employees is extremely important to business function, so it was an essential component of the models developed here. Population counts provide one measure of this important aspect of business, and in this project, total population within three distance bands (2, 10, and 50 km) was used to produce variables that can test for the effects of market area. Final models presented in this chapter use either total population within 50 km (which accounts for overall market area) or total population within 2 km and total population between 2 and 50 km (the first for local density and the second for overall market area). Market area variables were produced using block-level population totals from the decadal census and interpolated for years in between. The creation of the American Community Survey in the mid-2000s presents a tradeoff between fine spatial resolution (census blocks provide very detailed information about the distribution of people in small areas) and better-than-decadal temporal resolution, but in order to maintain a consistent dataset, census block totals were used for all years.

To create the market area dataset, block-level total population totals and boundary files for 1990, 2000, and 2010 available from the National Historical GIS (Minnesota Population Center 2011) were downloaded. Block polygons were converted to a raster grid with 50 m pixels snapped to the grid used for density and distance datasets; almost all blocks occupy numerous pixels. Block population totals were divided by the number of pixels each block occupies, and the resulting density value (measured in people per raster cell) was assigned to the block raster, producing a statewide map of block-level average population density. To convert this into a count of people within a certain distance, the ArcGIS Focal Statistics tool was used to count total value of all cells within 2 km, 10 km, and 50 km of each cell using a moving window technique (ESRI 2015a, b). Business establishment locations were

Formula Bar

Name	Agriculture, forestry, fishing, and hunting	Mining	Utilities	Construction	Manufacturing	Wholesale trade	Retail trade	Transport and warehousing	Information	Finance, insurance, real estate	Professional and business services	Education, health, and social assistance	Arts, ent., accommodation, and food services	Other services, except government
Agriculture, forestry, fishing, and hunting	103993	105	0	1399	280940	1547	2244	108	1	35	1799	916	8214	82
Mining	2520	58402	43079	10742	560306	46	49.00	1580	339	4297	1563	435	1286	592
Utilities	4955	4430	3076	2371	66293	5092	11171	6625	4414	71743	10765	25415	11448	4378
Construction	2884	5713	4749	158	16279	1577	2980	4683	2733	128332	2153	2970	2387	3884
Manufacturing	87134	52838	25686	281090	2042570	45665	59431	198525	105380	61146	163507	190020	155655	63114
Wholesale trade	26447	8437	4813	47406	300870	41132	25790	37865	29959	15763	28149	39472	26013	10642
Retail trade	220	230	378	82613	13762	813	7650	5505	406	7181	2694	1687	7992	5409
Transport and warehousing	14259	13205	18308	20930	160719	66389	71188	119692	19398	29061	53577	25000	16158	6490
Information	588	1225	1438	4530	26382	18091	20467	5818	241694	59498	79696	30260	10221	10078
Finance, insurance, real estate	21044	19499	10768	31460	85593	101708	147946	75213	68530	884455	246119	316319	93082	83370
Professional and business services	6245	45803	19550	46229	445816	206347	166840	65236	149996	440548	566526	251468	141266	45125
Education, health, and social assistance	495	0	148	20	85	1230	9106	75	927	50	817	27265	1975	2562
Arts, ent., accommodation, and food services	666	1197	2656	2444	23416	9312	6376	3398	35975	48182	69327	31235	29291	4222
Other services, except government	1156	734	737	4900	18173	18919	12264	5413	10735	33889	35044	36953	12107	7354
Government	47	6	454	26	6026	10472	6640	14479	2358	15152	9282	6726	6355	1824

Fig. 2 USA input-output table (source US BEA)

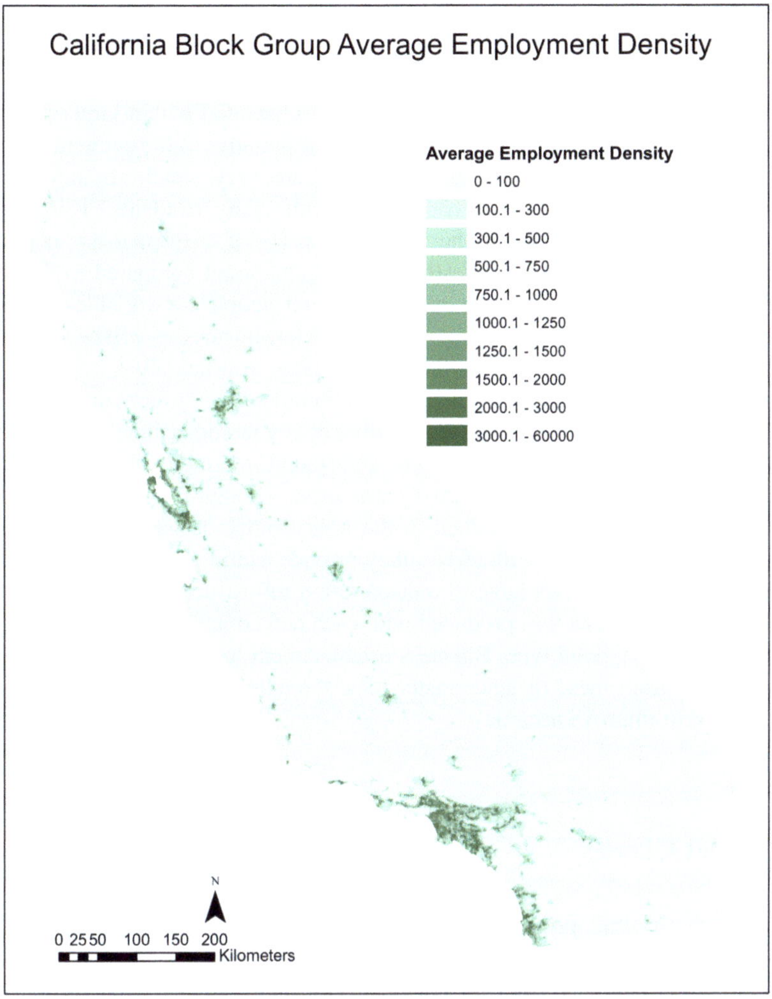

Fig. 3 Average employment density at the US census block-group level in 2012

then used to extract values from the nine market area raster files (three distance bands for each of three census years). Market area for a business location in a specific year between 1990 and 2010 was generated with a linear interpolation of the values from the two nearest census years (e.g., 2004 population was equal to 0.6 times the 2000 population plus 0.4 times the 2010 population). For 2011–2013, the annual population growth rate from the 2000s was projected forward. To make the resulting values more useful in a model, the 2 km radius population totals were divided by 10,000 and the 50 km radius and 2–50 km annulus population totals were divided by 1,000,000.

This method of estimating market areas entails two notable shortcomings: population is not evenly distributed within each census block and population totals are unlikely to change over time in a strictly linear fashion. Unevenness of human activity and natural attributes over space can be substantial within larger polygons (and assuming that zone-wide values are fully representative of the entire region they cover is called ecological fallacy), but blocks are very small, so they capture population density quite accurately in areas with many residents. This method underestimates local density in the populated parts of very sparsely populated blocks, but the scale of the underestimation is quite small compared to the range of density statewide (on the order of one hundred people for variables that have ranges of tens of thousands to several million) and few businesses will be affected by the error, so the problem will not affect model results substantially.

Access to transportation infrastructure is an important factor for many businesses. For this project, Euclidian distance to a number of key facilities, including freeways, small commercial airports, major airports, and freight/intermodal facilities (California Department of Transportation 2016) was used. Though these variables do not provide the level of detail contained in the accessibility datasets used in the Los Angeles case study (Davis et al. 2016), they provide useful general measures of the degree of access businesses have to transportation infrastructure. For each of these facilities, a raster dataset was produced with each cell containing the distance to the nearest feature of a given type. Business establishment locations were then used to extract values from these distance raster files; the extraction process used bilinear interpolation to improve accuracy.

6 Target Classes

The analysis in this project focused on four specific business types: retail, manufacturing, professional services, and health care. These businesses were chosen because they represent a range of business structures and are expected to prefer different types of environments. Retail businesses are included because retail establishments are generally very clearly delineated—each is an individual store. Retail businesses should respond to geographic variables because different locations provide different degrees of access to customers, real estate costs, and local competition. Retail businesses belong to NAICS categories 44 and 45. California retail businesses exhibited steady, but slow growth over the study period (Fig. 4). Manufacturing businesses were chosen because they are a classic focus of study in economic geography, and they are likely to respond to different locational cues than retail firms. Like retail establishments, manufacturers require access to the transportation network—in their case to moving inputs and products rather than to bring customers—and may prefer less-dense environments than retail businesses do. Manufacturing businesses belong to NAICS categories 31, 32, and 33, and the total number of establishments was very stable over time (Fig. 5).

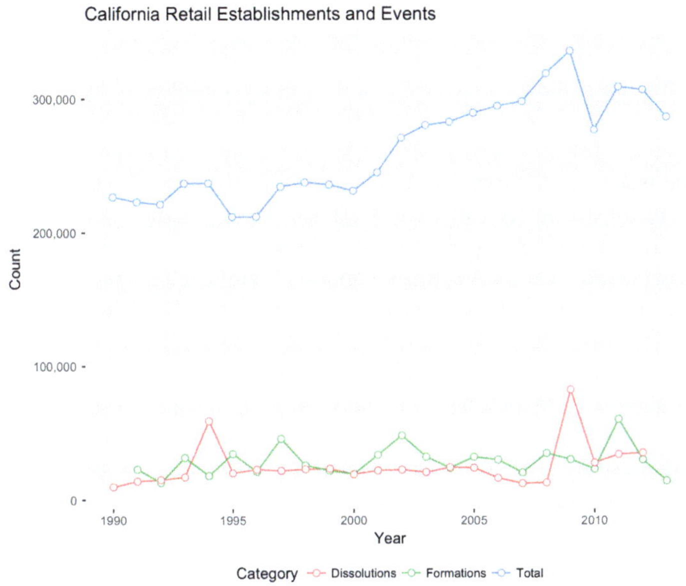

Fig. 4 Retail establishment events

Fig. 5 Manufacturing establishment events

7 Modeling Methods

While the reasons behind the success or failure of each individual firm are too varied and complex to fully capture in a general model, the modeling framework adopted in this study does allow to investigate the various attributes of business locations that make some places more favorable to businesses than others. Since these locational attributes are often directly subject to state and local planning decisions, they are particularly important to study. By analyzing how access to infrastructure, local land use, and local market areas affect the likelihood of individual businesses surviving or failing, a clearer picture of how to plan the economy of sustainable communities can be gained.

Probit regression was used to analyze the sensitivity of each event to a wide variety of internal and external variables of each business establishment. The reasoning for using probit regression and estimation details were provided in Greene (2003). Ravulaparthy (2013) provides a detailed review of the background of pooled estimation in this context. The results were presented in the same format as in linear regression models. In terms of interpretation, a positive significant coefficient indicates that the coefficient's associated variable is contributing positively to the occurrence of an event and negative indicates the opposite. A non-significant coefficient indicates no correlation between its associated variable and the event under study.

The panel models in this study were estimated with random effects. These are essentially a firm-level error term that contains unmeasured causes of variation between businesses that explains part of their success or failure. Random effects may be significant or insignificant depending on other variables included, but it is important to consider variability between members of the panel as well as between all observations because multiple observations of the same business are not independent.

The model development process started with a model that contains a large set of variables considered to be important, including most industry-specific densities. The model was then iteratively adjusted, removing some non-significant variables at each step until a final model was obtained that was both parsimonious and sensible. Internal attributes included at the start were business size (categorical as 1, 2, 3–8, 9–20 employees included as dummy variables, with businesses with more than 21 employees serving as the reference class), natural log of sales, efficiency (sales per employee), and a few measures of firm age. The initial market characteristics included were population within 2 km (divided by 10,000) and population between 2 and 50 km (divided by 1,000,000). When the model indicated that local density was not significant, both variables were replaced with total population within 50 km, again divided by one million. Log of kernel density for all 16 business category land uses were used in the initial testing for each class and insignificant variables were removed through an iterative process. Transportation access variables tested included distance to airports, distance to freeway, and distance to intermodal processing locations.

Sales per employee is an imperfect measure of the efficiency of a business because it ignores the differences in labor and input costs experienced by different businesses, but it should be relatively consistent within a given business category (for an alternate approach to business efficiency, see Ravulaparthy and Goulias 2015). To improve the usefulness of this variable and remove extreme cases, the analysis excluded establishments that bring in more than $400,000 in revenue per employee per year; depending on the random sample taken, this eliminated roughly 1.6% of retail business establishment observations and 2.2% of manufacturing business establishment observations.

The panel models included variables for which data existed at multiple time steps (such as the land use densities, population totals, and some firm internal attributes) or that were generally static over time (access to major infrastructure like airports and freeways, as well as some permanent firm attributes). In a parallel cross-sectional study for the Los Angeles area, these authors also experimented with the use of accessibility indicators used in regional simulation (Davis et al. 2016).

When coefficients are similar in size and direction between the two models, then the variable should be interpreted as predicting turnover rates rather than an overall shift in a sector's locational preferences. If a variable has a significant positive coefficient in the formation model and a similarly sized positive coefficient in the dissolution model, then it indicates that businesses in that sort of environment or with that characteristic will generally have shorter life spans but that those locations are also considered suitable sites for new business formation. If a variable has a negative effect on both formation and dissolution, it indicates lower turnover rates. Variables with a positive effect on firm birth and negative or no effect on death indicate areas of growth for a given category of firm, whereas variables with a positive effect on death and negative or no effect on birth indicate areas of decline. It is also worth noting that, in general, formation and dissolution are codetermined with the overall success of a business establishment, which is difficult to measure and is not directly addressed by this model.

7.1 Panel Models

For the panel models, the study focused on retail and manufacturing firms because these classes are similar in size and were more stable over the study period than were professional services and health care, both of which experienced substantial spikes in growth at the end of the period. These businesses were also expected to prefer different types of environments since retailers may prefer dense, mixed-use environments with access to customers and manufacturing firms are likely to perform just as well in more sparsely developed areas, so long as those are well served by transportation infrastructure.

As noted above, business creation, success, and failure are codetermined processes that may best be modeled simultaneously, but even these relatively simple models reveal many things about the ways in which business locations support or harm businesses. Birth and death models should be analyzed side-by-side to make it possible to distinguish between variables that cause change in the overall locations of businesses over time and those that predict higher turnover in certain locations than others. These models include only observations from the years 1995–2012; the study period was limited because establishment ages were not known for certain of those that began before 1990 and because excluding observations that were more than two decades old should improve the relevance of results to today.

In order to save processing time and keep within the memory requirements imposed by the statistics program used, random samples from the dataset were extracted and models were estimated on these samples. Given that these samples represent more than 10% of all observations for retail and over 20% for manufacturing, sampling error was small, and indeed repeating model estimation on new samples returned nearly identical results. Highly significant coefficients ($p < 0.01$) remained significant across all sample replications and the only variables to experience substantial changes were the insignificant ones included for comparison. Random effects were significant only in the birth models, possibly because the nonlinear time-trend variables used in the dissolution models masked the autocorrelation of firm events. The inclusion of random effects should not impact coefficient estimates one way or the other. The order in which variables are discussed in text matches the order in which these variables appear in the accompanying tables of coefficients.

7.2 Retail

In general, the models used for retail indicated that retail businesses fare best in dense, accessible urban environments in close proximity to entertainment and finance firms.

This project focuses primarily on the locational aspects of firm success, but internal attributes must be considered as well. The internal aspects of retail establishment formation and dissolution found in the panel models are generally unsurprising.

The four dummy variables for employee count that were used to indicate business size all have very significant and, with one exception, positive coefficients in both models (when compared to the reference class—businesses with more than 20 employees); this indicates that small businesses are generally more likely to fail and that new businesses are generally smaller than established firms (businesses are likely to start small).

As discussed in the section on general methods, sales per employee is an imperfect measure of a firm's efficiency, but it has a significant effect in these models. Less efficient businesses are more likely to fail since presumably they are less profitable. The negative coefficient in the formation model indicates that low efficiency is a useful predictor of a business's newness, suggesting both that it may take a few years of existence for a new retail business to reach its full market

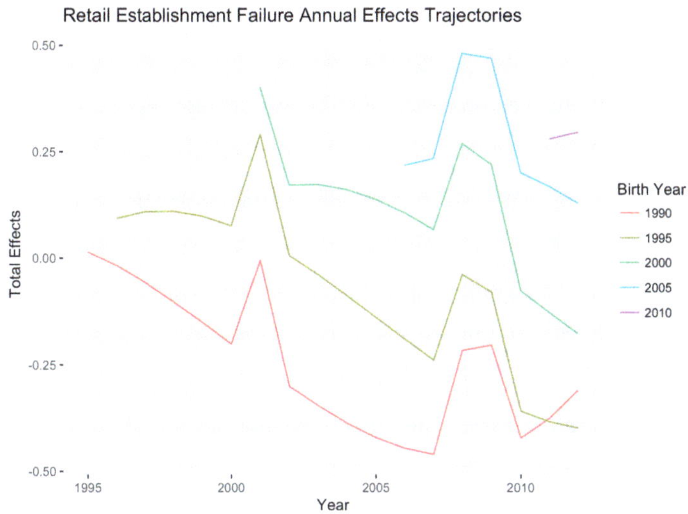

Fig. 6 Time effects on death of a retail business establishment

potential and that less efficient businesses are winnowed out within a few years, and are, therefore, disproportionately present among newer establishments.

Independent businesses experience higher rates of turnover than do those affiliated with a multi-establishment firm. The coefficient on the variable indicating whether an establishment is a standalone entity is positive for both the birth and death model, which means that they tend to have short life spans.

The effects of time into this model were incorporated in a number of ways. A simple linear time trend was included, as is an indicator for recession years (as discussed in the data processing section, it takes a value of 1 for the years 2001, 2008, and 2009 and is 0 for all other years). Additionally, the death model included firm age as well as the square and cube of age, which essentially represent a polynomial estimate of the change in a business's success over its life span. The combined effect of these five time variables over the study period is shown in Fig. 6, with each curve representing the effect on a business that opened in a different year. The combined effect of the time variables suggests that businesses are most vulnerable to fail in their first few years, after which their odds of failure decrease and eventually stabilize. There may be a significant uptick in failure odds for retail establishments over 20 years old. Recessions significantly increase the odds of business failure and significantly decrease the number of new firms that start.

The two market area variables that were included were chosen to indicate both a firm's local environment and its overall access to potential employees and customers. By assessing the formation and dissolution models jointly, it was found that increasing local density decreases turnover. Disproportionately, many businesses choose to locate in regions with high density, presumably because access to large numbers of customers is important for the success of retail firms, but the effect of regional density on failure rates was weaker and not clearly significant. The local population density effect may be related to the land use employment density variables, but it also may indicate that retail businesses benefit from having access to a very local customer base, despite this likely being associated with high rents.

As discussed in the data processing section, an attempt was made to capture the effects of collaboration/agglomeration and competition by using kernel density surfaces of employment counts in various sectors. These findings here can be taken as indicators of which sorts of mixed-use environments were best for retail businesses.

Retail businesses compete with each other for customers and real estate, leading to high retail formation and dissolution in areas with lots of retail employees. Retail also appears to compete to some degree with wholesalers though the effect on failure rates was much weaker. Retail establishments located near professional services firms also experienced higher rates of turnover; since these firms are often located in low-density suburban environments, this may indicate that these areas represent somewhat marginal sites for retail. Manufacturing does not have a significant effect on retail.

Health care businesses appear to exert an exclusively negative impact on retailers, as health care density predicts higher failure rates and lower birth rates. One potential explanation of this effect is that since health care companies often cluster around major hospitals, their rapid expansion over time may have pushed most other land uses out of these areas.

Retail firms benefit from areas with more entertainment businesses since these may attract shoppers and entice them to spend more time in the area. Finance firms and public administration density also appeared to benefit retail firms; the local presence of employees with high paying or stable jobs provides retail businesses with a stable customer base beyond what is offered by local residents.

Transportation infrastructure is a key component of business success, and for retail businesses, access to freeways was found to be the most significant predictor of all the variables tested. Retail establishments farther away from freeways were more likely to fail and new retail businesses were less likely to be located far from freeways. Retail businesses also prefer to locate near airports though this does not have a consistent effect on their success rate.

7.3 *Manufacturing*

These models indicated that manufacturing businesses generally were less likely to survive in mixed-use environments but had similar requirements for access to transportation infrastructure as retailers do.

The internal characteristics of manufacturing businesses mainly had the same effects as those of retail firms. Small-scale manufacturers generally performed worse, as did less efficient ones. In contrast to retail, standalone manufacturers generally experienced less turnover. The capital costs involved in opening a new factory are likely larger than those experienced by retailers, which may reduce the

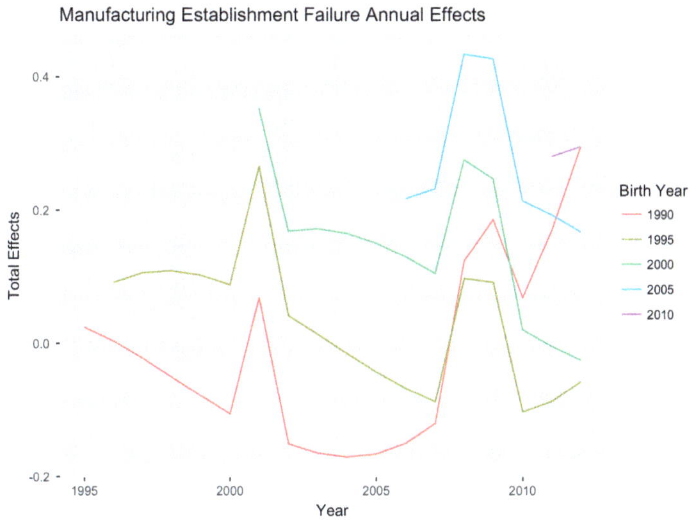

Fig. 7 Time effects on death of a manufacturing business establishment

flexibility of smaller manufacturers. Larger, multi-establishment manufacturing firms may also be more flexible for other reasons since they can more easily move operations to other parts of the United States or overseas.

The combined effect of time-trend variables on manufacturing firms was generally similar to the results found for retailers, but the effect of age cubed was much stronger (see Fig. 7), which means business failure probabilities were low and stable only briefly before beginning to rise again. A study that specifically addressed obsolescence would be better suited to investigate the causes of this increasing failure rate, but it was hypothesized that it may result from technological change and obsolescence of machinery and the increasing automation of manufacturing processes since these do not impact retail in the same way (Table 2).

The effects of market size on manufacturing firms were also generally similar to those experienced by the retail sector. Firms in dense areas experienced much lower turnover, though the effect was stronger on the formation side, indicating new manufacturing companies prefer to avoid extremely dense areas. Manufacturers also showed a slight preference for dense regions, presumably because access to large, diverse labor pools is important, though these effects were not consistently significant.

Whereas local competition increased turnover among retailers, manufacturing firms were less likely to fail when surrounded by other manufacturing firms. This difference may be explained by the absence of competition for customers due to specialization and complementarity, reliance on shared infrastructure, and zoning laws that seek to keep factories separate from other land uses. In fact, manufacturing firms appeared more likely to fail in any kind of mixed environment, as they performed worse in areas with high density of agriculture, retail, or transportation. In addition, information technology density was hypothesized to be beneficial for

Table 2 Panel data probit regression for births (formation) and deaths (dissolution)

Retail business establishment	Formation		Dissolution	
Variables	Coefficient	P-value	Coefficient	P-value
(Intercept)	−1.7945		−1.7781	
One employee	0.5418	<0.0001	0.1731	<0.0001
Two employees	0.5895	<0.0001	0.1917	<0.0001
3–8 employees	0.4470	<0.0001	0.2981	<0.0001
9–20 employees	−0.0783	0.0012	0.0633	0.0006
Sales/employees	−0.0196	<0.0001	−0.0116	<0.0001
Standalone business	0.3122	<0.0001	0.1527	<0.0001
Year—1990	−0.0230	<0.0001	0.0124	<0.0001
Recession year	−0.0279	0.0004	0.2462	<0.0001
Establishment age (years)			0.0285	<0.0001
Age squared			−0.0091	<0.0001
Age cubed			0.0003	<0.0001
Population within 2 km (per 10,000)	−0.0150	<0.0001	−0.0081	<0.0001
Population between 2 and 50 km (per million)	0.0062	<0.0001	0.0010	0.1891
Manufacturing density	0.0003	0.9299	0.0033	0.2379
Retail density	0.0274	<0.0001	0.0203	<0.0001
Wholesale density	0.0354	<0.0001	0.0065	0.0708
Prof. Service density	0.0272	<0.0001	0.0295	<0.0001
Health care density	−0.0214	<0.0001	0.0037	0.1435
Entertainment and food service density	−0.0287	<0.0001	−0.0166	<0.0001
Finance density	0.0013	0.8023	−0.0078	0.0668
Public Admin. Density	−0.0138	<0.0001	−0.0040	0.0001
ln(distance to freeway)	−0.0155	<0.0001	0.0065	0.0009
ln(distance to commercial airport)	−0.0141	0.0012	0.0012	0.7301
Rho (panel variance component)	0.3993		0.0000	

P-Values are reported here to show the exact Type I error. The lower the value the more significant a regression parameter is

manufacturing because research and development operations and manufacturing are likely to collocate, and this was borne out by the model. Over the study period, the high birth rate and low death rate of manufacturing firms situated in areas with information technology workers meant that these firms became more common over the study period and a strengthening relationship between information technology and manufacturing ensued.

Manufacturers generally performed best in areas with good access to freeways and airports (odds of failure increased with distance and new starts were concentrated at lower distance). Proximity to freight processing centers decreased turnover rates for manufacturing firms though the effect was much stronger for new starts (Table 3).

Table 3 Panel data probit regression for births (formation) and deaths (dissolution)

Manufacturing	Formation		Dissolution	
Variables	Coefficient	P-value	Coefficient	P-value
(Intercept)	−1.7752		−1.5653	
One employee	0.8896	<0.0001	0.1252	<0.0001
Two employees	1.0223	<0.0001	0.1471	<0.0001
3–8 employees	0.6352	<0.0001	0.1354	<0.0001
9–20 employees	0.1664	<0.0001	−0.0041	0.7252
Sales/employees	−0.0172	<0.0001	−0.0077	<0.0001
Standalone business	−0.0551	<0.0001	−0.1029	<0.0001
Year—1990	−0.0244	<0.0001	0.0126	<0.0001
Recession year	−0.0557	<0.0001	0.1989	<0.0001
Establishment age (years)			0.0228	<0.0001
Age squared			−0.0076	<0.0001
Age cubed			0.0003	<0.0001
Population within 2 km (per 10,000)	−0.0110	<0.0001	−0.0033	0.0148
Population between 2 and 50 km (per million)	0.0026	0.0400	−0.0013	0.1450
Agriculture density	−0.0100	<0.0001	0.0040	0.0048
Utilities sector density	−0.0028	0.0255	−0.0020	0.0334
Manufacturing density	−0.0237	<0.0001	−0.0208	<0.0001
Retail density	0.0192	0.0001	0.0151	0.0001
Transportation and warehousing density	−0.0052	0.1098	0.0079	0.0018
Information sector density	0.0056	0.0778	−0.0055	0.0273
Prof. Service density	0.0451	<0.0001	0.0270	<0.0001
Other service density	−0.0212	0.0002	0.0114	0.0094
ln(distance to freeway)	−0.0054	0.0664	0.0053	0.0157
ln(distance to commercial airport)	−0.0078	0.1036	0.0085	0.0156
ln(distance to freight processing)	−0.0269	<0.0001	−0.0059	0.0434
Rho (panel variance component)	0.3688		0.0000	

P-Values are reported here to show the exact Type I error. The lower the value the more significant a regression parameter is

7.4 *Leaving California*

Relatively few businesses relocate between states, but according to these models, relocation events share major commonalities with firm dissolution events, which suggests there are similar characteristics among the factors that cause businesses to leave the state and the locations that push businesses towards failure.

Among retail businesses, standalone firms were much more likely to relocate out of state, as were very large (>20 employees) and very small stores (one or two employees). Firms in retail-dense areas were less likely to leave, whereas retail firms in low-density/suburban environments were more likely to leave the state. Long-distance relocation of retail firms was generally rare (since it entails abandoning any long-term customers), the higher rates of relocation among retail firms in unusual locations indicated that there were unmeasured variables that may be playing an important role (e.g., extreme specialty stores and mail-order companies may be less tied to specific locations and thus more willing to leave).

Manufacturing firms that leave the state are disproportionately likely to be large, efficient firms (may be because they can afford substantial relocation costs). Firms in areas with less manufacturing were more likely to leave the state, which matches the birth/death model result that manufacturers prefer areas that specialize in manufacturing (e.g., technology parks). In contrast to what the death model showed, firms far from freight processing locations were more likely to leave the state, as were manufacturers located far from freeways, though this result was not clearly significant. These results indicated that manufacturing firms were more likely to leave relatively remote areas in California and may be relocating to find more suitable locations.

7.5 Relocation Within California

Though some business relocations involved moves between states, relocations within California were much more common. For these moves, our data provided a more complete picture because detailed information about the origin and destination environments was available. Unfortunately, this symmetry of information could not be extended to businesses that did not relocate, so these comparisons were made without the use of an econometric model. To eliminate autocorrelation, this analysis included only the final move recorded for each business establishment in the dataset. In this section, it is briefly investigated if the ways in which businesses that relocate within the state were different from those that did not, as well as the ways in which origin and destination locations differed from each other.

Table 5 compares the mean characteristics of businesses that moved with those of the entire dataset. The Total Observations row contains the number of business establishment—years recorded in the dataset. Slightly less than 1% of retail business observations include a relocation, and slightly more than 2% of manufacturing business observations do. For all variables, the difference between the move sample mean and the population are highly statistically significant, but this is partly an effect of the large sample size, and these differences are generally fairly small compared to the overall differences between firms. For both business types, the firm internal characteristics of movers generally match the population as a whole, but medium-sized, highly efficient firms were significantly more likely to move. Standalone retail establishments were somewhat more likely to relocate, but standalone manufacturers were slightly less likely to do so. For both business types, establishments in heavily populated regions (movers have a higher population within 50 km than do non-movers) with a good access to transportation infrastructure (the three distance to transportation variables are lower for the movers) were more likely to relocate, but the effects of local land use (population within 2 km, total employment density, and same-class density) were mixed. Retail businesses in highly dense environments

Table 4 Panel data probit regression for relocation out of California

Leaving California	Retail		Manufacturing	
Variables	Coefficient	P-value	Coefficient	P-value
(Intercept)	−2.3567		−2.3346	
One employee	0.0123	0.7534	−0.1114	<0.0001
Two employees	−0.0020	0.9601	−0.1300	<0.0001
3–8 employees	−0.0646	0.0852	−0.0905	<0.0001
9–20 employees	−0.0902	0.0411	−0.0232	0.3380
Sales/employees	0.0000	0.9556	0.0001	0.2024
Standalone business	0.1106	<0.0001	−0.2280	<0.0001
Year—1990	0.0165	<0.0001	0.0090	<0.0001
Recession year	0.0436	0.0137	−0.0010	0.9557
Establishment age (years)	0.0012	0.9134	0.0449	<0.0001
Age squared	0.0006	0.5962	−0.0040	0.0004
Age cubed	0.0000	0.2488	0.0001	0.0018
Population within 2 km (per 10,000)	−0.0623	<0.0001	−0.0424	<0.0001
Population between 2 and 50 km (per million)	−0.0033	0.1733	−0.0014	0.5670
Agriculture density			−0.0112	0.0018
Utilities sector density			−0.0101	<0.0001
Manufacturing density	−0.0027	0.7148	−0.0179	0.0055
Retail density	−0.0413	0.0010	−0.0229	0.0163
Wholesale density	−0.0082	0.3899		
Transportation and warehousing density			−0.0123	0.0466
Information sector density			0.0135	0.0289
Prof. Service density	0.0986	<0.0001	0.0783	<0.0001
Health care density	−0.0008	0.9060		
Entertainment and food service density	−0.0010	0.9137		
Other service density			0.0079	0.4631
Finance density	−0.0071	0.5268		
Public Admin. Density	−0.0067	0.0217		
ln(distance to freeway)	0.0212	0.0001	0.0073	0.1866
ln(distance to commercial airport)	−0.0203	0.0277	0.0004	0.9597
ln(distance to freight processing)			0.0340	<0.0001

were less likely to relocate (their mean density values are lower), whereas manufacturers in high density areas were more likely to relocate (Tables 4 and 5).

Comparing the origins and destinations of business relocation reveal some slight location preferences but also reveal the idiosyncratic nature of relocation decisions. Tables 6 and 7 contain mean values for origins and destinations of moves, as well as the percent of relocating businesses that increased the value of a given variable. Both retail and manufacturing businesses generally moved to areas of less-dense land use;

Table 5 Relocation origins versus all locations

Within California	Retail		Manufacturing	
Variables	Move origins	All locations	Move origins	All locations
Total observations	60,336	6,302,151	44,978	2,713,860
One employee	21.1%	22.1%	12.8%	19.9%
Two employees	26.8%	26.1%	14.2%	18.7%
3–8 employees	39.9%	39.2%	36.4%	33.1%
9–20 employees	8.2%	7.5%	18.6%	13.7%
21+ employees	4.1%	5.0%	18.1%	14.6%
Standalone business	87.1%	84.5%	84.0%	85.9%
Efficiency	$108,775	$98,320	$119,427	$99,401
Population within 2 km	32,076	32,418	29,025	28,532
Population within 50 km	5,505,318	5,073,490	6,128,219	5,603,230
Total employment density	2475	2617	2953	2820
Density of same class	259	267	580	454
Distance to freeway	2.2	3.1	1.7	2.6
Distance to commercial airport	19.8	22.6	17.9	20.9
Distance to freight processing	23.5	27.4	19.7	24.2

Table 6 Retail relocation origins versus destinations

Variables	Origin	Destination	% that Increased
Population within 2 km	32,076	28,806	43.4
Population within 50 km	5,505,318	5,366,531	54.6
Total employment density	2475	2067	41.3
Manufacturing density	262	236	43.7
Wholesale density	156	132	43.3
Retail density	259	217	41.0
Transportation and warehousing density	55	47	43.6
Information sector density	109	91	42.8
Professional services density	421	356	43.1
Health care density	223	184	42.5
Entertainment and food service density	210	172	41.7
Other service density	135	110	40.4
Finance density	248	206	42.5
Public Admin. Density	157	116	43.5
Educational services density	109	89	42.7
Distance to freeway	2.2	2.4	51.0
Distance to commercial airport	19.8	20.8	52.7
Distance to freight processing	23.5	24.7	52.9

Table 7 Manufacturing relocation origins versus destinations

Variables	Origin	Destination	% that Increased
Population within 2 km	29,025	25,282	42.4
Population within 50 km	6,128,219	5,816,715	50.4
Total employment density	2953	2421	41.3
Manufacturing density	580	513	44.4
Wholesale density	271	220	43.7
Retail density	243	196	41.1
Transportation and warehousing density	80	70	44.5
Information sector density	121	98	43.1
Professional services density	489	400	42.5
Health care density	176	144	42.9
Entertainment and food service density	189	152	41.7
Other service density	133	108	40.7
Finance density	264	204	42.6
Public admin. Density	135	94	44.4
Educational services density	104	81	42.8
Distance to freeway	1.7	2.0	51.9
Distance to commercial airport	17.9	19.5	53.5
Distance to freight processing	19.7	21.8	53.2

this result holds on average across all measures of employment density for both business types, but more than 40% of individual businesses moved to areas with increased density.

Population density tells a slightly different story—while businesses generally located to areas with less-dense population in their immediate vicinity and regional density decreases on average, a majority of businesses moved to higher-density regions. This effect likely represents a combination of two types of moves to suburban developments (whether malls or industrial parks): some businesses move from very dense areas to less-dense areas in order to decrease property rental costs and a larger number move from very low-density areas in order to increase access to customers and labor. Businesses also appear willing to sacrifice some access to transportation infrastructure when they move, but these results are much closer to 50–50.

Relocation decision-making patterns also changed over the course of time, as demonstrated by Figs. 8 and 9. These figures display the changes in origin and destination characteristics over time, smoothed using a spline technique both for easier visualization and to reduce the noise caused by year-to-year variation. Darker grey parts of the graph indicate the confidence interval around the moving mean value. It is important to note that these mean values are estimated quite precisely because they are calculated off of a sample of several thousand moves each year, but the origin and destination values from individual relocation events are quite varied.

Density around move destinations never surpasses the density around move origins, which indicates that businesses were more likely to relocate to lower-

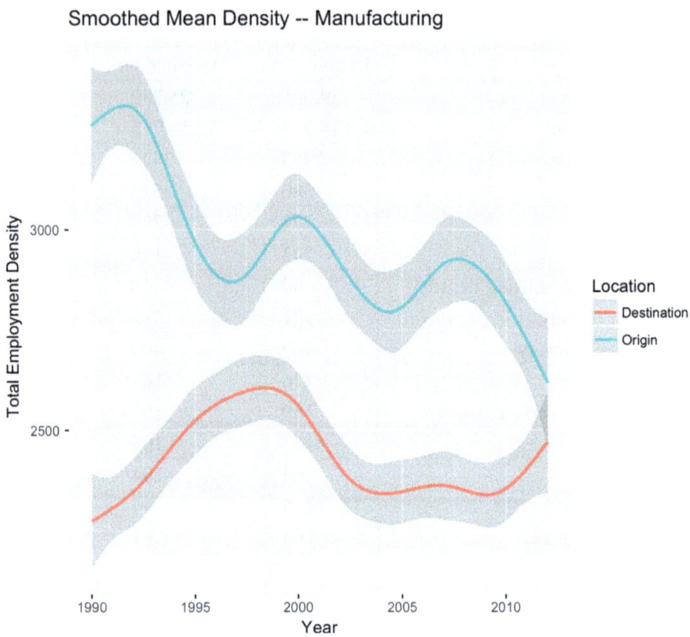

Fig. 8 Retail establishment relocation densities origins vs. destinations

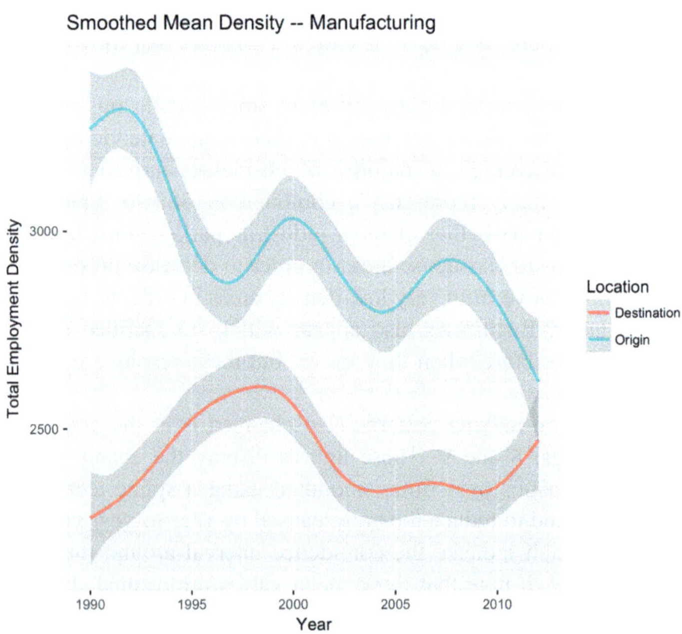

Fig. 9 Manufacturing establishment relocation densities origins vs. destinations

density areas throughout the study period, but the relationship between the means did change over time. Retail firms became much more likely to move to higher-density areas after the last recession (2010 to present), perhaps reflecting the relative strength of urban areas, which were less badly damaged by the collapse of the housing market than were many suburbs. Both figures indicate that economic expansions (mid-to-late 1990s and 2011-present) favored moves to denser areas. While the characteristics of manufacturing moves' origins is highly variable over time, the origin densities of retail firms remained remarkably stable over the study period, with a gradual increase that generally parallels the overall increase in density of California over time.

Relocation is a more likely decision than dissolution is. The formation and dissolution models can say that some areas make businesses more likely to fail or experience high degrees of business turnover, but the relocation results suggest that the dissolution results cannot be uniformly applied to the relocation process; relocation decisions change over time and are highly variable between businesses. Overall, businesses generally relocate to less-dense environments, which may indicate that relocation represents a type of failure—if businesses cannot succeed in high-competition dense environments, they may relocate to less-dense ones in order to stay open, whereas businesses that are successful in low-density environments may be more likely to expand by opening new branches rather than by relocating and abandoning a successful location. However, this relocation-as-failure interpretation is challenged by the overall high efficiency values of firms that chose to relocate.

8 Conclusions

Answers to the main research questions of this study are summarized below followed by limitations and next steps.

In what ways does access to transportation infrastructure affect the success, failure, and relocation of businesses?

Business establishments of all types rely on access to transportation infrastructure, but businesses with good access to infrastructure are not uniformly more likely to succeed. Transportation impacts the initial location decisions of most firms, as well as influencing success later on, and this increased demand drives up real estate costs and sometimes increases failure rates. Businesses are less likely to relocate to less-accessible areas than to less-dense ones, but local land use density appears to play a more significant part in relocation decisions than access to transportation does. Data limitations and the large scope of this project mean it may not be possible to extract the importance of specific transportation policies and projects on business establishment success on a statewide scale.

In what ways does local land use affect the success, failure, and relocation of businesses? Which types of mixed-use environments are beneficial to business establishment success?

Density provides lots of opportunities for collaboration but also for competition. Denser environments generally experience higher levels of turnover due to general competition and high real estate and labor costs. Mixed use is clearly beneficial in some cases, but not in others. Certain specific pairings are particularly beneficial–retail establishments are less likely to fail when they are located in environments that also provide customers with entertainment and dining opportunities; manufacturing businesses may perform better when in close proximity to information technology firms. Others are not—many business types perform worse in areas with high same-class density since direct competition drives down profits and small professional services firms perform best in areas with very low land use/employment density across all categories. Interestingly, manufacturing firms seem to do best in limited-use developments. Manufacturing turnover is lowest in areas with high density of manufacturing employees, and other land uses (including residential) likely prefer it that way as well since it keeps the noise and air pollution generated by manufacturing companies somewhat more contained in specific areas.

How do the effects of land use and transportation accessibility vary across different business types?

Manufacturing and retail showed similar preferences across all the models, but transit accessibility was notably important for retail and not at all important for manufacturing. Somewhat surprisingly, manufacturing showed no especially strong preference for access to freight processing centers, possibly reflecting the isolation of many of these centers in massive ports and inland train depots. Access to freeways is generally beneficial for all types of business establishments. As hypothesized, health care and professional services do not show any particularly strong preferences about transportation access, which indicates that their location decisions play a smaller role in their success/failure than do those made by retail and manufacturing firms.

How do the factors that predict the formation of new business establishments relate to the factors that predict the dissolution of businesses?

Many of the variables examined here had a strong positive bearing on firm birth but also on firm death (or were negative in both models). Rather than representing a net cost or benefit to a type of business, these variables serve as predictors of business turnover rates. It is beyond the scope of this study to investigate whether turnover has any substantial effect on a region's overall economic health, but the consistently high rate of turnover in dense areas indicates that this event may not be a disastrous occurrence. Turnover likely has more negative consequences for individual employees. This may be especially true in low-wage sectors like retail, where business closures mean changes in schedule and the need to learn new products, as opposed to higher-paying business types. Still, these areas also have higher rates of firm establishment, and the constant creation of new opportunities may be beneficial to workers.

Although this study succeeded in exploring important facets of business establishment evolution in California, a main limitation emerges from the lack of extensive accessibility indicators that are sensitive to the level of service offered by different transportation modes. This was done for a small region in Southern California but not for the entire State (Davis et al. 2016). A second limitation of the overall analysis is the lack of network-based centrality indicators statewide and for each year of the NETS business establishment inventory. This would have allowed a more direct correlation between business establishment events and central location measured using the infrastructure instead of the business density. Although it is feasible to expand the analysis here using network-based centrality indicators, this type of work may be more appropriate for city-level analysis (Ravulaparthy and Goulias 2015). A third limitation is due to the focus of this study on four types of businesses, and it may be worth testing the same models for all types of business establishments. Finally, it should be noted that business establishment events and not firm evolution and correlation of events with business practices were analyzed.

The present findings support the California land use change initiatives that are pushing for increased density and diversity of businesses and residences. For example, areas of high entertainment and food services industries have relatively low turnover in retail businesses creating diverse neighborhoods. Similarly, high density residential areas also have low turnover. This points to ordinances cities could possibly enact to strengthen the mix of land uses and create stable neighborhoods. In contrast, manufacturing does not lead to a very informative set of models to guide local land use policies. However, access to infrastructure is very important as one would expect and this could be an incentive to attract manufacturing plants.

Acknowledgement Funding for this research was provided by the California Department of Transportation (CALTRANS) through the University of California Transportation Center (UCTC).

References

Bodenmann, B. R., & Axhausen, K. W. (2010). *Synthesis report on the state of the art on firmographics*. Zurich: Institute for Transport Planning and Systems, ETH.

California Department of Transportation. (2016). *Caltrans GIS Data Library*. Retrieved February 14, 2016, from http://www.dot.ca.gov/hq/tsip/gis/datalibrary/

Chen, Y., Ravulaparthy, S., Deutsch, K., Dalal, P., Yoon, S., Lei, T., et al. (2011). Development of indicators of opportunity-based accessibility. *Transportation Research Record: Journal of the Transportation Research Board, 2255*, 58–68.

Davis A. W., Lee J. H., McBride E., Ravulaparthy S., & Goulias K. G. (2016). *Business establishment survival and transportation level of service*. Final Report. University of California Santa Barbara. Department of Geography, GeoTrans Laboratory, Contract Number: 65A0528. Submitted to the California Department of Transportation, May 2016.

de Bok, M., & van Oort, F. G. (2011). Agglomeration economies, accessibility and spatial choice behavior of relocating firms. *The Journal of Transport and Land Use, 4*(1), 5–24.

Elegar, I., & Miller, E. (2006). Conceptual model of location of small office firms. *Transportation Research Record: The Journal of Transportation Research Board, 1977*, 190–196.

ESRI. (2015a). *Kernel Density. ArcMap 10.3 Online Help*. Retrieved from http://desktop.arcgis.com/en/arcmap/10.3/tools/spatial-analyst-toolbox/kernel-density.htm

ESRI. (2015b). *Focal Statistics. ArcMap 10.3 Online Help*. Retrieved from http://desktop.arcgis.com/en/arcmap/10.3/tools/spatial-analyst-toolbox/focal-statistics.htm

Greene, W. H. (2003). *Econometric analysis* (5th ed.). Upper Saddle River.

Hunt, J. D., & Abraham, J. E. (2005). Design and implementation of PECAS: A generalized system for the allocation of economic production exchange and consumption quantities. In Lee Gosselin & Doherty (Eds.), *Foundations of integrated land-use and transportation models: Assumptions and new conceptual frameworks*. Elsevier.

Hunt, J., Khan, J., & Abraham, J. (2003). *Microsimulating firm spatial behavior*. In Proceedings of 8th International conference on computers in urban planning and urban management, Sendai.

Kumar, S., & Kockelman, K. M. (2008). Tracking size, location, and interactions of businesses: Microsimulation of firm behavior in Austin, Texas. *Transportation Research Record: Journal of the Transportation Research Board, 2077*, 113–121.

Lei, T., Chen, Y., & Goulias, K. (2012). Opportunity-based dynamic transit accessibility in Southern California: Measurement, findings, and comparison with automobile accessibility. *Transportation Research Record: Journal of the Transportation Research Board, 2276*, 26–37.

Manzato, G. G., Arentze, T. A., Timmermans, H. J. P., & Ettema, D. (2010a). Exploration of location influences on firm survival rates with parametric duration models. *Transportation Research Record: The Journal of Transportation Research Board, 2245*, 124–130.

Manzato, G. G., Arentze, T. A., Timmermans, H. J. P., & Ettema, D. (2010b). Location and accessibility mediated influences of office firm closure rates: A proportional hazard model. *International Journal of Urban Sciences, 14*(1), 1–15.

Maoh, H., & Kanaroglou, P. (2007). Business establishment mobility behavior in urban areas: A microanalytical model for the City of Hamilton in Ontario, Canada. *Journal of Geographical Systems, 9*, 229–252.

Maoh, H., & Kanaroglou, P. (2009). Intrametropolitan location of business establishments: Microanalytical model for Hamilton, Ontario, Canada. *Transportation Research Record: The Journal of Transportation Board, 2133*, 33–45.

Maoh, H., & Kanaroglou, P. (2013). Modeling firm failure: Towards building a firmographic microsimulation model. In F. Pagliara et al. (Eds.), *Employment location in cities and regions. Advances in spatial sciences*. Berlin: Springer.

Miller, E. J., Farooq, B., Chicgcuanco, F., & Wang, D. (2010). *Microsimulating urban spatial dynamics: Historical validation tests using ILUTE model system*. Presented at the Workshop in Urban Dynamics, Marbella, Chile March 22–25, 2010.

Minnesota Population Center. (2011). *National Historical Geographic Information System: Version 2.0*. Minneapolis, MN: University of Minnesota. Retrieved from http://www.nhgis.org

Moeckel, R. (2009). Simulation of firms as a planning support system to limit urban sprawl of jobs. *Environment and Planning B, 36*, 883–905.

Ravulaparthy, S. K. (2013). *Spatial perspectives in business establishment behavioral modeling: A case-study analysis in Santa Barbara County*. PhD Dissertation, University of California Santa Barbara.

Ravulaparthy, S. K., & Goulias, K. G. (2011). *Forecasting with dynamic microsimulation: Design, implementation, and demonstration*. Final report on review, model guidelines, and a pilot test for a Santa Barbara County application. University of California Transportation Center (UCTC) Research Project. Geotrans Research Report 0511-01, May, Santa Barbara, CA.

Ravulaparthy, S. K., & Goulias K. G. (2014) (in press). Characterizing the composition of economic activities in central locations: A graph-theoretic approach to urban network analysis. *Transportation Research Record: Journal of the Transportation Research Board*.

Ravulaparthy, S. K., & Goulias, K. G. (2015). *Exploring locational influences on technical efficiency of business establishments in Santa Barbara County using stochastic frontier modeling framework 2!*. In Transportation Research Board 94th Annual Meeting (No. 15-3498).

Ravulaparthy, S. K., Selby, B., Kupam, A., Jeon, K., Nippani, S., & Livshits, V. (2017). *Spatial firm demographic microsimulator: Development and validation for Phoenix and Tucson Mega-Region*. Paper presented at the 97th Annual Transportation Research Board Meeting, Washington, D.C.

Strauch, D., Moeckel, R., Wegener, M., Gräfe, J., Mühlhans, H., Rindsfüser, G., & Beckman, K. J. (2005). Linking transportation and land-use planning: The microscopic dynamic simulation model ILUMASS. In P. M. Atkinson, G. M. Foody, et al. (Eds.), *Geodynamics*. Boca Raton: CRC Press.

U.S. Bureau of Economic Analysis. (2014). Use Table (Before Redistribution, Producer Prices). Retrieved January 14, 2016, from http://www.bea.gov/industry/xls/io-annual/IOUse_Before_Redefinitions_PRO_1997-2014_Sector.xlsx.

van Wissen, L. J. G. (2000). A microsimulation model of firms: Applications of concepts of demography of the firm. *Papers in Regional Science, 79*, 111–134.

Waddell, P., Ulfarsson, G. F., Franklin, J. P., & Lobb, J. (2007). Incorporating land-use in metropolitan transportation planning. *Transportation Research Part A: Policy and Practice, 41*, 382–410.

Walls, D. (2007). *National Establishment Time Series Database: Data Overview*. Presented at 2007 Kauffman Symposium on Entrepreneurship and Innovation Data.

Investigating the Internal Compromise Between Wife and Husband's Commute Time Changes in Residential Relocation

Milad Ghasri and Taha Hossein Rashidi

Abstract This study investigates how employed members in a household compromise their commute times when deciding about the location of household's living place. The hypothesis of an internal negotiation between household breadwinners is statistically tested using a longitudinal survey of "Household Income and Labour Dynamics in Australia" (HILDA). The residential relocation history of dual-worker households is extracted from HILDA and according to the variations in commute time before and after relocation a reciprocal model specification is proposed. The reciprocal model is then transformed to a linear model, and the test of overall significance is conducted to validate the hypothesis. It is found that the household structure and wife's job type have significant impacts on how breadwinners adjust their commute times in response to their partner's conditions.

1 Introduction

Commute time for mandatory trips is a key indicator in transport and urban planning. The importance of commute time in transport planning is quite apparent as commuting patterns have direct impacts on network performance indicators such as total system travel time in the network and travel time between origin and destinations. The importance of commute time in urban planning comes from the long-term impact of travel time on people's decision regarding their home and workplace locations that form the land use pattern in urban areas and also determine commute distance and vehicles kilometers traveled for work trips (Clark et al. 2003). Therefore, commute distance affects and is affected by land use structure. On the one hand, households' home search behavior is tightly related to available transportation infrastructure and urban structure. On the other hand, work and school trips, which

M. Ghasri (✉) · T. H. Rashidi
Research Centre for Integrated Transport and Innovation (rCITI), School of Civil and Environmental Engineering, University of New South Wales (UNSW), Sydney, NSW, Australia
e-mail: m.ghasri@unsw.edu.au; rashidi@unsw.edu.au

H. Briassoulis et al. (eds.), *The Practice of Spatial Analysis*,
https://doi.org/10.1007/978-3-319-89806-3_15

contribute greatly to congestion in metropolitan areas, are highly affected by the location of homes and workplaces (Bayer et al. 2005). Furthermore, other activities of households such as shopping and recreational activities are highly affected by home and workplace location (Waddell 2002).

As commute time is a critical variable in both transportation and urban planning, there is a growing interest in developing integrated travel demand and land use models with a special focus on modeling residential and job location to estimate commute distance/time/cost in a more precise way (Salvini and Miller 2005; Pendyala et al. 2012; Rashidi et al. 2011; Rashidi and Mohammadian 2011; Farooq and Miller 2012). In an integrated travel and land use framework, information about travel time/cost is commonly exchanged between transportation and land use components in a sequential fashion or in a jointly modeled fashion (Wegener 2004a). Integrated transportation/land use frameworks appear to gradually receive wide recognition from the communities of urban planners and transportation engineers, while they are still in their infancy. According to Wegener (2004b), who has listed 20 urban land use/transport models in his review of urban modeling, the spatial input-output TRANUS (De la Barra 1989) and MEPLAN (Echenique et al. 1990), the MUSSA model (Martinez 1996) and METROSIM by Anas (1994) are integrated land use/transport models based on urban economics and econometric theories. Almost in all these integrated models, the connection between land use and transportation components is provided through commute distance/time.

In this vein, there are ample studies on modeling the relationship between commute distance/time and urban structure. In classical economic studies, residential location models are typically based on the trade-off between commuting costs and housing costs (Alonzo 1964). The general assumption in these models is that households prefer to live closer to their workplace (Clark and Burt 1980). People's tendency to reduce their commute distance provides the basis for modeling residential location (Alonzo 1964) and workplace location (Van Ommeren et al. 1997). Changing job location can also result in reducing commute distance (Van Ommeren et al. 1997; Van Ommeren 1998; Rouwendal 1999). Studies that have considered the decision of residential relocation and job change endogenously suggest that as commute distance increases, households are more likely to either change their residential location or accept a new job offer. Apart from the trade-off between living costs and transport costs, a wide range of topics related to commute distance is addressed in the literature, including the difference between commute distance for suburban employees and central city employees (Cervero and Wu 1997; Gordon et al. 1989a, b; Wachs et al. 1993), difference between men's and women's commute distance (Tkocz and Kristensen 1994) and the impact of commute distance on residential relocation and workplace change (Deding et al. 2009).

However, most classical studies consider a monocentric model of the city, which no longer is appropriate for modern urban regions. Since the development of the classical monocentric models of urban structure a profound spatial transformation has taken place. On the one hand, the monocentric structure of urban areas is giving up its place to polycentric structures (Berry and Kim 1993). Furthermore, the number of two-worker households is increasing which adds to the complexity of the relationship between commuting cost and housing cost (Clark et al. 2003). This is because while the two workers of the household live together, they typically do not have similar job locations. Therefore, selecting a residential location closer to one's job location might worsen the commuting condition for the other worker in the household.

Households with more than two workers introduce another level of complexity to the decision-making process as commute time of each household member matters. Complexity of group decision-making in households with two workers has attracted a few researchers while it requires further analysis. Since relocation affects the commuting distance and time of both workers, the question is how household members compromise their own commute time with their partner's commute time. There are several attempts in developing modeling frameworks for two-worker households in polycentric cities. Their findings suggest that two-earner households select their residential location in a way that makes commuting distance shorter for women (White 1977); Freedman and Kern 1997). The location of household residence is highly affected by women's earning opportunities and their commuting cost (Freedman and Kern 1997). Besides, households with children experience more challenges to reach an agreement as motherhood responsibility is critical (Fan 2015). Wife's commute time is not just affected by children's circumstances, but by social and, to a lesser extent, economic factors such as the casual attitude of women towards job, as they are often the secondary wage earner. This attitude towards wife's job can be considered as a critical variable determining their involvement in household-related issues that require them to be closer to home (Jun and Kwon 2015).

The dissimilar travel behavior of men and women in two-worker households and its complex impact on residential relocation is already acknowledged in the literature; however, the internal discussion between household members about the impact of residential relocation on their commute time is not addressed to the best of these authors' knowledge. The hypothesis of this study is that in a residential relocation process, workers in a household negotiate their changes in commute time to achieve or maintain a desirable commute time. In order to investigate the validity of this hypothesis, a mathematical relationship between the workers commute time changes is developed and its overall significance is tested. This study uses the longitudinal survey of "Households Income and Labour Dynamics in Australia" (HILDA) to estimate the parameters of the proposed model. Since the longitudinal structure of HILDA allows for tracking changes in commute time before and after relocation, it provides a suitable dataset to study the compromise made by household members.

Understanding how household members compromise their commute time is helpful in modeling the housing relocation process more accurately. When the internal compromise between household members is formulated, it can be utilized to adjust the choice set of households when deciding about their next living place. Residential location choice is a challenging decision to model mainly because of the large number of alternatives in a households' choice set. Several protocols are developed to reduce the choice set size including randomly selecting a subset of alternatives, using a primary model, or using a priory information, to name a few (Fatmi et al. 2016). One of the outcomes in this study is a mathematical formulation between changes in workers' commute time in dual-worker households. Therefore, it provides supplementary information for generating a more reasonable choice set in modeling residential location choice.

The rest of the paper is organized as follows. First, some basic descriptive statistics of the utilized dataset and the process of extracting it from HILDA are presented. Section 3 discusses the proposed model specification for capturing the commute time trade-off. In Sect. 4, the model parameters estimated are presented and the impacts of the explanatory variables are discussed. The conclusions section summarizes the highlights of the study.

2 Data

The data for estimating the parameters of the proposed model and for examining its overall significance were extracted from the longitudinal survey of Households Income and Labour Dynamics in Australia (HILDA). HILDA provides socio-demographic attributes of individuals, as well as, information on households' residential relocations on a yearly basis. This survey started in 2001 with nearly 17,000 individuals included in each wave of the survey. This study utilizes available information in the first 12 waves of HILDA between 2001 and 2013. The data processing started with extracting all dual-worker households. Then, for those households with a record of residential relocation, the change in commute time was calculated. If the household had multiple residential relocations, only the latest relocation was included in the dataset. Commute times before and after relocation were self-reported average travel time to/from the place of paid work during a week. Note that households' relocations do not necessarily occur in the same wave (year).

Once the dataset was processed, it encompassed details of 409 households. The variable of interest is commute time and its changes after relocation.

2.1 Explanatory Variables of the Model

Table 1 shows descriptive statistics of the variables included in the model. The Table presents the socio-demographic attributes of couples before their residential relocation (except for "commute time after relocation"). Four life-course events—"marriage," "separation," "giving birth to a child," and "job change"—were examined. Accordingly, four binary variables were defined which get the value of 1, if a person had experienced the life-course events in the year preceding to their relocation. Job type and education level are the other categorical variables that are transformed to binary format to be included in the model.

Descriptive statistics for husbands' and wives' are presented separately. According to Table 1, the average commute times before and after relocation are

Table 1 Descriptive statistics of examined variables

	Husband		Wife	
Variables	Mean (sd)	Percentage	Mean (sd)	Percentage
Commute time before relocation	4.68 (3.9)		3.74 (3.54)	
Commute time after relocation	4.96 (4.07)		3.89 (3.69)	
Changes in commute time	0.28 (4.07)		0.16 (3.67)	
Personal income	55499 (41340)		38785 (20148)	
Age	36.23 (10.18)		34.48 (9.65)	
Life-course events				
Marriage		6%		6%
Separation		2%		1%
Birth to a child		6%		6%
Job Change		24%		23%
Job type				
Managers		18%		9%
Professionals		30%		34%
Technicians and trades workers		17%		6%
Community and personal service workers		7%		11%
Clerical and administrative workers		8%		29%
Sales workers		4%		7%
Machinery operators and drivers		8%		1%
Laborers		8%		3%
Education level				
Postgrad—masters or doctorate		6%		5%
Bachelor or honors		24%		28%
Others		70%		67%

higher for men compared to women. This observation aligns with findings of previous studies (Tkocz and Kristensen 1994; White 1977).

Plots in Fig. 1 illustrate the relationship between percentage of change in husbands' commute time and wives' commute time. This figure shows that changes in commute time for both workers are less than 500% for the majority of households. At first glance, substantial changes, such as 500%, seem questionable. However, further investigation showed that the average commute time for people with a high percentage of increase in their commute time is below 15 min per week. Thereby, it is their short commute time before relocation that causes a very large percentage of increase in commute time.

In a broader perspective, when extreme cases of commute time changes are considered, a reciprocal relationship between changes in commute times can be discerned. Accordingly, large commute changes happen for one of the workers.

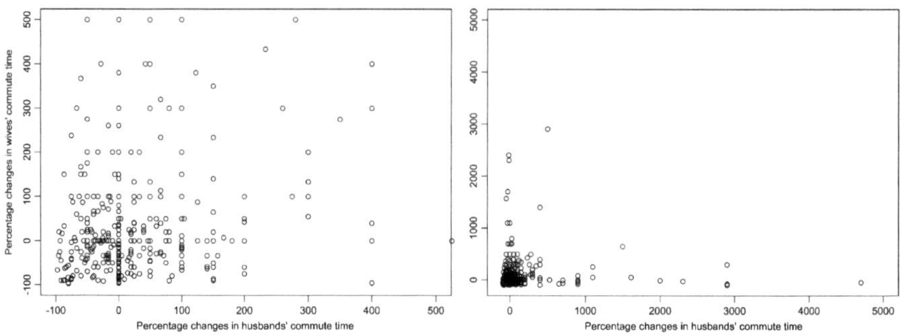

Fig. 1 Percentage changes in commute time for husband and wife

Table 2 Results of statistical test for husbands' and wives' commute time difference

Dataset	Husbands' commute time	Wives' commute time	Difference in commute time	Standard error	No. of observations	Test statistic	Critical statistic (t^*)
Before relocation	4.84	3.98	0.86	4.46	409	3.900	1.655
After relocation	5.18	4.05	1.12	4.76	409	4.773	1.655
Aggregated dataset	5.01	4.01	0.99	4.61	818	6.151	1.648

2.2 Differences Between Wives' and Husbands' Commute Time

According to Table 1, the average value of commute time for husbands is larger than that for wives, both before and after relocation. To confirm a significant difference, a statistical test was conducted where the null hypothesis was that "husbands' commute time is not greater than wives' commute time." The test was applied to three datasets: before residential relocation, after residential relocation, and the aggregation of both datasets. In each case, the difference between commute times was calculated and a one-tailed test for the difference being less than zero was examined. The results are summarized in Table 2. In all three cases, the null hypothesis was rejected at the 95% confidence level; therefore, the alternative hypothesis is accepted that husbands have an average higher commute time compared to wives in two-worker households.

This finding is already acknowledged in the literature (Tkocz and Kristensen 1994; White 1977; Freedman and Kern 1997; Jun and Kwon 2015). The explanations behind this observation can be attributed to sociological and economic reasons. From the economic perspective, women's jobs are more flexible since they have a more casual approach towards employment (White 1977; Fan 2017), which implies that women would find a job close to the living place of the household. From the

Table 3 Results of statistical test for commute time difference before and after relocation

Dataset	Before relocation	After relocation	Difference in commute time	Standard error	Test statistic	Critical statistic (t^*)
Husband	4.84	5.18	0.34	3.48	1.981	1.974
Wife	3.98	4.05	0.08	3.42	0.456	1.974
Aggregated dataset	4.41	4.61	0.21	3.45	1.731	1.964

sociological perspective, women are considered as the primary responsible person for family service roles (Jun and Kwon 2015). Therefore, households tend to live closer to the women's workplace.

2.3 *Changes in Commute Time After Residential Relocation*

Similar to the previous section, the hypothesis of meaningful changes in commute time after relocation was tested. Comparing mean values of commute time after relocation indicated an increase in commute times. The reason behind generally higher commute times after relocation might be due to increasing travel time. To further investigate these changes in commute time after relocation, the dataset was categorized into three groups of female workers, male workers, and the aggregation of both. The null hypothesis of this test was that the commute times before and after relocation are not different. Table 3 summarizes the results. The hypothesis of unchanged average commute time before and after relocation was rejected for the husbands' category. However, for wives, or for the aggregated category, the hypothesis of the mean value of commute time to be unchanged could not be rejected. This finding suggests that men are more willing to accept higher commute times which can be explained in two ways. First, men are more willing to adjust to the new circumstances of the family and to compromise their commute time in order to reduce their partners' commute time. Second, men show greedier behavior towards appealing opportunities and might accept large commute times to try new job opportunities.

3 Methodology

The main contribution of this study is to propose and validate a mathematical formulation between changes in commute times after residential relocation for dual-worker households. In order to formulate the relationship between changes in couples' commute time, two issues were taken into consideration: first, gender on its own should not play a role unless there is solid evidence available, and second, the formulation should be supported by the observed pattern in data. According to

Eq. (1), a reciprocal relationship between changes in the two workers' commute times is expected, especially for large changes. Moreover, according to the first issue, changes in wives' and husbands' commute times should have equal roles in terms of their importance. Equation (1) shows the assumed relationship between commute time changes in this study. In this equation, $\mathrm{PC_w}$ and $\mathrm{PC_h}$ denote percentage change in commute travel time for wife and husband, respectively, and α_w, α_h, and γ are the model parameters. As seen, substituting $\mathrm{PC_w}$ for $\mathrm{PC_h}$ does not make any difference in this formulation. In other words, the proposed reciprocal relationship is not systematically biased towards a specific gender.

By applying a few transformations, Eq. (1) was transformed into a liner one. The parameters α_w and α_h are the vertical and horizontal asymptotes of the model. In the present case, commute time cannot have a negative value; therefore, the percentage changes in commute time cannot be less than -1. So, both α_w and α_h parameters are set to 1. γ is the vertical dilation factor of the function which determines the smoothness of the curvature. In this study, γ is associated with the households' attributes. Since negative values of γ result in unrealistic cases, an exponential functional form was assumed for γ which is shown in Eq. (2). In this equation, x is the vector of explanatory variables, β is the vector of coefficients, and ϵ is the stochastic error term of the model.

$$(\mathrm{PC_w} + \alpha_\mathrm{w}) = \frac{\gamma}{(\mathrm{PC_h} + \alpha_\mathrm{h})} \tag{1}$$

$$\gamma = e^{\beta X + \epsilon} \tag{2}$$

After replacing γ in Eq. (1) and defining a new variable as $Y = (\mathrm{PC_w} + 1)(\mathrm{PC_h} + 1)$, Eq. (3) is obtained. One key feature of this functional form is that under the assumption that the error terms ϵ follow a normal distribution, the parameters of the model can be estimated by using the OLS method (Gujarati 2009).

Apart from the ease in estimating the parameters of the model, the transformed version of the relationship between percentage changes in commute time allows to conduct a standard statistical test to verify the hypothesized relationship in Eq. (1). The F test is used to examine the overall significance of Eq. (3).

$$\ln\left((\mathrm{PC_w} + 1)(\mathrm{PC_h} + 1)\right) = \beta X + \epsilon \tag{3}$$

4 Results

The Estimated paremeters of Eq. (3) based on the data collected are discussed below.

Table 4 Model estimation results

Variable	Estimated coefficient	*t* value	Pr(>|*t*|)
Intercept	1.816	5.10	0.00
Husband's commute time (HCT)	−0.254	−5.15	0.00
Wife's commute time (WCT)	−0.302	−4.98	0.00
Squared value of husband's commute time	0.011	3.49	0.00
Squared value of wife's commute time	0.015	3.17	0.00
Number of children younger than 5	−0.128	−0.97	0.33
House value (for owners)	1.92E-07	0.98	0.33
Couple without children	−0.507	−1.66	0.10
Couple with children under 15	−0.360	−1.15	0.25
Husband's job category: manager	−0.413	−2.32	0.02
Husband's job category: sales worker	−0.719	−2.16	0.03
Wife's income	3.90E-06	1.06	0.29
Wife's job category: clerical and administrative	0.340	2.25	0.03
Wife's job category: machinery and operator and driver	−1.117	−1.44	0.15
Wife's job category: labors	0.471	2.96	0.00
			$R^2 = 0.25$

4.1 Overall Goodness-of-Fit

Table 4 summarizes the model estimation results. The model passed the F test which indicates that the hypothesis of no linear relationship between the explanatory variables and the dependent variable is rejected. This finding suggests that the hypothesized relationship between changes in commute times after relocation cannot be statistically rejected.

However, the model suffers a low value of coefficient of determination (R^2). The poor goodness-of-fit can be explained by the complex nature of the decision. Although this study hypothesized that the decision on workers' commute time changes after relocation is more dependent on households' characteristics rather than on external factors (such as housing market and transportation system changes), the complex mechanisms of this decision cannot be easily formulated mathematically. For instance, as it was discussed in the Introduction, changes in commute time and the internal negotiation between household members is highly dependent on the reason of residential relocation. The changes in commute time of households who intended to move closer to one of the workers workplace is different from those of households who are moving to a more affordable living place. In the first case, one of the commute times is to be decreased, as it was the reason of relocation, but in the second case, both commute times might increase.

This study did not intend to estimate commute times after relocation. The present modeling exercise was helpful in verifying an internal negotiation between household workers on their changes in commute time after relocation and illustrating the impact of certain socio-demographic attributes on this compromise. However, modeling commute time after relocation requires further information and more holistic modeling approaches, as commute time and its variations depend on a wide range of factors, including socio-demographic attributes of households, job market, housing market, and transportation system.

4.2 Vertical Dilation Factor

Based on Eq. (2), several variables were incorporated in the model to parameterize the vertical dilation of the reciprocal function. As the value of the vertical dilation factor approaches zero, the curve gets closer to the origin of the system of coordinates. In contrast, larger values of the dilation factor indicate a smoother transition between the two variables. Figure 2: illustrates the impact of the dilation factor on the shape of the curve. For extremely low values of γ, the relationship between changes in husbands' and wives' commute time represents a *complementary* relationship. In this case, for a fixed change in the value of one worker's commute time, a wide range of commute time changes is possible for the other worker. That is to say, the decision of changes in commute time for one of the workers is fixed to some extent, and the changes for the other worker will be decided accordingly.

In contrast, when γ approaches larger values, changes in commute time represent a *substitution* relationship. The trade-off between changes in commute time is more

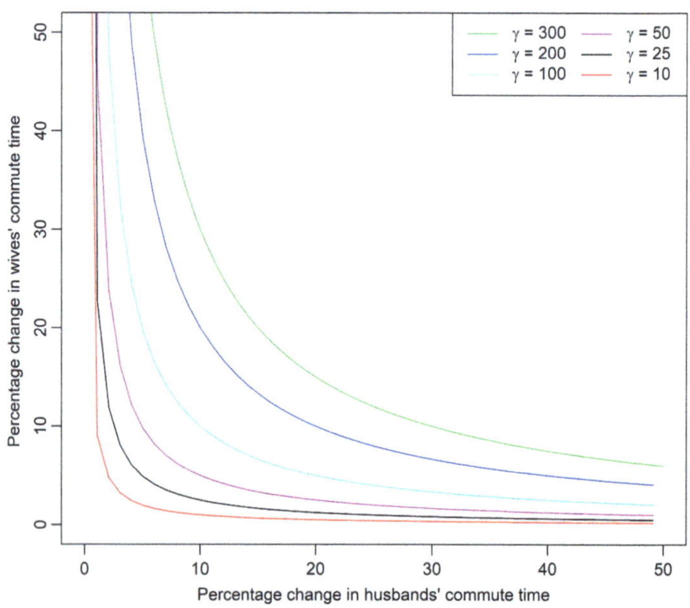

Fig. 2 The impact of vertical dilation factor

tangible in this scenario. In this case, at any point on the curve, one worker can exchange his/her commute time with the other person to improve the other worker's condition.

4.3 Impact of Current Commute Times

Interpretation of the estimated coefficients in Table 4 is based on the discussion on the vertical dilation factor. Based on the definition of γ, positive coefficients increase the dilation factor and vice versa. According to Table 4, the estimated coefficients of the current commute times for both workers are negative. Therefore, larger values of the current commute times result in complementary rather than substitution relationships. Two justifications are possible for this observation. First, when one of the workers has a large commute time, the residential relocation is possibly planned to reduce the unbearable commute time. In this case, the other worker will adjust his/her commute time accordingly. The second justification is for cases where residential relocation has purposes other than reducing one of the workers travel time. In this case, the worker with larger commute time feels less sensitive towards changes in commute time, therefore, allowing the other worker's condition to determine his/her new commute time. Note that, the squared values of commute times are also statistically significant in the model. This suggests a nonlinear, and perhaps more complex, impact from current commute time on the compromise between the changes.

4.4 Impact of Household Structure

Household structure plays a vital role because it reflects household members' commitments (Fan 2015). According to Table 4, the number of children below 5, as well as the binary variable indicating if household structure is "coupled with children under 15" have negative coefficients. Similar to current commute time, increasing the value of these variables results in stronger complementary relationships. This observation can be justified by the fact that, in general, having children adds to the responsibilities of partners. These can include additional house chores related to children, spending time with children for educational and entertainment purposes, or drop off at and pick up from childcare or school. All these responsibilities make the trade-off less flexible. A possible scenario can be a wife willing to have a short commute time to take the responsibility of children at home. In this case, it is the husband who will adjust his commute time accordingly.

4.5 Impact of Job Type

Different job types have different requirements and limitations. Some jobs have fixed working hours and require physical presence at the work location, while others are flexible and allow for teleworking. Therefore, workers' job type can have an impact on the decision of commute time compromise.

Based on the model estimation results, if husband is a manager or sales worker the decision on commute time compromise becomes a unilateral decision. This can be explained in two ways. First, higher responsibilities in these jobs and their specific requirements, such as fixed working hours, allow less leeway for commute time compromise. Second, the managerial jobs and sales workers have higher income; therefore, the husband can be considered as the main earner of the household which gives higher priority to his commute time changes.

On the other hand, job categories of "labor" and "clerical and administrative" for wives have a positive coefficient in the model. This indicates that employed women in these categories of jobs are more flexible and willing to compromise their commute time.

5 Discussion

The findings of this study can help in developing and appraising policies in two directions. Firstly, the quantitative tool developed to evaluate how workers commute time changes after residential relocation can be utilized to identify households' residential location after moving houses and to estimate changes in commute times due to relocation. Secondly, the study revealed the factors that have an impact on the balance of workers' commute times in dual-worker households. Other studies have shown that commute time has a direct impact on people's life satisfaction (Olsson et al. 2013). Therefore, encouraging households to maintain reasonable commute times and avoid extreme cases of commute times would have a positive impact on life satisfaction in society. The present findings suggest that if workers in a household have unbalanced commute times before relocation, they are more likely to have unbalanced commute times after relocation as well. This is because higher values of commute time before relocation shifts the relationship towards a complementary relationship where having extreme commute times after relocation is more likely. Similarly, having children under 15 produces a more complementary relationship between changes in commute time. Providing facilities around workplace locations, developing neighborhoods with mixed-land use, where childcare and schools are close to workplaces, are not unusual and can help households to plan their daily tasks more easily and achieve a more balanced commute time.

6 Conclusion

Commute time is a complex notion and depends on a variety of factors including household decisions on residential and job relocation. However, the negotiation between household members regarding their commute time changes in a residential relocation situation and their willingness to compromise their commute time to improve the condition of their partners is primarily an internal decision of households. This study supported this hypothesis by developing a mathematical formulation of changes in commute times in dual-worker households and conducting an overall significance statistical test.

Tracking changes in commute times after residential relocation showed that only one of the workers experiences large increases in commute time. This pattern represents a reciprocal relationship between changes in commute times in two-worker households. In this model, observed heterogeneity is addressed in the vertical dilation of the model which determines to what extent the changes in workers' commute time has a complementary or substitution relationship. In a complementary relationship, variations in one of the workers commute time does not improve or worsen the commuting condition for the other worker, whereas in a substitution relationship, a small increase in one workers' commute time will result in a high increase in the other workers' commute time and vice versa.

The vertical dilation parameter is found to depend on current commute time, family structure, and workers' job types. According to the results, for households with larger commute times, for those who have children, and for workers who are employed in managerial jobs, changes in one worker's commute time is more independent from changes in the other worker's commute time. That is to say, the relationship between changes in commute times in these families is closer to a complementary relationship. In this case, workers' commute times are less negotiable, and household members would prioritize their personal concerns over their partner's commuting condition.

The modeling framework of this study can help in predicting households' preferences for acceptable ranges of commute distance/time which, then, can be used to determine an acceptable set of zones the households will consider for relocating. In other words, knowledge about how family members decide about their commute time may enable modelers to identify a manageable set of regions in which the family will choose to live in. From another perspective, a second relationship between changes in commute times can be established based on households' budget and their expenditure on commuting. Setting up two equations for commute times after relocation enables modelers to identify the households' next dwelling location. Therefore, this study can be extended in two directions: first by improving the predictive capability of the model by investigating more explanatory variables and classifying households based on the reason of relocation, and second by investigating households' budget and their expenditure on commuting.

References

Alonzo, W. (1964). *Location and land use*. Cambridge, Mass.: Harvard Univ. Press

Anas, A. (1994). *METROSIM: A unified economic model of transportation and land-use*. Williamsville: Alex Anas & Associates.

Bayer, P., Ross, S., & Topa, G. (2005). Place of work and place of residence: Informal hiring networks and labor market outcomes. National Bureau of Economic Research.

Berry, B. J., & Kim, H. M. (1993). Challenges to the monocentric model. *Geographical Analysis, 25*, 1–4.

Cervero, R., & Wu, K.-L. (1997). Polycentrism, commuting, and residential location in the San Francisco Bay area. *Environment and Planning A, 29*, 865–886.

Clark, W. A., & Burt, J. E. (1980). The impact of workplace on residential relocation∗. *Annals of the Association of American Geographers, 70*, 59–66.

Clark, W. A., Huang, Y., & Withers, S. (2003). Does commuting distance matter?: Commuting tolerance and residential change. *Regional Science and Urban Economics, 33*, 199–221.

De la Barra, T. (1989). *Integrated land use and transport modelling. Decision chains and hierarchies*.

Deding, M., Filges, T., & Van Ommeren, J. (2009). Spatial mobility and commuting: The case of two-earner households. *Journal of Regional Science, 49*, 113–147.

Echenique, M. H., Flowerdew, A. D., Hunt, J. D., Mayo, T. R., Skidmore, I. J., & Simmonds, D. C. (1990). The MEPLAN models of Bilbao, Leeds and Dortmund. *Transport Reviews, 10*, 309–322.

Fan, Y. (2015). Household structure and gender differences in travel time: Spouse/partner presence, parenthood, and breadwinner status. *Transportation, 44*, 1–21.

Fan, Y. (2017). Household structure and gender differences in travel time: Spouse/partner presence, parenthood, and breadwinner status. *Transportation, 44*, 271–291.

Farooq, B., & Miller, E. J. (2012). Towards integrated land use and transportation: A dynamic disequilibrium based microsimulation framework for built space markets. *Transportation Research Part A: Policy and Practice, 46*, 1030–1053.

Fatmi, M. R., Chowdhury, S., & Habib, M. A. (2016). *A two-stage panel model of residential search and location choice decisions*. Transportation Research Board 95th Annual Meeting, 2016.

Freedman, O., & Kern, C. R. (1997). A model of workplace and residence choice in two-worker households. *Regional Science and Urban Economics, 27*, 241–260.

Gordon, P., Kumar, A., & Richardson, H. W. (1989a). The influence of metropolitan spatial structure on commuting time. *Journal of Urban Economics, 26*, 138–151.

Gordon, P., Kumar, A., & Richardson, H. W. (1989b). The spatial mismatch hypothesis: Some new evidence. *Urban Studies, 26*, 315–326.

Gujarati, D. N. (2009). *Basic econometrics*. Tata McGraw-Hill Education.

Jun, M.-J., & Kwon, K.-H. (2015). Why dual-earner households in Seoul live closer to the wife's workplace than the husband's? *Urban Policy and Research, 33*, 217–232.

Martinez, F. (1996). MUSSA: Land use model for Santiago city. *Transportation Research Record: Journal of the Transportation Research Board, 1552*, 126–134.

Olsson, L. E., Gärling, T., Ettema, D., Friman, M., & Fujii, S. (2013). Happiness and satisfaction with work commute. *Social Indicators Research, 111*, 255–263.

Pendyala, R. M., Konduri, K. C., Chiu, Y.-C., Hickman, M., Noh, H., Waddell, P., Wang, L., You, D., & Gardner, B. (2012). Integrated land use-transport model system with dynamic time-dependent activity-travel microsimulation. *Transportation Research Record: Journal of the Transportation Research Board, 2303*, 19–27.

Rashidi, T. H., & Mohammadian, A. K. (2011). A dynamic hazard-based system of equations of vehicle ownership with endogenous long-term decision factors incorporating group decision making. *Journal of Transport Geography, 19*, 1072–1080.

Rashidi, T. H., Mohammadian, A., & Koppelman, F. S. (2011). Modeling interdependencies between vehicle transaction, residential relocation and job change. *Transportation, 38*, 909–932.

Rouwendal, J. (1999). Spatial job search and commuting distances. *Regional Science and Urban Economics, 29*, 491–517.

Salvini, P., & Miller, E. J. (2005). ILUTE: An operational prototype of a comprehensive microsimulation model of urban systems. *Networks and Spatial Economics, 5*, 217–234.

Tkocz, Z., & Kristensen, G. (1994). Commuting distances and gender: A spatial urban model. *Geographical Analysis, 26*, 1–14.

Van Ommeren, J. (1998). On-the-job search behavior: The importance of commuting time. *Land Economics, 74*, 526–540.

Van Ommeren, J., Rietveld, P., & Nijkamp, P. (1997). Commuting: In search of jobs and residences. *Journal of Urban Economics, 42*, 402–421.

Wachs, M., Taylor, B. D., Levine, N., & Ong, P. (1993). The changing commute: A case-study of the jobs-housing relationship over time. *Urban Studies, 30*, 1711–1729.

Waddell, P. (2002). UrbanSim: Modeling urban development for land use, transportation, and environmental planning. *Journal of the American Planning Association, 68*, 297–314.

Wegener, M. (2004a). Overview of land-use transport models. In *Handbook of transport geography and spatial systems* (Vol. 5, pp. 127–146).

Wegener, M. (2004b). Overview of land use transport models. In *Handbook of transport geography and spatial systems.* Emerald Group Publishing Limited.

White, M. J. (1977). A model of residential location choice and commuting by men and women workers. *Journal of Regional Science, 17*, 41–52.

Part IV
Travel-Related Air Pollution Analysis

Aerosol Optical Thickness Mapper (AOT-Mapper): A Geo-Information Software for AOT Mapping at Urban Scale Using Landsat TM or ETM$^+$ Satellite Images

Nikolaos Soulakellis, Themistoklis Kontos, Nikolaos Sifakis, and Christos Iossifidis

Abstract Aerosol pollutants can engender, via light scattering and absorption, optical atmospheric effects on the signal recorded by high spatial resolution Earth Observation satellite sensors. Under certain conditions, these effects can be isolated on Landsat TM or ETM$^+$ satellite imagery, allowing tracking and mapping aerosol pollution at local and at regional scale. This study aimed at the development of a geo-information software, the aerosol optical thickness mapper (AOT-mapper). The AOT-mapper is a multi-module object-oriented software, specifically designed and developed for aerosol optical thickness monitoring and mapping on the basis of high spatial resolution satellite imagery in urban environments. It comprises the following processing modules: satellite image import, geometric correction of images, radiometric calibration and AOT calculation, and AOT visualization. The AOT-mapper has been successfully tested for mapping the aerosol load distribution over the city of Athens, Greece.

1 Introduction

Thematic mapping based on satellite imagery information facilitates analysis, modelling, and forecasting of both natural and anthropogenic phenomena, and of the important interactions between the natural and cultural domains, as well as

N. Soulakellis (✉)
Department of Geography, University of the Aegean, Mytilini, Lesvos, Greece
e-mail: n.soulakellis@aegean.gr

T. Kontos
Department of Environment, University of the Aegean, University Hill, Mytilini, Greece
e-mail: kontos@aegean.gr

N. Sifakis · C. Iossifidis
Institute for Space Applications and Remote Sensing, National Observatory of Athens, Athens, Greece
e-mail: sifakis@space.noa.gr

H. Briassoulis et al. (eds.), *The Practice of Spatial Analysis*,
https://doi.org/10.1007/978-3-319-89806-3_16

communication of results (Taylor 1994). Atmospheric pollution, produced from the interaction of human activities with the atmospheric environment, is of key importance from the environmental, socio-economic, and human health point of view. This interaction is more direct and intense in the case of urban air pollution. Thematic maps of air pollution, derived from satellite image processing, can be used to bridge the gap between air pollution simulation models and detailed ground-based measurements from local monitoring stations at various scales (Soulakellis et al. 2004). High spatial resolution (HSR) satellite images covering urban areas during air pollution episodes could provide spatial, timely, and reliable information on air pollution for assessing associated indicators, such as the aerosol optical thickness (AOT or τ_a). AOT can be retrieved by satellite measurements and corresponds to the atmospheric aerosol loading. The problem of urban aerosol load monitoring has attracted interest as scientific evidence has shown the adverse health effects of aerosols, particularly of small particles (e.g. Lipfert 1994; Pope et al. 1995). The AOT, which is defined as the linear integral of the extinction coefficient due to small airborne particles, can be considered as an air pollution indicator appropriate for aerosol assessment in urban areas because: (1) it is sensitive to a number of emissions and pollutant concentrations, (2) even if it is not limited to ground level aerosol concentrations it can be correlated to aerosol concentrations especially when relative humidity is less than 60% (Colls 1997; Cachorro and Tanre 1995, 1997), and (3) it is linked to visibility, which is the air quality indicator most people readily understand.

Various algorithms pertaining to AOT assessment by means of HSR satellite imagery have already been developed and applied to European urban sites, e.g. to Toulouse (France) by Sifakis and Deschamps (1992), to Nantes (France) by Wald and Baleynaud (1999), to Athens (Greece) by Sifakis et al. (1998) and Retalis et al. (1999), to Brescia-Italy by Soulakellis et al. (2004). There is, however, still a gap between existing algorithms and software that would consecutively perform all the required processing tasks. The purpose of the work presented herein was the development and implementation of a software, named "AOT-mapper", which allows the users to perform all necessary processing steps in order to calculate the AOT and visualize the results over urban areas via Landsat TM/ETM^+ images.

2 Method

2.1 Theoretical Background

Satellite Earth Observation systems (EO), apart from the retrieval of atmospheric parameters required in weather forecasting, namely pressure, temperature, humidity, and tracking of weather systems (retrieved mainly by meteorological satellites), can also be exploited to provide either direct or ancillary information on air pollution. EO systems have started showing their advantages in air pollution mapping at a variety of scales due to: (a) their coverage of extended territories in a homogeneous way,

(b) their variety of resolutions allowing to obtain comprehensive spatial data from the local to the global scale, and (c) a series of features of the electromagnetic radiation (EMR) measurements in the context of atmospheric investigations. Furthermore, contrary to the conventional sampling means that allow point measurements at ground level only, EMR used in satellite observations allows for a more representative coverage between two points along the observation path and informs on not-easily accessible areas of the globe (e.g. mountainous areas, polar regions). Despite the previous advantages, there are certain theoretical limitations in the use of EO systems in atmospheric observations. These are due to the fact that remote observations are feasible only if interaction mechanisms between EMR and the atmosphere are strong enough to be observable, isolated enough from other contaminating signals and sensitive enough to detect variations in the parameter of interest. Another limitation concerning satellite observations in urban atmospheres is the incompatibility between high spatial resolution required for observations in such big scales and very high spectral resolution required for detecting single atmospheric species.

Landsat satellites present a series of advantages in air pollution tracking compared to other high spatial resolution (HSR) satellites due to their stable orbit, large swath, and largest number of spectral bands (Wen et al. 1999; Moran et al. 2001). Sifakis (1987) classified AOT originating from industrial activities over Toulouse using SPOT images while Tanré et al. (1988) estimated Saharan AOT using Landsat-5 TM images. Several studies worldwide have continued using this approach for the calculation of AOT over urban areas. Sifakis and Deschamps (1992) developed the DTA algorithm while Sifakis et al. (1998) developed the SMA (Satellite Mapping of Aerosol) algorithm, specifically for use with Landsat data.

2.2 Algorithm Description

The SMA algorithm quantifies AOT using two physically independent principles, namely the "contrast reduction" and the "screening effect" both due to the presence of aerosols on the satellite images. The AOT-mapper software, which is presented in this article, has been based on this algorithm in order to retrieve AOT according to the following method. The EO imaging data, such as Landsat and SPOT, provide (usually 8-bit) digital values or numbers (DNs), which are functions of five components: (a) the target's real reflectance (ρ), (b) the technical characteristics of the sensor (gain A and offset B coefficients), (c) the observation geometry (solar angle θ_s, observation angle θ_v, azimuth angle z), (d) the illumination conditions (solar irradiance E_s), and (e) the interactions between EMR and the atmospheric constituents, quantified by the total optical thickness (τ). The total τ can be broken down into: (1) the optical thickness due to the natural atmosphere (τ_g), and (2) the optical thickness due to the pollution load (AOT or τ_a), which is assessed in this study. The objective of the AOT-mapper software is to isolate the pollution component and retrieve AOT through the following procedure. DNs are first calibrated by applying the A and B coefficients, which allows passing to at-satellite radiance values (R^*). R^* is then normalized by solar irradiance conditions and by observation geometry to obtain at-satellite reflectance values (ρ^*) as

$$\rho^* = \pi R^* / \cos(\theta_s) E_s \tag{1}$$

At this stage, AOT and the remaining observation parameters (θ_v and z) have been isolated as residual adjustment components between ρ^* and real reflectance values (ρ). If it is ensured that θ_v, z, and τ_{atm} are constant, two equations are needed to assess two unknown variables, namely AOT and ρ. The two equations can be provided by a couple of satellite images (say "image 1" and "image 2") of the same area so that $\rho_1 = \rho_2$. If the two images are acquired close in time, then z can be considered practically constant and, in the case of Landsat data, θ_v will be 90°, since images are always acquired vertically. Finally, the composition of the natural atmosphere may also be considered as invariant (τ_{atm} = constant). Based on a single scattering approximation, and according to Koschmieder's contrast reduction law (Middleton 1952), the difference of AOT in the two images can be assessed according to the ratio of contrasts (C) of ρ^* between the two images:

$$C(\rho_1)/C(\rho_2) = \exp(\tau_{a2} - \tau_{a1}) \tag{2}$$

$$\Rightarrow \ln[C(\rho_1)/C(\rho_2)] = \tau_{a2} - \tau_{a1} \tag{3}$$

Since "image 1" has been chosen as pollution free, which means $\tau_{a1} \approx 0$, then the retrieved AOT difference will correspond to AOT of "image 2" (τ_{a2}).

The SMA algorithm also engages a second procedure, based on the "screening effect" (Sifakis et al. 1992), as detector for possible artefacts in AOT variation induced by temporal changes in ρ (i.e. between "image 1" and "image 2"). The physical principle of this second effect is totally distinct from "contrast reduction" since the spectral region in which it takes place, that is the thermal infrared, corresponds to the energy emitted by the Earth (T) not by the energy reflected (ρ). This spectral area (between 800 and 14,000 nm) is little disturbed by scattering mechanisms. It is, however, sensitive to absorption mechanisms by aerosols. An increase of pollutant concentrations in the troposphere would decrease the observed at-satellite radiative temperatures through attenuation. At the same time, a reduction in the incoming solar radiation reaching the ground corroborates to the previous mechanism through reduction of the surface temperature. The following Eq. (4) is applied to the thermal band in the two images to estimate the observed radiative temperature variations (ΔT^*) at pixel level:

$$\Delta T^* = T_1^* + T_2^* = T_{E1} - T_{E2} + \Delta T_{a1} - \Delta T_{a2} + \Delta T_{g1} - \Delta T_{g2} \tag{4}$$

where

T_1^*, T_2^* are at-satellite radiative temperatures measured on two images

T_{E1}, T_{E2} are the surface radiative temperatures at the moments of image acquisition

ΔT_{a1}, ΔT_{a2} are local radiative temperature variations due to aerosols

ΔT_{g1}, ΔT_{g2} are radiative temperature variations due to changes in gas concentration.

If there is no noticeable difference in atmospheric pressure between the 2 days examined, then $\Delta T_{g1} = \Delta T_{g2.}$ Intrinsic radiative temperatures of the surface (T_{E1} and T_{E2}) are subject to variations in season and sun exposure between the two images. Assuming that $[T_{E1} - T_{E2}]$ is constant throughout the land area covered by the satellite scene, any residual at-satellite radiative temperature difference detected by the SMA code at pixel level can be attributed to attenuation engendered by aerosols. The SMA code combines the two procedures described previously by considering their mathematic intersection; the temperature-attenuation procedure based on the "screening effect" in the thermal infrared allows cross-checking the results of contrast reduction in the visible. The code finally maintains AOT values only for the areas where the presence of aerosols has been confirmed by both procedures.

The main prerequisite for the application of SMA is the availability of a pollution-free image. This image can be selected on the basis of available pollution data or by application of textural indices to the available images (Sifakis et al. 1992). AOT values can be then retrieved over the common cloud-free parts of the images and refer to atmospheric columnar values from the ground up to the satellite. Nonetheless since aerosol sources are on the surface, emitted particles remain in the troposphere and most particularly in its lowest part in less than 1.5 km of altitude (Lenoble 1993).

3 Software Description

The aerosol optical thickness mapper (AOT-mapper) is a multi-modular software (Fig. 1) especially designed and developed in order to meet the real needs and requirements for accurate aerosol monitoring and mapping using high spatial resolution satellite data. Its main goal is to provide a user-friendly geo-information software for pollution monitoring and mapping at both local and regional levels. More specifically, the AOT-mapper enables the accurate calculation and mapping of aerosol optical thickness via Landsat TM and ETM images by performing the following procedures:

1. Database management
2. Geo-registration
3. AOT calculation
4. AOT visualization

Taking advantage of the capabilities offered by commercial GIS and advanced computational software packages (i.e. ARC/GIS, ARC/MAP), the AOT-mapper software has been built in order to enhance its ease-of-use and its transferability from one operational area to another. Its different modules, however, have been developed on two programming languages, namely C and Visual Basic. The entire programming environment is object oriented in order to ensure an uncomplicated

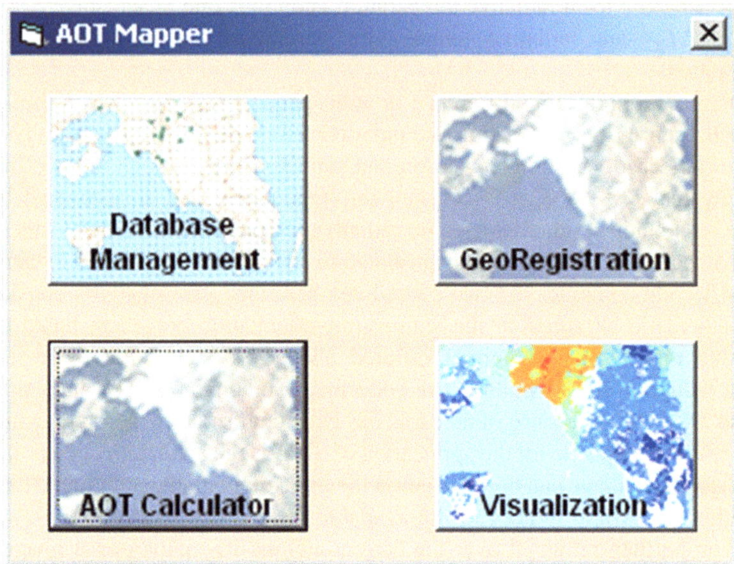

Fig. 1 Four modules of the AOT-mapper software as they appear in the user interface

upgrading of the software in subsequent releases. The AOT-mapper software comprises the following four modules:

Module 1 (Database management): This module enables the effective and user-friendly import, validation, and update of all the geographical datasets required for AOT calculation, mapping, and visualization. As such, it enables managing and storing not only of satellite images and AOT maps derived but also ancillary information in both geographical and tabular formats. The most important role of that module is that it acts as the *"main input gate"* of the AOT-mapper software for all required datasets. It concentrates, in one single module, all data import functionalities either from own databases ("inside" information—created by the rest modules) or external databases ("outside" information). This module can import data in any of the following standard formats such as ASCII data, *.xls, *.dbf, *.dat, GIS, and IMG. The second important role of this module is that it is the main information server for the rest of the AOT-mapper modules. Having a predefined database scheme, the other modules can directly access these datasets without having to perform any format conversion either manually or automatically (i.e. through the use of ad hoc "scripts" and pre-processors).

Module 2 (Geo-registration): This module enables the accurate geometric correction of satellite imagery (Fig. 2) required for AOT calculation. Despite the fact that the users of the AOT-mapper can import an image which has been already geo-registered by some other image processing software (e.g. ERDAS, ENVI), the existence of this module is crucial mainly for those users who are unfamiliar with other software.

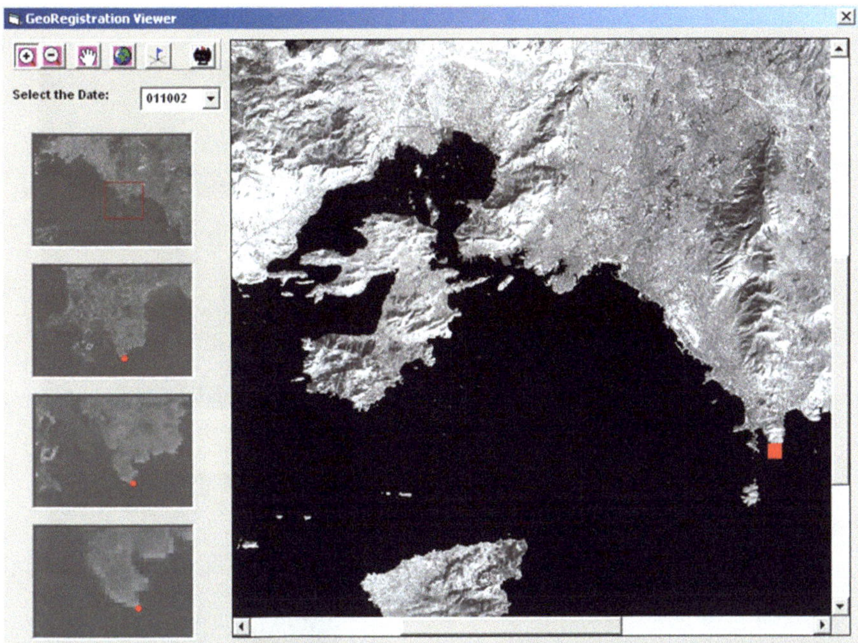

Fig. 2 Snapshot of the geo-registration module's user interface

Module 3 (AOT Calculator): This module enables the calculation of AOT values by processing high resolution Landsat TM and ETM satellite images (Fig. 3). The module implements updated versions of the existing SMA algorithm. This module is fully automatic since there is no need for the user to intervene in order to perform the necessary calculations. The software reads automatically all necessary input information from the satellite image header files and directly implements the appropriate algorithms. More specifically this module performs the following two main processing steps:

1. *Radiometric Counts to Radiance*: It performs counts to radiance conversion of the satellite data by reading all the required parameters, i.e. gain and offset values, sun elevation angle, and day of the year from the leader files of the images (Fig. 4) and
2. *AOT Calculation*: It performs the AOT calculation by applying the appropriate algorithm to the radiometrically corrected satellite images. This process is taking place after the radiometric correction of a new image, without the interaction of the end-user. The final output (in terms of geo-referenced AOT values) is stored automatically in the AOT-mapper database for its cartographic representation by the AOT visualization module.

Module 4 (AOT Visualization): This last module of the AOT-mapper software acts as "main output gate" of the results obtained (Figs. 4 and 5). The effective visualization

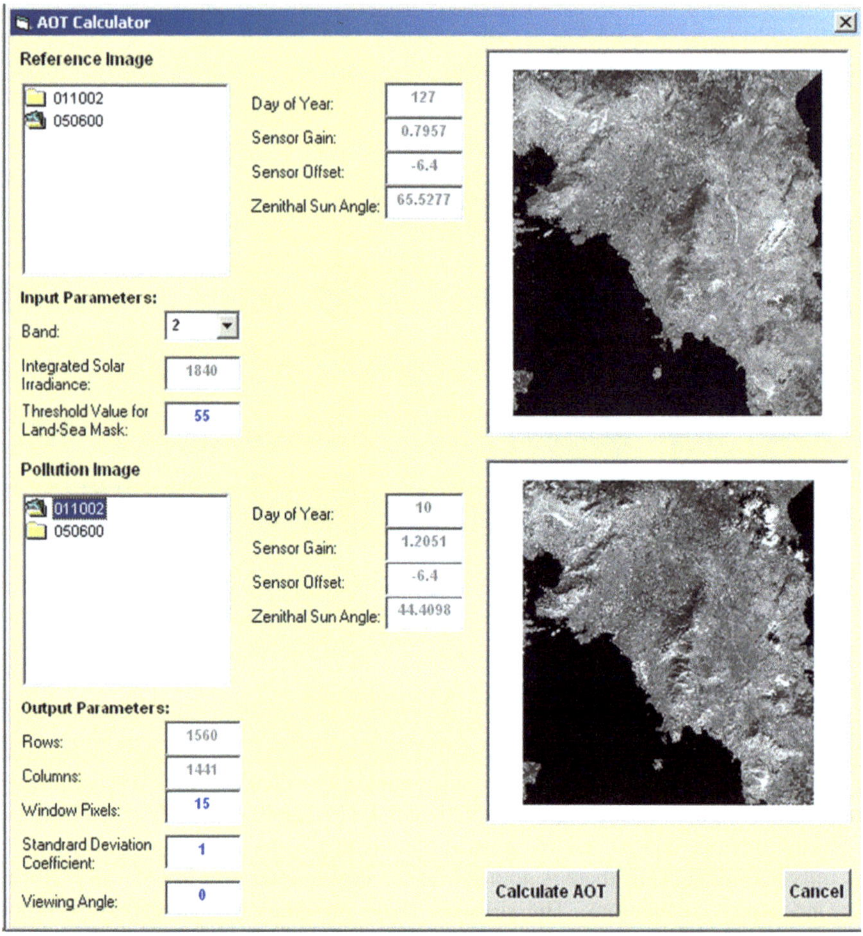

Fig. 3 Snapshot of the AOT calculator module's user interface

of the results, in terms of AOT values, is critical for a better understanding of air quality of the area of interest and, consequently, for decision-making concerning pollution abatement. The visualization was implemented through the development of suitable software for:

- Querying the database, mainly based on the date, and selecting data for mapping (using Visual Basic).
- Interactive cartographic visualization of the selected items (using Visual Basic and Arc Objects).

The need covered by this module is mainly to retrieve data and display them in a meaningful manner. These data are available in and retrieved from the database that has been developed and described in module 1. The elements involved have certain interrelations, which should be shown cartographically, in order to make meaningful

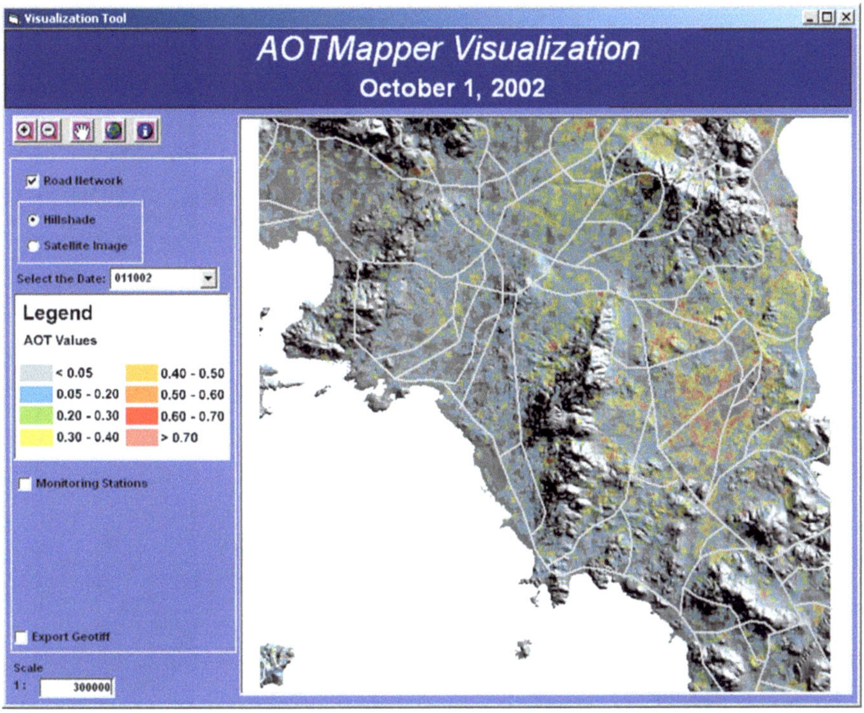

Fig. 4 AOT visualization module showing the AOT values over Athens city on 1st of October 2002

visual correlations on the map. The AOT visualization module (Figs. 4 and 5) offers to the users the following functionalities:

(a) *Zoom in/out to the desirable scale,* which permits the user to choose the appropriate scale for the visualization of the AOT values.
(b) *AOT values selection,* which permits the user to select the date desirable for visualization.
(c) *Information retrieval,* which permits the user to activate a thematic layer and interact with the map in order to retrieve the exact AOT value for a desirable point of the area.

During the development processes of this module, special attention was given to the interaction between user and visualization. This means that for all datasets there is a default palette for the AOT values. This palette (intervals of AOT values and their corresponding colour) cannot be changed by the user in order to avoid misconceptions during the visual interpretation of the obtained results among different dates.

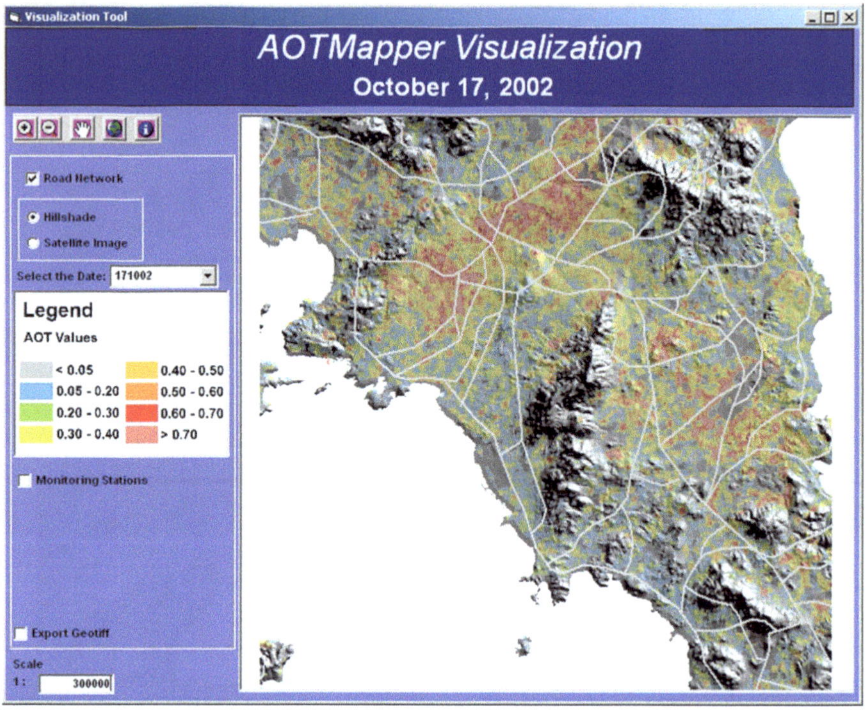

Fig. 5 AOT visualization module showing the AOT values over Athens city on 17th of October 2002

4 Application and Results

4.1 Area Description

The AOT-mapper was tested on the Athens urban area. Athens, the capital of Greece, has a population of approximately four million inhabitants, 15% of whom are over 65 years old, confined in an area of less than 500 km^2. The Athens urban area lies in a basin, having a NE-SW orientation, surrounded by three mountains and the sea (see Fig. 6). The climate of Athens is typically Mediterranean with wet mild winters and hot dry summers. The prevailing wind direction is north-northeast at the end of summer, in autumn, and in the winter and south-southwest in spring and the beginning of the summer. Because of her topography, population density, and climate, Athens experiences frequent air pollution episodes. The main air pollution sources are automobile traffic, industrial activity, and heating (during winter).

During the last two decades, air quality in Athens became an important research subject. Many researchers have focused on the collection and analysis of air pollutant measurements and meteorological parameters derived by the ground monitoring network (e.g. Kalabokas et al. 1999). The 17 monitoring stations operated by the

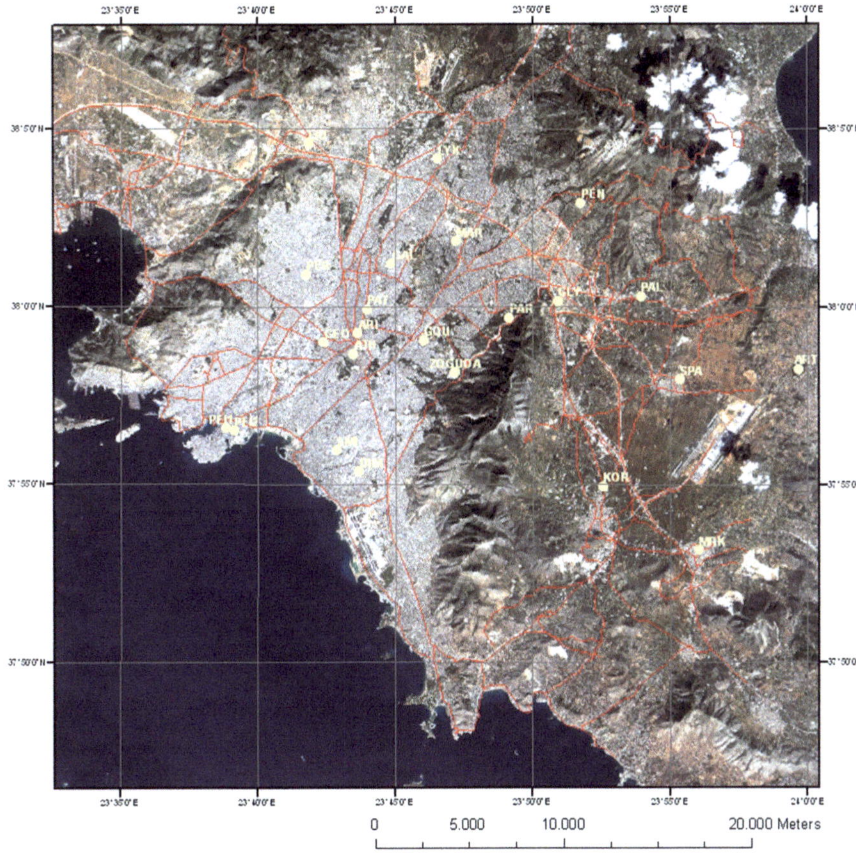

Fig. 6 Landsat TM image map (false colour composite, date: 17/10/2002) of Athens city showing synoptically the physical geography of the area, the road network, and the air pollutant ground monitoring stations

Ministry of the Environment measure regularly O_3, NO_X, CO, SO_2, and black smoke and, since 2001, PM_{10} and BTEX concentrations and provide a quite comprehensive time-series dataset for analysis aiming to better understand air pollution temporal variation over the city. In parallel, many researchers have focused on the development and operational use of numerical simulations (Kallos et al. 1993; Pilinis et al. 1993; Moussiopoulos et al. 1995; Ziomas et al. 1998; Kalabokas et al. 2000; Varotsos et al. 2001; Sotiropoulou et al. 2004; Bossioli et al. 2007). Finally, research activity is also focusing on the exploitation of EO data for obtaining spatially continuous information on air quality over Athens (Sifakis 1995; Sifakis et al. 1998; Retalis et al. 1999). Air quality monitoring and mapping from satellite images is based on several geo-information software, i.e. geographic information systems, digital image analysis software, and individual software, to perform all required processes.

4.2 Results and Discussion

The AOT-mapper software, comprising all the necessary algorithms and tools, has been applied to the Athens greater area using Landsat-5/TM and Landsat-7/ETM+ images. The satellite data were acquired on June 5, 2000, October 1, 2002, and October 17, 2002. The first of the three dates was selected as reference image; this date was ranked as the 225th most polluted day of 2000 with pollutant levels very low and can, therefore, be considered as the "cleanest" day among all. The processing was performed according to the following four steps:

- The first step was the development of the geographical database of Athens by importing the two satellite images, required for the calculation of the AOT values, and auxiliary geographical information, i.e. road network, administrative boundaries, and DEM, required for visualization purposes. This step was performed successfully using the database management module of the AOT-mapper software.
- The second step was the geometric correction of the two imported satellite images and was successfully by using the geo-registration module of the AOT-mapper software.
- The third step was the calculation of AOT values using the two imported and geometrically corrected satellite images and was performed successfully by the use of the AOT calculator module of the AOT-mapper software.
- The fourth step was the examination of the results obtained by the calculation of AOT values over the city of Athens on October 1, 2002. The examination was performed by the use of the AOT visualization module of the AOT-mapper software.

The two AOT maps derived were first visually examined and then compared to available ground-based pollution data. Considering both maps, a first general remark is a cluster of clouds covering the area northeastern of the city centre. These clouds were present on the reference image and the AOT-mapper successfully detected and delineated them. The AOT map of October 1, 2002 (Fig. 4) revealed significantly lower pollution levels compared to those of October 17, 2002 (Fig. 5). In particular, the AOT map of October 1 showed low to moderate air pollution over the city centre and AOT values ranged from 0.1 to 0.4 inside the urban area. However, in the eastern and northeastern part of the city as well as in the vicinity of the international airport (east of the central part of the image) some areas presented higher AOT values ranging from 0.5 to 0.8. The AOT map of October 17, 2002 (Fig. 5) revealed higher pollution load west and north of the city centre and AOT values ranged from 0.6 to 0.8. On the same map, the pollution cloud has a southwest to northeast orientation, which is quite typical for days when a land-sea breeze system develops (Sifakis 1995). Similar to the previous map, air quality east of the Athens basin, particularly around the international airport, appeared particularly degraded with AOT values reaching 0.8.

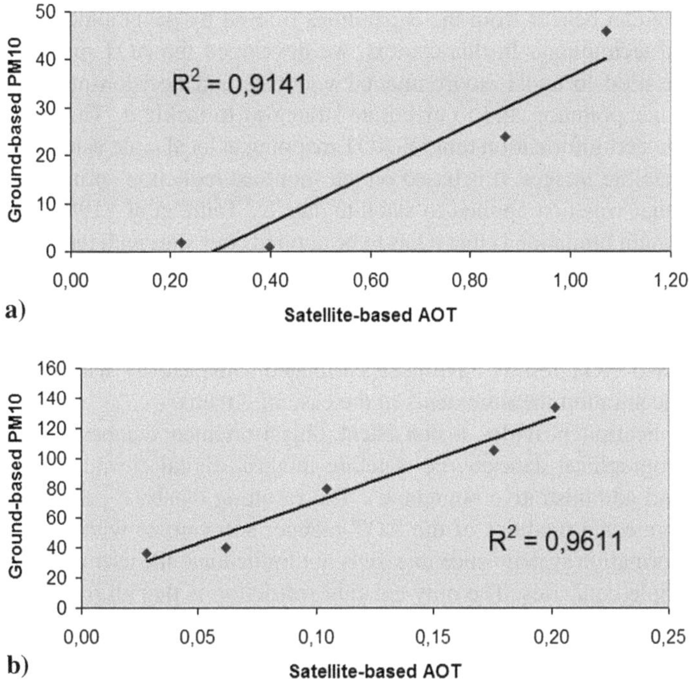

Fig. 7 Aerosol optical thickness derived by means of the AOT-mapper vs. PM10 concentrations measured at ground level by the Earth air quality control network. Graph (**a**) corresponds to three stations located at the periphery of the city centre and are surrounded by vegetation while graph (**b**) corresponds to three station located in city centre

The above findings agree with pollution data from the Earth monitoring network as the 1st of October ranked as the 170th most polluted date in 2002 and the 17th of October ranked as the 73rd most polluted date in 2002. A quantitative correlation to pollution levels in terms of aerosol concentrations was also carried out. Landsat-based AOT were correlated to PM10 concentrations measured at ground level by the Earth air quality control network ($R^2 = 0.91$). This correlation was considerably higher when AOT was correlated to PM10 only from the stations located in the city centre ($R^2 = 0.96$) (i.e. Aristotelous, Goudi, and Pireas) (Fig. 7a, b). A possible explanation may be the existence of different land cover classes in the two cases: dense urban tissue in the city centre vs. less dense urban tissue mixed with vegetation in the second case. Mis-calibration between the measuring devices at the various stations is also possible.

5 Conclusions

Urban air quality greatly depends on aerosol load which, in turn, is directly linked to an optical indicator that can be retrieved by satellite sensors; the aerosol optical thickness (AOT). AOT retrieval and mapping by means of high spatial resolution satellite data can benefit from the capabilities offered by geographical information science and technology. In this context, we developed the AOT-mapper software that can be used to assist environmental scientists and decision-makers to better understand air pollution and to elaborate strategies to tackle it. The AOT-mapper software is a geo-information tool for AOT mapping at local scale using Landsat TM or ETM^+ satellite images. It is based on the "contrast reduction" principle (Middleton 1952) that was first applied to satellite data by Tanré et al. (1988) and Sifakis (1987). Its main limitation is that it has to be applied over sites with temporally stable reflectance (Liang et al. 2001), which is often (but not always) the case in urban areas. The software was tested on the urban area of Athens over a limited but representative series of Landsat satellite data for AOT retrieval and mapping. Satellite-based AOT results have been correlated satisfactorily with ground-based aerosol concentration measurements in the case of Athens.

The application provides a consistent object-oriented database including all primary geographical datasets, i.e. satellite images, digital elevation model, road network, and administrative boundaries. The resulting database can be used either with the connected modules of the AOT-mapper software or with any other geographic information system since its size is not limited and the user does not need to create multiple databases. The only existing restriction is that all imported datasets should have the same geographic projection. At the same time, the AOT-mapper database handles AOT values retrieved by the AOT calculation module. The present version of the software can explore this information only by the AOT visualization module but it can be further used by any other GIS software without performing any format transformation.

The set of resulting geometrically corrected satellite images can be further used in order to provide accurate spatial information concerning land use, land cover, and also their changes through time. This information can be exploited in air quality-related studies since land cover can be related to potential emissions sources. Finally, a set of spatially continuous AOT values at different dates was derived by application of the AOT-mapper software. This dataset of AOT values is suitable not only for studying the spatial distribution of air pollution over the city of Athens on the specific dates the satellite passes but also for spatiotemporal monitoring and mapping.

The AOT-mapper software has been developed with object-oriented technology, and it is open for further improvements in the near future. For example, other high spatial resolution satellite sensors, such as SPOT or IRS, can be taken into consideration and tested accordingly. In parallel, a spatial analysis module will allow users to apply the most appropriate spatial analysis methods for the correlation of AOT values with ground-based data as well as with numerical simulation results.

Acknowledgements This research work was partially funded by the European Commission in the framework of the ICAROS-NET project under contract IST-2000-29264 (DG INFSO). Thanks go to the Earth monitoring network for providing us with all necessary data on air quality over the Athens area.

References

Bossioli, E., Tombrou, M., Dandou, A., & Soulakellis, N. (2007). Simulation of the effects of critical factors on ozone formation and accumulation in the Greater Athens Area. *Journal of Geophysical Research - Atmospheres, 112*, D02309, 19pp. https://doi.org/10.1029/2006JD007185.

Cachorro, V., & Tanre, D. (1995). Particle mass loading estimation from extinction data: Application to satellite determination. In *Proceedings of SPIE 2311, Atmospheric Sensing and Modelling*. doi: https://doi.org/10.1117/12.198586.

Cachorro, V., & Tanre, D. (1997). The correlation between particle mass loading and extinction: Application to desert dust aerosol content estimation. *Remote Sensing of Environment, 60*(2), 187194.

Colls, J. (1997). *Air pollution: An introduction* (341p). London: E & FN SPON.

Kalabokas, P. D., Viras, L. G., Repapis, C. C., & Bartzis, J. G. (1999). Analysis of year record (1987-1997) of air pollution measurements in Athens, Greece. Photochemical air pollutants. *Global Nest: The International Journal, 1*, 169–176.

Kalabokas, P. D., Viras, L. G., Bartzis, J. G., & Repapis, C. C. (2000). Mediterranean rural ozone characteristics around the urban area of Athens. *Atmospheric Environment, 34*(29–30), 5199–5208.

Kallos, G., Kassomenos, P., & Pielke, R. (1993). Synoptic and mesoscale circulations associated with air pollution episodes in Athens, Greece. *Boundary-Layer Meteorology, 62*, 163–184.

Lenoble, J. (1993). *Atmospheric radiative transfer*. Hampton: Deepak Publishing.

Lipfert, F. W. (1994). *Air pollution and community health: A critical review and data sourcebook*. New York: Van Nostrand Reinhold.

Liang, S., Fang, H., & Chen, M. (2001). Atmospheric correction of Landsat ETM+ land surface imagery—Part I: Methods. *IEEE Transactions on Geoscience and Remote Sensing, 39*(11), 2490–2498.

Middleton, W. (1952). *Vision through the atmosphere*. Toronto: University of Toronto Press.

Moran, M. S., Bryant, R., Thome, K., Ni, W., Nouvellon, Y., Gonzalez-Dugo, M. P., Qi, J., & Clarke, T. R. (2001). A refined empirical line approach for reflectance factor retrieval from Landsat-5 TM and Landsat-7 ETM+. *Remote Sensing of Environment, 78*, 71–82.

Moussiopoulos, N., Sahm, P., & Kessler, C. (1995). Numerical simulation of photochemical smog formation in Athens, Greece—A case study. *Atmospheric Environment, 29*(24), 3619–3632.

Pilinis, C., Kassomenos, P., & Kallos, G. (1993). Modeling of photochemical pollution in Athens, Greece. Application of the RAMS-CALGRID modeling system. *Atmospheric Environment, 27B*, 353–370.

Pope, A. C., Thun, M. J., Namboodir, M. M., Dockery, D. W., Evans, J. S., Spezer, F. E., & Heath, C. W. J. R. (1995). Particulate air pollution as a predictor of mortality in a prospective study for US adults. *American Journal of Respiratory and Critical Care Medicine, 151*, 669–674.

Retalis, A., Cartalis, C., & Athanassiou, E. (1999). Assessment of the distribution of aerosols in the area of Athens with the use of Landsat Thematic Mapper data. *International Journal of Remote Sensing, 20*(5), 939–945.

Sifakis, N. (1987). *Tentative mapping of air pollution using SPOT satellite images*. M.Sc. thesis, University Paris-7, 30 pp. +photos/maps, doc: GDTA-Toulouse, ENSG-Paris, CNES-Toulouse, University Paris-6, Technical Chamber of Greece.

Sifakis, N. (1995). Remote sensing of veils of atmospheric pollution and degradation of the environment in the region of Athens. *Photo-interprétation: images aériennes et spatiales, 33*, 220–225/233–236.

Sifakis, N., & Deschamps, P. Y. (1992). Mapping of air pollution using SPOT satellite data. *Photogrammetric Engineering and Remote Sensing, LVIII*, 1433–1437.

Sifakis, N., Bildgen, P., & Gilg, J. P. (1992). Use of the band 6 of Thematic Mapper to localise atmospheric-pollution clouds; the case of Athens (Greece). *Pollution Atmosphérique, 133*, 96–107.

Sifakis, N., Soulakellis, N. A., & Paronis, D. (1998). Quantitative mapping of air pollution density using Earth observations: A new processing method and application on the urban area of Athens. *International Journal of Remote Sensing, 19*, 3289–3300.
Sotiropoulou, R. E. P., Tagaris, E., Pilinis, C., Andronopoulos, S., Sfetsos, A., & Bartzis, J. G. (2004). The BOND project: Biogenic aerosols and air quality in Athens and Marseille greater areas. *Journal of Geophysical Research, 109*, D05205.
Soulakellis, N., Sifakis, N., Tombrou, M., Sarigiannis, D., & Schäfer, K. (2004). Estimation and mapping of aerosol optical thickness over the city of Brescia—Italy using diachronic and multiangle SPOT 1, 2 & 4 imagery. *Geocarto International Journal, 19*(4), 57–65.
Tanré, D., Deschamps, P. Y., Devaux, C., & Herman, M. (1988). Estimation of Saharan aerosol optical thickness from blurring effects in Thematic Mapper data. *Journal of Geophysical Research, 93*, 15955–15964.
Taylor, D. R. F. (1994). Preface. In J. Denègre (Eds.), *Thematic mapping from satellite imagery: A guidebook*. Published on behalf of the International Cartographic Association by Pergamon, UK.
Varotsos, C., Kondratyev, K. Y., & Efstathiou, M. (2001). On the seasonal variation of the surface ozone in Athens, Greece. *Atmospheric Environment, 35*(2), 315–320.
Wald, L., & Baleynaud, J. M. (1999). Observing air quality over the city of Nantes by means of Landsat thermal infrared data. *International Journal of Remote Sensing, 20*(5), 947–959.
Wen, G., Tsay, S. C., Cahalan, R. F., & Oreopoulos, L. (1999). Path radiance technique for retrieving aerosol optical thickness over land. *Journal of Geophysical Research Atmospheres, 104*(24), 31,321–31,332.
Ziomas, I. C., Tzoumaka, P., Balis, D., Melas, D., Zerefos, C. S., & Klemm, O. (1998). Ozone episodes in Athens, Greece. A modelling approach using data from the MEDCAPHOT-TRACE. *Atmospheric Environment, 32*(12), 2313–2321.

Mapping Air Pollution Health Risk: An Application of Canada's AQHI

Matthew Adams, Denis Corr, and Weeberb Requia

Abstract Human exposure to elevated air pollution has many negative health outcomes. Communicating elevated air pollution concentrations with an air quality (health) index is one an approach to reduce population exposure. The indices translate a cocktail of air pollutants to a single value that can be understood by the general public. People can use the index to avoid activities that will elevate their exposure. Most indices report the health risk for an entire city or large area as a single value. Research into air pollution spatial variability shows that major variations can occur within cities and neighborhoods, so air quality index information, while valuable, may mislead citizens when they estimate their own risk. This chapter describes the development of a neighborhood level, real time, internet enabled air pollution map that can be used by citizens to become aware of their localized air quality health risks and then take appropriate actions.

1 Introduction

Air pollution health risk is a global issue that varies in both space and time. The spatiotemporal variability is driven by many factors that can include: (1) Meteorological conditions, for example, temperature inversions that stagnate the mixing of air in the troposphere; (2) The presence or absence of solar radiation that controls the generation of secondary pollutants, such as tropospheric ozone; (3) Human-activity patterns, including vehicular congestion that tends to occur during the morning and afternoon rush-hour periods; or (4) Wind-driven regional transport of air pollutants,

M. Adams (✉)
Department of Geography, University of Toronto Mississauga, Mississauga, ON, Canada
e-mail: md.adams@utoronto.ca

D. Corr
Corr Research, Hamilton, ON, Canada

W. Requia
McMaster University, McMaster Institute for Transportation and Logistics, Hamilton, ON, Canada

H. Briassoulis et al. (eds.), *The Practice of Spatial Analysis*,
https://doi.org/10.1007/978-3-319-89806-3_17

which moves pollution generated in one region across geographical and political boundaries. Population exposure is affected by changes in concentration; however, an individual's exposure must consider their daily activity patterns, which may include cycling to work along a polluted urban corridor. The combination of ambient pollution concentration and personal activity define an individual's air pollution exposure, and subsequently their health risk for associated diseases (Park and Kwan 2017). Overall a person's exposure and subsequent air pollution health risk will vary throughout time (i.e., seasons, day of the week, or hour of the day) based on their location and a multitude of environmental and anthropogenic factors that are largely outside of their control.

We would argue that the optimal solution to the global environmental health burden caused by air pollution is regulation and enforcement, including limits placed on emitters, and improvement and substitution of technologies to reduce emissions. These potential solutions are difficult to implement and may not align with other political directions, such as economic development incentives. One non-optimal solution that has been introduced to potentially reduce citizens' risk of air pollution exposure, is the communication of air quality health risk information. In Canada, the system that has been adopted for this purpose is the Air Quality Health Index (AQHI), which translates a cocktail of air pollutants to a single value (from 1 to 10 and 10+) and can be easily communicated to the public (Steib et al. 2008). The translation of the multiple pollutant values to a single value is beneficial to the public because it conveys the cumulative risk of pollution. The Canadian AQHI is only one system that integrates air pollution health risk; Plaia and Ruggieri (2010) reviewed 18 different air quality or air pollution indices that have been developed since 1999 for different countries and regions.

Across Canada, the AQHI is calculated for each city or major area. This provides a generalized view of air pollution, but for areas where air quality is known to vary across space and with time, it does not address the risk for all individuals. The epidemiological community has shifted away from modelling city level air pollution risk, and insists, instead, that local level air pollution estimates are required to improve the accuracy and precision of their models (Baxter et al. 2013). In this chapter, the development of a spatial pollution risk communication tool is discussed, which provides a spatially and temporally refined field of AQHI values.

2 Methodology

2.1 Study Area

Hamilton, Ontario, Canada has historically, and at least regionally, been known for higher air pollution concentrations, which has led to significant research in the area (Jerrett et al. 2001; Wallace et al. 2009; Adams et al. 2012; Kanaroglou et al. 2013; Adams and Kanaroglou 2016a; Adams and Requia 2017; Requia et al. 2017). In addition to academic research, Hamilton has a strong presence of community and municipally driven air pollution initiatives, which include monitoring air pollution concentrations and community outreach (Public Health Services City of Hamilton 2015). Examining the research in the city of Hamilton, it is apparent that spatial and temporal variability occurs (Adams and Kanaroglou 2016b), which is unlikely to be captured using the current approach of only a single AQHI value for the entire spatial extent of the city.

The City of Hamilton has a unique geography that leads to complex and highly variable air pollution conditions, both in time and space. First, a 100-meter escarpment passes through the city dividing it into an upper and a lower city with an urban forest along the escarpment. The lower city experiences temperature inversions, which cause increased air pollution concentrations because of reduced atmospheric mixing (Wallace et al. 2010). Second, the city is developed around an industrial core, which generates a separate heat island in addition to the normal central city effect. When winds blow from the west (primary), this pollution does not pass into the city; however, easterly (secondary) winds move industrial air pollutants into the city (Kanaroglou et al. 2013). Third, the city is located at the western end of Lake Ontario, one of Canada's Great Lakes, thus Hamilton experiences significant lake effect winds, including onshore easterly winds during certain weather conditions. In addition to this unique geography, a major source of air pollution in the city is the dense transportation network that surrounds and transects the city (Adams and Kanaroglou 2016b). Traffic is exacerbated on the highway network due to the fact that this area is the only route to the USA around the western end of Lake Ontario. These geographical conditions and the main freeways are visualized in (Fig. 1).

2.2 Data Collection

Mobile air pollution monitoring was conducted in the study area for 12 years, from 2005 to 2016. Full details on the data collection methodology can be found in Wallace et al. (2009). The mobile unit's operation was overseen by the Ontario Ministry of the Environment and Climate Change (MOECC) and Corr Research, who conduct all calibration, maintenance, and operation processes. The MOECC applies the same procedures for mobile monitoring, as applied in the regional fixed location government monitoring network, which they also operate. Generally, mobile pollution monitoring campaigns were conducted to maximize spatial coverage during both easterly and westerly wind conditions.

2.3 Data Processing

The mobile data collection produces a spatially and temporally discontinuous data set, i.e., not all locations were observed at all points in time. This discontinuity occurs when mobile campaigns are conducted because the monitoring unit is regularly relocated during monitoring, which is very different from traditional

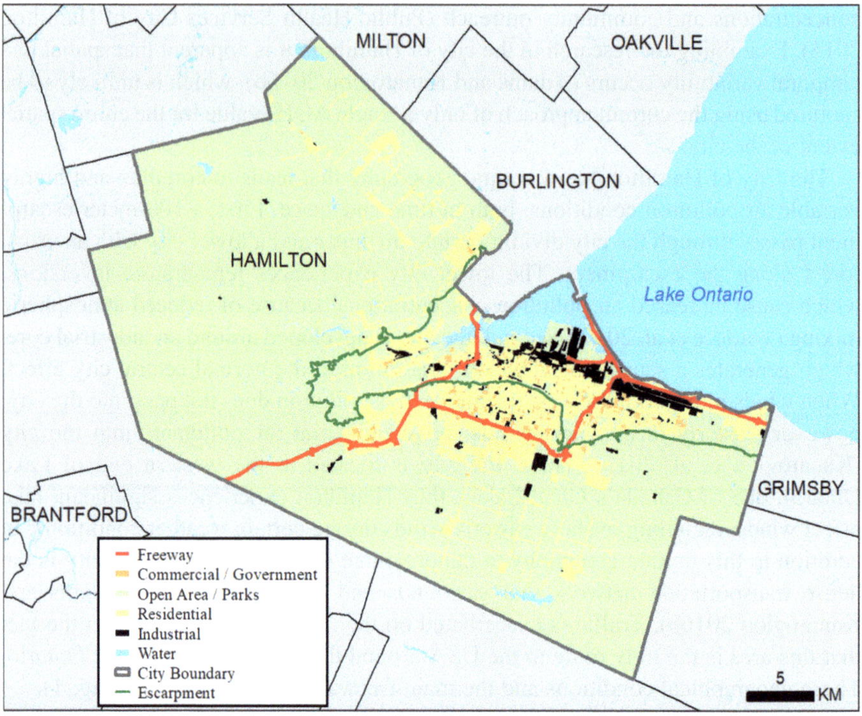

Fig. 1 Study area map, City of Hamilton

monitoring approaches that install a monitor at a fixed location. To overcome this discontinuity, the temporal signal was removed from the data to isolate the spatial signal in the mobile air pollution data. Removal of the temporal signal required standardizing the mobile data to a continuous fixed location air pollution monitor, which was centrally located in the study area. Formula (1), modified from Adams and Kanaroglou (2015), was applied to remove the temporal signal.

$$O_S = O_{ST} \div \log_e(S_T + e) \quad (1)$$

where O_S is the mobile air pollution value with the temporal component removed, O_{ST} is the observed mobile air pollution value, S_T is the value observed at the fixed location monitoring unit during the same time as the mobile air pollution data collection, and e is Euler's number, which is included to ensure that when the log of the stationary value is calculated it will be a positive value above one. Following this adjustment, the spatial differences between all observations are left and not the variation caused by collection during different times of the year. The stationary air pollution data monitor was located in downtown Hamilton, which is centrally located within the study area.

Meteorological data collected by Environment Canada in Hamilton were examined to check for primary and potentially secondary prevailing wind directions. The mobile pollution data were separated for further processing based on identified wind regimes. This was motivated because it is known that in Hamilton, temperature inversions can occur during easterly winds (Wallace et al. 2010), which have shown to generate significant air pollution concentrations. The increased pollution occurs because of an industrial complex along the northern edge of the study area, that is responsible for the generation of air pollution, along with freeways bounding the city (Adams et al. 2012; Adams and Kanaroglou 2016b).

Data Aggregation

Air pollution is a continuous spatial process that is well suited for mapping. The mobile air pollution data collection generated point observations; however, presenting individual points is an unintuitive approach for the map reader, nor is it representative of the real-world phenomena because air pollution is a continuous process. It was necessary to aggregate the point information to spatial units that can be mapped to represent the continuous spatial phenomena. The AQHI maps generated in this work, are web-maps that update each hour.

The adjusted mobile data point values were aggregated within polygons covering the study area. Data were separated by wind regimes prior to aggregation. To assess locations where future data collection should focus, the number of observations and the variance of the values within each polygon were mapped. The spatial units were hexagons with a distance of 1000 m from the centroid of top edge to the centroid on the parallel bottom edge. Data collection in this region is an ongoing process, and this simple aggregation process is designed for rapid updating, which is in contrast to previous work that predicted spatial variation in the region by fitting neural network models in a land-use regression framework (Adams and Kanaroglou 2016b).

2.4 Air Pollutants and Air Quality Health Index

Providing the public with health risk information usually requires a translation from the raw values, which in this case are air pollution concentrations, to a value that can be easily interpreted. For air pollution risk communication, this is usually achieved with an air quality index (Plaia and Ruggieri 2010). Many indices exist and the adoption of a specific index is commonly determined by a regional or national government. Across Canada air quality risk is provided by the Air Quality Health Index (Steib et al. 2008). This index relies on pollution concentration estimates of nitrogen dioxide (NO_2), ground-level ozone (O_3), and particulate matter 2.5 microns or smaller in aerodynamic diameter ($PM_{2.5}$). Two formulae were developed for Canada based on seasonal temperature variations, which included the cool season (October–March) and the warm season (April–September). The cool season formula is

$$\mathrm{AQHI}_{\mathrm{cool}} = 10/6.43^* \left(100^* \left(e^{(0.000457*\mathrm{NO}_2)} - 1 + e^{(0.000462*\mathrm{PM}_{2.5})} - 1\right)\right).$$

The warm season includes an extra component of ground-level ozone (Steib et al. 2008),

$$\mathrm{AQHI}_{\mathrm{warm}} = 10/6.43^* \left(100^* \left(e^{(0.00101*\mathrm{NO}_2)} - 1 + e^{(0.000621*\mathrm{PM}_{2.5})} - 1 + e^{(0.00104*\mathrm{O}_3)} - 1\right)\right)$$

The appropriate formula is applied in this work based on the time of year.

2.5 Temporal Signal

During the data processing the temporal signal was removed from the mobile data values, which were then aggregated based on their meteorological and spatial contexts. Following the removal of the temporal component from the data, the spatial factor (multiplier) is left. Then, for any given time, the current air pollution values from a local MOECC operated AQHI station can be used to calculate the current temporal AQHI value across the study area. Within Canada, the AQHI is updated on an hourly basis, using a rolling 3-h average of each pollutant to calculate the AQHI. In the present application of this mapping approach, the same 3-h rolling average is applied to calculate the spatially refined AQHI.

3 Results

3.1 Mobile Data Description

The mobile data set included 16,194 observations. These were collected once per minute during mobile air pollution monitoring campaigns, lasting for 1 s. The average spatial distance between mobile monitoring points was 451 m, which equates to an average travel speed for the mobile monitoring vehicle of 27 km/h.

The mobile air pollution data summary statistics are included in Table 1. The higher number of missing observations in the ozone data occurs because monitoring

Table 1 Summary statistics of mobile air pollution data

	$PM_{2.5}$ (μg/m^3)	NO_2 (ppb)	O_3 (ppb)
Minimum	LDL	LDL	LDL
Median	12.5	13.0	23
Mean	14.7	16.6	21.65
Max	731	498	238
Missing data	1136	650	10,183

LDL Lower than detection level

for this pollutant did not occur until 2012. The average change in pollution concentrations between consecutive air pollution observations during campaigns were 2.05 μg/m^3, 7.16 ppb, and 1.27 ppb for $PM_{2.5}$, NO_2, and O_3, respectively. The maximum differences between consecutive observations for $PM_{2.5}$, NO_2, and O_3 were 653 μg/m^3, 493 ppb, and 212 ppb, respectively.

3.2 *Data Processing*

Temporal Signal Removal

The temporal signal of each datum was removed using air pollution data collected at a fixed location air pollution monitor. The summary statistics for the adjusted values are presented in Table 2. The temporal removal utilized the natural logarithm of the fixed location air pollution datum, with the intent to reduce extreme adjustments when the fixed location observation was observing concentrations at the extreme tail of the distribution. Air pollution is known to follow a log-normal distribution (Ott 1990).

Wind Regimes

Wind data were acquired from a local meteorological station operated by Environment Canada. These data were hourly observations that were collected as the average speed during the 2-min period ending at the time of observation. The data covered the same period from 2005 to 2016 and were collected 10 m above the ground. The average wind speed during this time was 7.5 km/h, with a maximum observed speed of 39 km/h.

We separated the air pollution observations based on two wind regimes that were observed in the data set. The two groupings were based on a split of the data along the north-south axis. The wind rose is presented in Fig. 2.

Table 2 Summary statistics for mobile data with temporal signal removed

	$PM_{2.5}$ (μg/m^3)	NO_2 (ppb)	O_3 (ppb)
Minimum	LDL	LDL	LDL
Median	0.0102	0.005	0.0678
Mean	0.0262	0.022	0.136
Max	0.994	1.38	2.08

LDL Lower than detection level

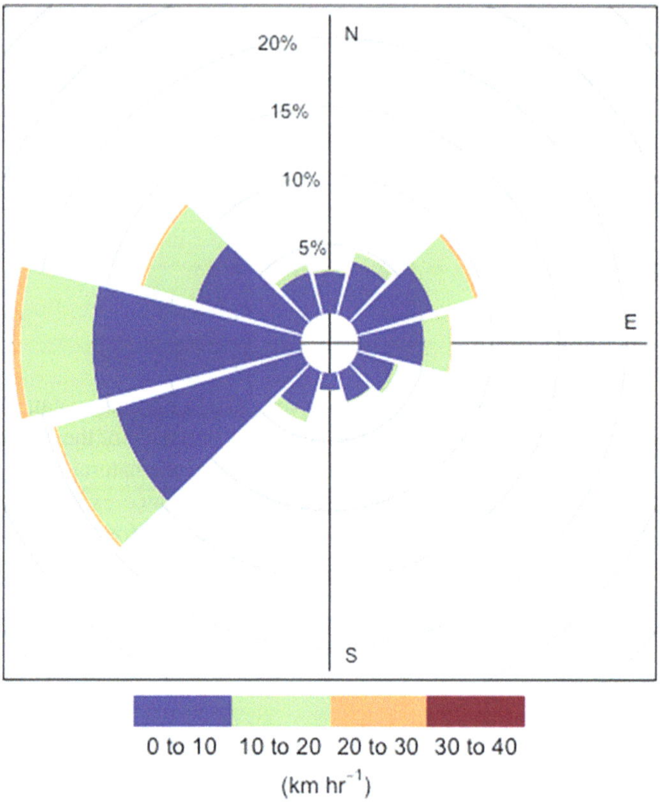

Fig. 2 Wind rose for Hamilton, Ontario from 2005 to 2016

3.3 *Spatial Data Aggregation*

The creation of the polygon data set that covered the spatial extent of the city resulted in 1256 hexagons that were not intersecting water bodies. Any hexagon located on top of water was removed from the analysis.

Following the separation of the mobile data based on the two wind regimes, mobile observations in the east regime represented 36% of the data and the remaining 64% were observed during westerly winds. The average wind speed during mobile air pollution data collection was 7.4 and 11.2 km/h during the east and west wind regimes, respectively.

Mobile air pollution observations did not occur within the spatial boundaries of all polygons, particularly those polygons that are located on the periphery of the city, which are rural and agricultural lands. Descriptive statistics are provided on the number of observations that fell within each polygon in Table 3. The number of observations for each wind direction is visually presented in Fig. 3.

Table 3 Summary statistics of aggregation of mobile pollution data to hexagons (temporal signal removed)

	East			West		
	$PM_{2.5}$	NO_2	O_3	$PM_{2.5}$	NO_2	O_3
Average count of observations (Excluding null polygons)	22	22	22	31	31	31
Null polygons	1004	1004	1004	956	956	956
Standard deviation (Excluding null polygons)	0.09	0.18	0.11	0.84	0.18	0.16

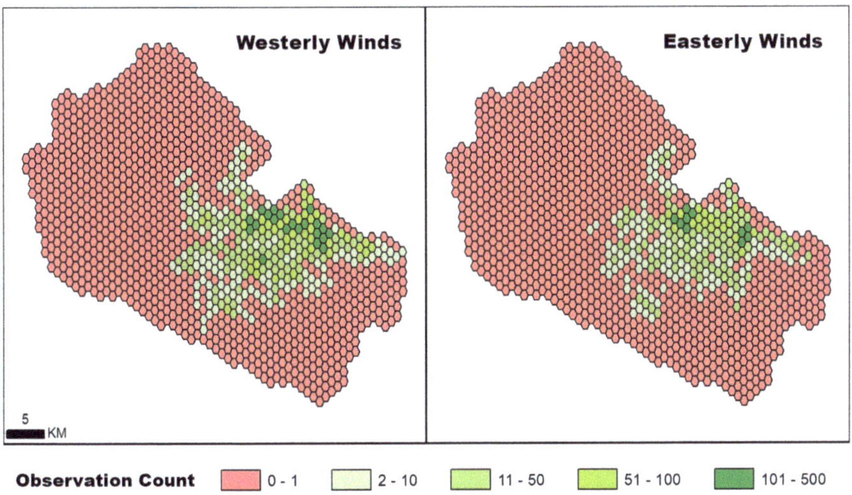

Fig. 3 Observation counts separated by wind direction

Polygons without observations were filled by inverse distance-weighted interpolation. This interpolation primarily occurred within polygons near the core of the city without road access or at the periphery of the city where rural and low-density suburban land use occurs. In Fig. 4, the spatial multipliers for each pollutant and wind direction are presented.

A realization of the AQHI field based on the long-term average for each of the three pollutants from the Hamilton downtown fixed location air pollution monitoring units is presented; the long-term average concentrations were 28.43 (μg/m^3), 10.62 (ppb), and 7.736 (ppb) for $PM_{2.5}$, NO_2, and O_3, respectively. As well, in Fig. 5, the AQHI calculated with the long-term 95th percentile concentrations, which were 51 (μg/m^3), 23 (ppb), and 19 (ppb) for $PM_{2.5}$, NO_2, and O_3, respectively, is presented. The color scale of the maps is the AQHI color scale from Environment Canada.

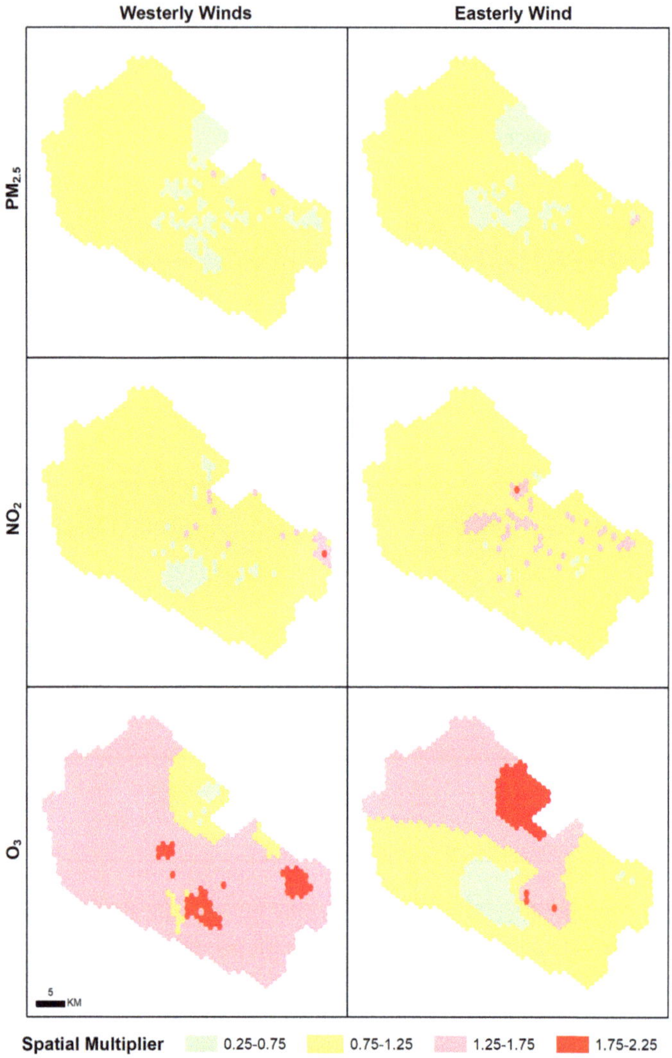

Fig. 4 Spatial multipliers for $PM_{2.5}$, NO_2, and O_3 separated by westerly and easterly winds

4 Discussion

The ideal approach for understanding the spatial and temporal variation of air pollution across a domain would be a dense network of fixed location continuous air pollution monitors. This idealized network would come at a great expense with current technology, and we are not aware of any such network. The challenge occurs because urban air pollution can vary locally, such as local significant increases near freeways or heavily trafficked urban corridors. Highway emissions of ultrafine

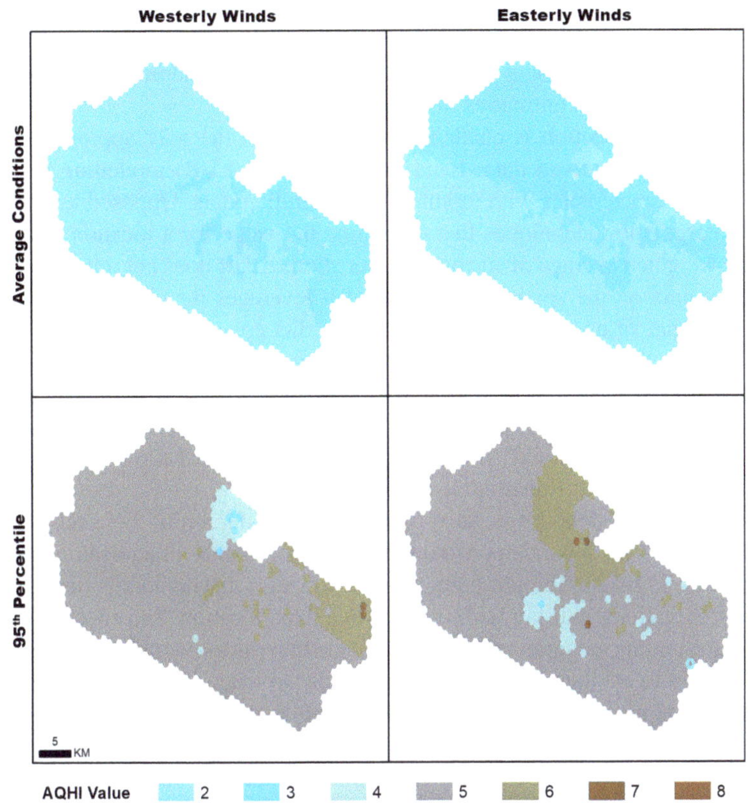

Fig. 5 Realizations of the AQHI field based on the long-term average and the 95th percentile of data collected at the fixed location long-term monitors

particulate matter have been found to recede to background concentrations within a few hundred meters of the freeway (Zhu et al. 2002). This chapter has presented one approach to modelling spatially and temporally variable concentrations that leverages mobile and stationary air pollution monitoring units.

The limitations of the present study include the data collection of mobile air pollution data. A large data set that has been collected for general and specific monitoring in the city of Hamilton over a long-term period was leveraged. Admittedly, this type of data is not easily collected due to the high-cost and time requirements. Second, the process followed requires fixed location continuous air pollution monitors for removing the temporal trend, and then for updating the spatial multipliers each hour to produce the spatially refined concentrations. In the present work, temporal and spatial stationarity is assumed; however, as urban environments evolve, air pollution sources and emission profiles will change, which may result in non-stationarity. Third, a suitable approach that can define the number of monitoring visits necessary at a location, to ensure a correct estimate has yet to be determined. When air pollution monitors were installed on Google Street View Cars, it was found that using 1-s samples, similar to this work, that 10–20 repeated samples were sufficient to capture the temporal variation and calculate a long-term average (Apte et al. 2017). Additional work is necessary to understand the uncertainty in the estimates based on discontinuous input data.

Mobile data collection is challenging, but it is not the only approach that can provide spatially resolved data. Land use regression (LUR) modelling associates urban form characteristics with monitored air pollution data. Opposed to traditional spatial interpolation techniques that determine the value for a location based on a distance-weighted average of surrounding locations, LUR does not rely on distances between points on the spatial plane, instead it leverages the characteristics of the urban form that sit upon the spatial plane to predict concentrations. Because of the short distances that air pollution varies, LUR allows for a better spatial prediction of concentrations, because it is not possible to have a dense enough monitoring network for traditional spatial interpolation techniques. A well-located monitoring network combined with LUR could produce the necessary information for space–time modelling (Adams and Kanaroglou 2016b).

When LUR models have been tested for transferability between different spatial regions, they have demonstrated poor transferability (Vienneau et al. 2010; Mukerjee et al. 2012). The present approach utilizes a discontinuous time-series of mobile observations and standardizes them to a fixed location. Since no other aspect of the surrounding geography is incorporated, the values would not provide any transferability to another region. Thus, using this approach it would be necessary to replicate all steps with new data collected by mobile monitoring or other means in a different region.

Another emerging approach that may negate the need or supplement mobile data collection is the observation of air pollution concentrations by remote sensing, which has been applied to estimate annual average pollution concentrations (Geddes et al. 2016) and even daily concentrations (Lee et al. 2016). Currently, new sensors are being developed, such as the TEMPO (Tropospheric Emissions: Monitoring Pollution) mission, which is designed to collect daily air pollution data. When this and similar satellites become active in the coming years, and with the new models to calibrate remote-sensing data, including chemical transport, land use, statistical, and hybrid models, it will provide additional spatiotemporal coverage for assigning air pollution risk.

5 Conclusions

Communicating localized potential health risks of air pollution should be adopted over non-spatial approaches to increase an individual's utility of the information. The system should be capable of identifying short-term events by constantly updating the information provided. Also, long-term health risks, which may cause even greater negative health outcomes, should also be communicated. This approach for the communication of an environmental health risk would help satisfy one of the seven cardinal rules for the effective communication of health risks that were suggested by the EPA: People have the right to have a voice and participate in decisions that affect their lives (Maibach and Covello 2016). Providing this risk information in a localized, real-time internet map format allows individuals to make informed choices on their daily activity in order to limit the personal air pollution exposures of themselves and their families.

References

Adams, M. D., & Kanaroglou, P. S. (2015). A method for reducing classical error in long-term average air pollution concentrations from incomplete time-series data. In P. S. Kanaroglou, E. M. Delmelle, & A. Paez (Eds.), *Edited Volume: Spatial analysis in health geography* (pp. 183–196). New York: Ashgate Publishing.

Adams, M. D., & Kanaroglou, P. S. (2016a). A criticality index for air pollution monitors. *Atmospheric Pollution Research, 7*, 482–487. https://doi.org/10.1016/j.apr.2015.11.004.

Adams, M. D., & Kanaroglou, P. S. (2016b). Mapping real-time air pollution health risk for environmental management: Combining mobile and stationary air pollution monitoring with neural network models. *Journal of Environmental Management, 168*, 133–141. https://doi.org/10.1016/j.jenvman.2015.12.012.

Adams, M. D., & Requia, W. J. (2017). How private vehicle use increases ambient air pollution concentrations at schools during the morning drop-off of children. *Atmospheric Environment, 165*, 264–273. https://doi.org/10.1016/j.atmosenv.2017.06.046.

Adams, M. D., De Luca, P. F., Corr, D., & Kanaroglou, P. S. (2012). Mobile air monitoring: Measuring change in air quality in the City of Hamilton, 2005–2010. *Social Indicators Research, 108*, 351–364. https://doi.org/10.1007/s11205-012-0061-5.

Apte, J. S., Messier, K. P., Gani, S., et al. (2017). High-resolution air pollution mapping with Google street view cars: Exploiting big data. *Environmental Science & Technology, 51*, 6999–7008. https://doi.org/10.1021/acs.est.7b00891.

Baxter, L. K., Dionisio, K. L., Burke, J., et al. (2013). Exposure prediction approaches used in air pollution epidemiology studies: Key findings and future recommendations. *Journal of Exposure Science & Environmental Epidemiology, 23*, 654–659. https://doi.org/10.1038/jes.2013.62.

Geddes, J. A., Martin, R. V., Boys, B. L., & van Donkelaar, A. (2016). Long-term trends worldwide in ambient NO2 concentrations inferred from satellite observations. *Environmental Health Perspectives, 124*, 281–289. https://doi.org/10.1289/ehp.1409567.

Jerrett, M., Burnett, R. T., Kanaroglou, P., et al. (2001). A GIS—Environmental justice analysis of particulate air pollution in Hamilton, Canada. *Environment & Planning A, 33*, 955–973. https://doi.org/10.1068/a33137.

Kanaroglou, P. S., Adams, M. D., De Luca, P. F., et al. (2013). Estimation of sulfur dioxide air pollution concentrations with a spatial autoregressive model. *Atmospheric Environment, 79*, 421–427. https://doi.org/10.1016/j.atmosenv.2013.07.014.

Lee, M., Kloog, I., Chudnovsky, A., et al. (2016). Spatiotemporal prediction of fine particulate matter using high-resolution satellite images in the Southeastern US 2003-2011. *Journal of Exposure Science and Environmental Epidemiology, 26*(4), 377–384. https://doi.org/10.1038/jes.2015.41.

Maibach, E., & Covello, V. T. (2016). Communicating environmental health. In H. Frumkin (Ed.), *Environmental health: From global to local* (pp. 769–786). San Francisco: Wiley.

Mukerjee, S., Smith, L., Neas, L., & Norris, G. (2012). Evaluation of land use regression models for nitrogen dioxide and benzene in Four US cities. *ScientificWorldJournal, 2012*, 865150. https://doi.org/10.1100/2012/865150.

Ott, W. R. (1990). A physical explanation of the lognormality of pollutant concentrations. *Journal of the Air & Waste Management Association, 40*, 1378–1383. https://doi.org/10.1080/10473289.1990.10466789.

Park, Y. M., & Kwan, M. P. (2017). Individual exposure estimates may be erroneous when spatiotemporal variability of air pollution and human mobility are ignored. *Health & Place, 43*, 85–94. https://doi.org/10.1016/j.healthplace.2016.10.002.

Plaia, A., & Ruggieri, M. (2010). Air quality indices: A review. *Reviews in Environmental Science and Bio/Technology, 10*, 165–179. https://doi.org/10.1007/s11157-010-9227-2.

Public Health Services City of Hamilton. (2015). Clean Air Hamilton 2014 Air Quality Progress Report.

Requia, W. J., Adams, M. D., Arain, A., et al. (2017). Spatio-temporal analysis of particulate matter intake fractions for vehicular emissions: Hourly variation by micro-environments in the Greater Toronto and Hamilton Area, Canada. *Science of The Total Environment, 599–600*, 1813–1822. https://doi.org/10.1016/j.scitotenv.2017.05.134.

Steib, D. M., Burnett, R. T., Smith-Doiron, M., et al. (2008). A new multipollutant, no-threshold air quality health index based on short-term associations observed in daily time-series analyses. *Journal of the Air & Waste Management Association, 58*, 435–450. https://doi.org/10.3155/1047-3289.58.3.435.

Vienneau, D., de Hoogh, K., Beelen, R., et al. (2010). Comparison of land-use regression models between Great Britain and the Netherlands. *Atmospheric Environment, 44*, 688–696. https://doi.org/10.1016/j.atmosenv.2009.11.016.

Wallace, J., Corr, D., Deluca, P., et al. (2009). Mobile monitoring of air pollution in cities: The case of Hamilton, Ontario, Canada. *Journal of Environmental Monitoring, 11*, 998–1003. https://doi.org/10.1039/b818477a.

Wallace, J., Corr, D., & Kanaroglou, P. (2010). Topographic and spatial impacts of temperature inversions on air quality using mobile air pollution surveys. *Science of The Total Environment, 408*, 5086–5096. https://doi.org/10.1016/j.scitotenv.2010.06.020.

Zhu, Y., Hinds, W. C., Kim, S., & Sioutas, C. (2002). Concentration and size distribution of ultrafine particles near a major highway. *Journal of the Air & Waste Management Association, 52*, 1032–1042. https://doi.org/10.1080/10473289.2002.10470842.

Index

H. Briassoulis et al. (eds.), *The Practice of Spatial Analysis*,
https://doi.org/10.1007/978-3-319-89806-3

B

MIX
Papier aus verantwortungsvollen Quellen
Paper from responsible sources
FSC® C105338

If you have any concerns about our products,
you can contact us on
ProductSafety@springernature.com

In case Publisher is established outside the EU,
the EU authorized representative is:
Springer Nature Customer Service Center GmbH
Europaplatz 3, 69115 Heidelberg, Germany

Printed by Libri Plureos GmbH
in Hamburg, Germany